A COMPARATIVE STUDY IN COLOUR PATTERNS AND BEHAVIOUR IN SEVEN ASIATIC BARBUS SPECIES

(CYPRINIDAE, OSTARIOPHYSI, OSTEICHTHYES)

A PROGRESS REPORT

BEHAVIOUR

AN INTERNATIONAL JOURNAL OF COMPARATIVE ETHOLOGY

SUPPLEMENT XIX

K. KORTMULDER

A COMPARATIVE STUDY IN COLOUR PATTERNS AND BEHAVIOUR IN SEVEN ASIATIC BARBUS SPECIES (CYPRINIDAE, OSTARIOPHYSI, OSTEICHTHYES)

A PROGRESS REPORT

LEIDEN
E. J. BRILL
1972

A COMPARATIVE STUDY IN COLOUR PATTERNS AND BEHAVIOUR IN SEVEN ASIATIC BARBUS SPECIES

(CYPRINIDAE, OSTARIOPHYSI, OSTEICHTHYES)

A PROGRESS REPORT

B. CONCHONIUS . B. NIGROFASCIATUS . B. TETRAZONA
B. STOLICZKANUS . B. CUMINGI . B. PHUTUNIO . B. GELIUS

BY

K. KORTMULDER

LEIDEN
E. J. BRILL
1972

ISBN 90 04 03516 8

PRINTED IN THE NETHERLANDS

CONTENTS

ERRATA

p. 126, fig. 44; p. 128, fig. 45; p. 131, fig. 46: The uppermost line of each section of the figures should be on the same height with: a♂

p. 232, table 10, *stoliczkanus* 3♂♂ groups: 3♂♂pl. n.l. *instead of* 3♂♂b. n.l.; *tetrazona* 3♂♂ groups: 3♂♂b. a.l. *instead of* 3♂♂pl. n.l.

p. 233, table 11, *conchonius* 2♂♂ groups: 2♂♂b. B *instead of* 2♀♀b. B; *nigrofasciatus* 2♂♂ groups: 2♂♂pl. A *instead of* 2♀♀pl. A.

p. 285, *top line should read*: its interactions with other systems (flight, sex, territorial tendencies *etc.*)

p. 329, line 15 from top: 3♂♂ *instead of* 3♀♀.

PREFACE

This book is an attempt to put together data I have collected during a year-long study of a number of Asiatic *Barbus* species in captivity. The central problems of this study are: (1) is there some relation between the colour markings of different species on the one hand and their behaviour on the other and (2) if there are such relations can these be interpreted as a result of mutual evolutionary adaptation of the colour patterns and the behaviour of a species to each other. Three steps are necessary to solve these problems. First, it is necessary to make a detailed description and comparison of the colour patterns of different species. Second, an analysis and comparison of the behaviour of these species must be made. Third, the effects of the colour patterns on the behaviour of conspecifics must be studied. The three sets of problems will be dealt with in the chapters I, III and IV respectively. In addition, chapter I provides evidence that the 5 species which are the main object of the study belong to one closely related subgroup of the genus; chapter II gives a description and a preliminary analysis of the behaviour elements of the species to serve as a basis for the comparison made in chapter III; chapter IV offers a general discussion of the problems involved and the evidence available.

This book is a progress report. It doesn't claim any completeness in the solution of the problems outlined above. Especially the evidence concerning the effects of the colour patterns is meagre still. Yet the data presented may be interesting in themselves and it may be worthwhile to present them in this form.

I am afraid this won't be an easy book to read. This is mainly so because a considerable quantity of information, the elements of which are interconnected in such a way that they might only be satisfactorily rendered in a multidimensional space, had to be squeezed into unidimensional script. By way of substitute for multidimensional presentation, part of the information is given in separate appendices at the end of the book. It is indicated at which points the appendices might be inserted into the main text. Other parts of the information are given in small type passages and foot-notes and extensive cross-references have been inserted. Further the reader may use as Ariadne's clue the sketch presented in the first paragraph of this preface. Finally, I have taken some pains to present things in such a way as to enable the reader to read separate chapters or sections as independent units even though this involved the creation of some redundancy in the complete text.

NOMENCLATORY

Nomenclatorial questions were discussed with Dr P. H. GREENWOOD, British Museum (Nat. Hist.) London. In view of the diversity of opinions as to a possible (and probably desirable) further sub-division of the genus, I decided to stick to the generic name *Barbus*. (Perhaps the kind of data discussed in the first chapter of this study may prove useful in a future revision of the genus). The specific names used in this report are the same as those in TUCKER's 1963 english edition of STERBA's "Süsswasserfische aus aller Welt". Full names of the species appearing in this report are listed below:

B. conchonius (Ham.-Buch. 1822)
B. cumingi Günther 1868
B. nigrofasciatus Günther 1868
B. tetrazona tetrazona (Bleeker 1855)
B. tetrazona partipentazona Fowler 1934
B. stoliczkanus Day 1871 1)
B. gelius (Ham.-Buch. 1822)
B. phutunio (Ham.-Buch. 1822)
B. titteya (Deraniyagala 1929)
B. arulius (Jerdon 1849)
B. filamentosus (Cuv. & Val. 1844)
B. binotatus Cuv. & Val. 1842
B. everetti Boulenger 1894
B. fasciatus Bleeker 1853
B. lateristriga Cuv. & Val. 1842
B. oligolepis (Bleeker 1853)
B. pentazona pentazona Boulenger 1894
B. pentazona hexazona Weber & De Beaufort 1912
B. semifasciolatus Günther 1868

MATERIAL AND METHODS

The fishes used in this study were (a) imported specimens caught in the wild as adults; (b) fishes bred in captivity by professional breeders; such fishes were usually purchased when sub-adult; (c) fishes bred in our laboratory; parents were from either category (a) or (b). Categories (a) and (b) were mainly obtained through dutch dealers.

Practically all *B. stoliczkanus* used for the behavioural studies belonged to category (a). A few imported *B. cumingi* were used in some of the experiments (chapter III, 2♂♂ 2♀♀ groups). For the rest all fishes used for the behavioural studies belonged to categories (b) or (c).

As to the fishes of category (b) we took some care to buy different groups of the same species from different dealers in order to prevent the material to be too homogeneous genetically. (Fishes of all five species are imported at regular times from the wild; so one may assume that the breeding stocks do not differ too much from the wild populations). Occasionally a group of category (b) was excluded from the experiments because we suspected them of too close inbreeding or loss of general vitality. Of the fishes of category (c) only first laboratory generations were used for the comparative bebavioural studies. We took some pains, if possible, to take parents from different purchased groups. For the morphological studies also 2nd and 3rd laboratory generations have been used.

Adults of most of the other species mentioned in this study, such as *B. phutunio*, *B. gelius*, *B. arulius* and *B. filamentosus* were direct imports exclusively. In none of the species any obvious differences were found between specimens of the three cate-

1) = *Barbus (Puntius) ticto stoliczkae* according to STERBA (*in litt.*, 1969).

gories, either in morphological characters or — qualitatively and roughly quantitatively — in behaviour. One reservation must be made as to category (a). In no case we were able to check ourselves whether these fishes were genuine wild specimens or whether they had been bred overseas. We have some confidence in our dealers, however.

Specimens of *B. conchonius, B. nigrofasciatus, B. stoliczkanus, B. tetrazona tetrazona* and *B. cumingi* were identified by comparison with specimens present at the British Museum (Nat. Hist.) London; the *stoliczkanus* were identified as such, though they were not identical to the museum specimens.

The following specimens were deposited at the British Museum: *B. tetrazona tetrazona* 1966.9.14: 1-8; *B. nigrofasciatus* 1966.9.14: 9-13; *B. cumingi* 1966.9.14: 14-16; *B. conchonius* 1966.9.14: 17-19; 20-22; *B. stoliczkanus* (labeled *B. ticto*) 1966.9.14: 27-34; *B. cumingi* x *B. stoliczkanus* 1966.9.14: 23-26. Except for the *stoliczkanus* specimens, all belonged to categories (b) or (c).

When not used for experiments the fish were mainly kept in ambisexual unispecific or mixed groups in stock tanks (125 × 40 × 50 cm either with asbestos-cement sides and back and a glass front or with iron framing and glass on four sides. The tanks were aerated with plastic lines and air stones; the water was made to circulate through filters. Sand covered the bottom and the tanks were well supplied with mixed vegetation. The temperature of the room was maintained at levels varying between 21°C and 25°C but constant for long periods of time. Light was supplied to the tanks from above by incandescent or fluorescent lamps each day from 08.15 to 22.15. A few small incandescent lamps, directed into the four corners of the room were lit from 08.00 to 08.30 and from 22.00 to 22.30. The large lamps illuminating the whole room were lit at 08.30 and extinguished at 22.00. This substitute twilight prevented sudden frights. During the early stages of the study also daylight had access to the room and no standardized light regime was maintained during that period. A dry food mixture was supplemented, as far as possible, with live tubifex, daphnia, chironomid larvae or artemia. Most of the experiments were done in the same room. For details of experimental tanks and conditions see chapter III, sections II-IV.

Behavioural "experiments" mainly consisted of observations of the behaviour of individuals in groups varying in size from 2 ♂♂ to 6 ♂♂ 6♀♀. Observation times were standardized more or less according to the different kinds of experiments (see chapter III for details). Barbs of all species studied are highly social animals and more standardized experiments involving one individual at a time — such as have been carried out successfully in *e.g.* sticklebacks, bitterlings and guppys — were found to be unpracticable in the barbs. For instance, males introduced into another male's tank in a glass container do anything but behave in a constant way and often show drastic colour changes in the procedure. Males do fight each other when separated by glass but — except a few *tetrazona* males that had been alone for weeks — no male barb could be induced to court a female introduced to him in a glass container of any size or shape or at any location in the tank. This was so even if we took the very female that the male had been courting intensively just before experiment. No barb — except again the isolated *tetrazona* males mentioned before — ever fought or courted models spontaneously. Barbs do not have any form of nest-building or parental behaviour (see, however, *B. phutunio*, p. 165ff.). Thus there remains little to experiment with a single individual, if one is interested in reproductive behaviour. We therefore had to content ourselves with studying social situations. It was only during the later stages of this study that a technique of training individual males to fight and court models became available. Only a few experiments with these males are mentioned in this report.

Colour intensities (red and black) were assessed according to a subjective scale in which full intensity is represented by the value 1 and absence of colour by 0. Between these extremes the values 0—¼, ¼, ¼—½, ½, ½—¾, ¾ and ¾—1 are distinguished.

Subjective as this scale may be, estimations of colour intensity by different persons used to yield a rather good concordance.

Interspecific hybrids were bred by means of a forced breeding technique, usually by confining 2 ♂♂ of one species with 2 ♀♀ of another in a tank for several weeks.

TERMINOLOGY

A few terms used in this report may require some specification. 'Tendency' is used in a loose everyday sense. When saying that an animal tends to behave in a certain way, I mean that it more or less often does so — if not at the moment under consideration then at least shortly afterwards. The terms 'stronger' and 'weaker' tendency refer roughly to the degree of likelihood of occurrence. 'Tendency' as used here differs from the definition proposed by HINDE (1955; 1966 p. 245-6) in that it always refers directly to observable behaviour. For instance, when saying that a fish tends to rise in the water during threat, I mean that it does rise often or intermittently. 'Potential tendencies' (see also BLURTON-JONES, 1968) belong to the realm of 'motivation' or 'drive' (see below).

'Motivation' is used in two different ways: (1) In a qualitative sense. For instance, when saying that a certain threat posture is aggressively motivated, I mean that the posture and aggressive behaviour elements (such as attack and bite) have some causal factors in common. (2) In a quantitative sense; variations in the frequencies and/or intensities of certain behaviour patterns may be due to variations in the specific motivations for these patterns. In this sense the term is synonymous to 'drive' as used by various authors. Motivation in both senses refers to internal states.

> There is, however, a methodological problem here. Motivation in the second sense may be measured through the elimination of variability in the external stimulus strength. On the other hand, motivation in the first sense may be assessed in two principle ways: (a) by analyzing correlations in the occurrence (frequencies, intensities *etc.*) of two or more behaviour patterns. In this case it often is very difficult to prove that the correlations found are caused by common internal factors and that they are not a mere reflection of (inevitable) correlations in the external stimuli. This question is left open in the present report. (b) by means of form analysis. In this case common motivation refers to partial sharing of motor coordination systems by two or more behaviour patterns. Of course, the internal systems concerned here are not identical to those implied in the quantitative definition of motivation.

Motivation in the quantitative sense may vary within one individual between different time periods, between individuals of the same species (under comparable conditions) or (more or less consistently) between individuals of different species. As to the last case — which forms one of the main themes of this report — it should be emphasized that a statement to the effect that a certain motivation is stronger in one species than in another does not refer to the nature of the internal systems themselves but only to the fact that individuals of the one species show more of a certain type of behaviour than do those of the other species under influence of (objectively) the same external stimuli.

'Behaviour elements' refer to smallest units of behaviour which can be considered as a whole. 'Components' refer to smaller units which together constitute an element. 'Behaviour patterns' is used in a more loose and general sense.

A 'sequence' of behaviour is defined as a limited number of behaviour elements that follow each other immediately. The total duration may vary but is mostly of the order of magnitude of seconds or tens of seconds. A 'bout' is a period varying from several minutes to hours in which behaviour patterns of a certain class (such as reproductive or feeding behaviour) are frequent. Bouts of the same type are separated from each other by periods of roughly the same order of magnitude or longer.

'Aggression' and 'fear' are used as a short-hand for the motivations underlying attacking and fleeing behaviour.

ACKNOWLEDGEMENTS

I wish to thank Prof. J. J. A. van Iersel for originally drawing my attention to a possible relationship between black marking patterns and degrees of aggressiveness in different species — on the basis of the assumption that the markings act as threat colours, for leaving me entirely free in the organization of this study and for valuable criticism of the manuscript. I am grateful to Prof. P. Sevenster for encouraging a broad biological approach, to Dr P. H. Greenwood for identifying my fish, to both for critically reading the manuscript; to Mrs D. C. M. A. Brederoo-Huygen for the histological examination of the hybrids, to Mr C. Noorlander for preparing the cross-sections of young fish; to Mr G. Piket for caring for the fishes during the later stages of the study, to Mr G. Veldhuyzen for constructing the 3 m tank and many a small gadget used in this study; to Mr C. Elzenga for preparing the graphs.

I am much indebted to several of my students who helped in caring for the fishes, conducted pilot experiments and experiments parallel to those mentioned in this study. At the risk of doing paltry justice to others I want to mention Mr M. M. J. van Balgooy, Mr H. de Graaff, Mr A. A. Mabelis, Miss C. J. van der Voet and Mr F. Putters.

CHAPTER I. MORPHOLOGY

'Variations on a theme'

INTRODUCTION

The main objects of this chapter are the five species: *B. conchonius, B. nigrofasciatus, B. tetrazona tetrazona, B. stoliczkanus* and *B. cumingi*. In this chapter some morphological characters of these species are described. Special attention is given to the colour patterns on the body and fins.

Section I introduces the five species with a general description of their size, shape and colour patterns. Some data on the geographical distributions and the natural habitats of the species are given in Section II. Section III presents a comparison of the colour patterns of the five species. Some inter-specific hybrids are introduced in Section IV. The ontogeny of some colour patterns is described in Section V. Possible homologies between the colour patterns of the five species are discussed in Section VI. Section VII deals with the systematic status of the five species. Hybrids between the species are discussed and the black marking patterns of the five species and their ontogeny are compared with those of some other species. Finally, Section VIII presents a brief description of two more species: *B. phutunio* and *B. gelius*.

SECTION I. DESCRIPTION OF THE FIVE SPECIES

In all five species the sexes are similar in size and shape. The females tend to have a more rounded belly, the exact shape depending on the amount of ripe eggs carried. The general shape is streamlined and varies from more rounded to more rhomboid according to the species. The ground-colour varies from greyish to silvery. All scales, except those at the ventral side are thinly rimmed with black. Every species has its specific pattern of black spots and/or cross-bars on the body and tail. These patterns are common to both sexes. Green and golden colours are also mostly identical in both sexes. Males in all species tend to be more red than females, but the degree of difference varies with the species. Juvenile males are coloured liked females.

Black and red colours are caused by chromatophores which, by expansion or contraction, bring about changes in colour intensity. Green and golden colours are caused by iridocytes and are invariable in intensity.

1. *B. conchonius.*

In the laboratory this species attained a standard length of 5-6 cm, a total length of 6½-7 cm and a height of about 2½ cm. The shape is rather rounded.

The mouth is terminal and rounded. The body is moderately compressed laterally.

There is a small round to rhomboid black spot on either side of the tail above the anal fin. Immediately in front of this spot and partly surrounding it there is a small, roughly semilunar, golden area. This makes the spot look more or less like an eye. In some individuals there is a small area behind the head just below the lateral line where the black scale pigmentation is slightly stronger than on the rest of the body. The iris of the eye is greyish with a small darker spot both dorsally and ventrally of the pupil. In the males the iris is often reddish. There is a green iridescence on the back.

The females have a grey to silvery ground colour. Their fins are transparent except for a very thin black pigmentation of the hard rays of the dorsal, anal and ventral fins. In the males the entire head, body and tail become brightly red during reproductive activity. At the same time the frontal and distal parts of the dorsal, anal and ventral fins are pitch black and the caudal fin and the proximal parts of the dorsal, anal and ventral fins are rosy. In some individuals also the pectoral fins, the entire ventral fins and part of the ventral surface of the body become black. During reproduction the tail spot fades and often disappears completely. Outside reproductive periods the bodies of the males are more or less rosy, the proximal and distal ends of the dorsal, anal and ventral fins are greyish and the tail spot is black. Once adult, the males never loose the red colour of the bodies and the black on the fins completely. This makes them always easily distinguishable from females, also outside reproductive periods.

2. *B. nigrofasciatus.*

In the laboratory this species attained a standard length of 4½-5 cm, a total length of 5½-6½ cm and a height of 2-2½ cm. One female, which lived in the laboratory to an age of about 9 years, had a standard length of 5.5 cm, a total length of 6.7 cm and a height of 2.7 cm. The shape is rather rounded. The mouth is terminal or slightly sub-terminal and a little sharper than in *conchonius*. The body is more laterally compressed than in *conchonius*.

There are three black cross-bars on both sides of the body; the first just behind the base of the pectoral fin, the second below the dorsal fin and the third across the tail above the anal fin. The third bars of both sides of the tail are connected with each other both at the dorsal and the ventral side. The second bars reach to the base of the dorsal fin at their dorsal ends and are not connected at the ventral side. The first bars are not connected either at the ventral or at the dorsal side. Further there is a, less distinct, narrow bar running from the nape to just above the eye at both sides. Some individuals have a very thin black line along the base of the caudal fin. This line consists of melanophores situated on the bases of the caudal fin rays. The line is much more distinct in sub-adults (see ontogeny, section V) but it disappears gradually in most adults. The iris of the eye is greyish with a small dark spot both dorsally and ventrally of the pupil. In the male the eyes may have a reddish tinge. There are no golden colours in this species. In the males there is a green iridescence on the back. This is lacking in most adult females. Young females may have a weak iridescense, but it is always much weaker than in the males.

The females have a greyish ground colour. The proximal parts of the dorsal and anal fins are black and there is a large rounded black spot on the basis of each ventral fin. The rest of the fins are transparent. In full-grown males the entire body and tail become brightly red during reproduction. Often the red is distinctly more intense on the head and immediately behind it than further caudad. In young adult males only the head is red. During reproduction the entire dorsal, anal and ventral fins are pitch black; the pectorals may be greyish and the tail fin may have a reddish or blackish hue. At the same time the black bars on the body often fade partly or completely but, especially if a male not only courts but fights other males as well (see chapter III, p. 130) often a pitch black spreads over the entire body and tail covering the red colours except those on the head. This black pattern all over the body is typical of this species and does not occur in this intensity in any of the other species used in this study. Once adult, the males never completely loose the red colour on the head and body and the black colours on the fins. This makes them always easy to distinguish from females, even outside reproductive periods.

3. *B. tetrazona tetrazona.*

In the laboratory this species attained a standard length of about 4 cm, a total length of about 5 cm and a height of about 2 cm. The general shape is more rhomboid than in the two fore-going species. The mouth is terminal and comparatively sharp. The body is strongly compressed laterally.

There are four black cross-bars on both sides. The first runs from the nape to the top of the eye on both sides and for a short distance below the eye. These two parts merge with the comparatively strongly developed dark spots at the dorsal and ventral side of the iris to form one continuous bar. The second bar runs just behind the pectorals. The bars at both sides of the body are connected with each other at the dorsal side, but not at the ventral side. The third bar runs across the tail above the anal fin. The bars at both sides are connected both dorsally and ventrally. The fourth bar runs along the base of the tail fin. Its melanophores are not, as in *nigrofasciatus*, situated on the fin rays, but on the scales at the base of the fin (see ontogeny, section V). The bars at both sides are connected both dorsally and ventrally. The contrast of all four stripes with the background is very sharp, much sharper than in the fore-going two species. When the black of the bars fades, a bright green layer of iridocytes, situated below the melanophores, becomes visible. The iris of the eye, apart from the two black areas mentioned above, is greyish. There are no golden colours in this species. There is no green iridescence.

Both sexes have a shiny silvery back-ground colour. The black rims of the scales, which may become visible depending on the type of behaviour displayed, are usually comparatively inconspicuous. The proximal parts of the dorsal and anal fins are black. The distal parts of these fins and the entire ventral fins are red. In some males there are some black spots in the ventral fins. The females have a transparent tail fin. In the males this fin has two distinct red streaks which run roughly from the base of the fin to both tips. The snout is often reddish in both sexes. Except for the shape of the belly, which is often more rounded in females, the colour of the tail fin is the only reliable external character for sex determination outside reproductive periods. Though males are on the average more red than females, the intensities of all other red patterns may over-lap considerably. During reproduction the black body markings of both sexes often fade considerably; the fourth bar often disappears completely in males. At the same time the red colours of the males become more intensive, the ground-colour of the body becomes even more silvery and a faint rosy hue may spread over the entire body. On the other hand, the red colours of the females become less intense and often the black colours in the dorsal and anal fins fade considerably. This leads to a more distinct difference between the sexes than outside reproductive periods, but yet the differences never get by far as dramatic as they are in *conchonius* and *nigrofasciatus* (see also chapter III).

4. *B. stoliczkanus.*

In the laboratory this species attained a standard length of about 4 cm, a total length of about 5 cm and a height of 1½-2 cm. The shape is slightly more rounded than in *tetrazona* but more rhomboid than in *conchonius* or *nigrofasciatus*. The mouth is terminal and about as sharp in as *nigrofasciatus*. The lateral compression of the body is about as strong as in *nigrofasciatus*.

There are two black spots on both sides. The first, just behind the base of the pectoral fin, is small, covering about three scales. It is situated in the same area which in some individuals of *conchonius* is slightly more strongly pigmented than the rest of the body p. 2). The second spot, on the tail above the anal fin, is large and circular. It is considerably larger than the tail spot of *conchonius*. Like in *conchonius*, there is a small semi-lunar golden area in front of it. Like in *tetrazona*, the contrast between the black body markings and the back-ground is very sharp. When the black of the tail spot fades, a deeper layer of green iridocytes becomes visible. The iris of the eye is greyish, in the males often tinged with red. Like in the other species, there is a small darker spot to the dorsal and to the ventral side of the pupil. There is no green iridescence on the back.

Both sexes have a rather silvery ground-colour. In the females all fins are transparent, apart from a slight black pigmentation on the hard rays of the dorsal, anal and ventral fins. In some individual females there may be a weak rosy hue in the dorsal fin. The dorsal fin of the males is bright red with a number of black dots or bands. During reproduction the ventral fins of the males may become reddish, in some individuals with a few small black spots or streaks. At the same time the black body markings of both sexes may fade considerably. The entire body of the males becomes slightly yellowish. The colours of the dorsal fins of the males usually remain bright also outside reproductive periods. The colour of the dorsal fin is thus the best external character for sex determination both during and outside reproductive periods.

5. *B. cumingi*.

In the laboratory this species attained a standard length of about 3½ cm, a total length of about 4½ cm and a height of about 1½ cm. The shape is distinctly rhomboid. The mouth is terminal and pointed. The body is moderately compressed laterally.

There are two large black markings on both sides. The first, just behind the base of the pectoral fin, is elongate in dorso-ventral direction and intermediate in size (relative to the body size) between the pectoral spot of *stoliczkanus* and the pectoral bar of *nigrofasciatus*. The second spot, on the tail above the anal fin, is round and even larger than the tail spot of *stoliczkanus*. The tail spots of both sides of the tail are connected with each other at the dorsal side through a grey band. There is no ventral connection. Juveniles have a distinct semi-lunar golden area in front of the tail spot, but this is absent in adults. In adults the area just in front of the tail spot is usually lighter than the rest of the body. Below the dorsal fin, where *nigrofasciatus* has its second bar, there is a broad area in which the scale pigmentation is distinctly stronger than on the rest of the body, but no distinct cohesive bar is formed. The contrast between the black body markings and the background is not very sharp but about the same as in *conchonius* and *nigrofasciatus*. When the black colour of the tail spot fades, the green colour of a deeper layer of iridocytes becomes clearly visible. The iris of the eye is greyish in females and slightly yellowish in males. Like in the other species, there is a small darker spot both to the dorsal and to the ventral side of the pupil. There is no green iridescence on the back.

Both sexes have a greyish back-ground colour. As to the colours of the fins two forms could be distinguished. In one the fins of the females are transparent, apart from a slight black pigmentation of the hard rays of the dorsal, anal and ventral fins and a few tiny black dots in the dorsal. In some males the dorsal fin is distinctly yellowish with a number of small black dots and a very thin black rim. These colours are less well developed in other males, though on the average stronger than in the females. In the other form the dorsal and ventral fins of both sexes are slightly reddish with a few tiny black dots in the dorsal. In both forms the colours of the dorsal fins are (apart from he general shape of the body) the best external character for sex determination in this species outside reproductive periods. However, even this character is not always reliable. During reproduction the black body markings of the males often fade considerably. At the same time the whole body becomes slightly yellowish (or reddish in the second form) and often darker than normal. The colours of the dorsal fin become slightly more intense and also the caudal, anal and ventral fins become slightly yellowish (reddish or more intensely so in the second form). As the females change very little in colour during reproduction, the differences between males and females become more distinct during reproduction. Even so, however, the differences are slight (see also Chapter III, p. 161). It is not clear what the status of the two forms is. As they do not seem to differ in characters other than those described, they were used indiscriminately.

SECTION II. DISTRIBUTION AND HABITAT

The knowledge of the geographical distributions and ecological conditions of the species described in section I is incomplete. The following data seem to be relevant.

1. *B. conchonius.*

According to STERBA (1963, p. 284) this species occurs in N.-India, Bengal, Assam. HAMILTON (1822, p. 317-318: *Cyprinus conchonius)* found it "in ponds of the N.E. of Bengal, and in the rivers Kosi and Ami, which are far to the west. It probably, therefore, is found in all the northern borders of the Gangetic provinces." The Ami he describes as "a stagnant river full of weeds". He further writes: "In Bengal it has clear coloured fins, without spots or dots, and is greenish above, and silver coloured below; but the specimens which I saw in Behar had blackish fins; and those which I observed farther west, in the Ami... had the belly stained with black, while the fins of the back and the tail were yellow, with blackish ends; yet I am sure that the fishes which I observed in the three provinces belonged to the same species." This may suggest a considerable geographical variation in the species. However, all groups of *conchonius* used for this study were very homogeneous as described in section I.

2. *B. nigrofasciatus.*

According to STERBA (1963, p. 288) this species occurs in S.-Ceylon, in shallow, quiet-flowing waters. Other authors agree that it occurs in slow-flowing streams in the lowland of Ceylon (DERANYAGALA, 1930, p. 16-17: *Puntius nigrofasciatus*; MUNRO, 1955; LADIGES, 1951, p. 70; MENDIS & FERNANDO, 1962, p. 116: *Puntius nigrofasciatus*; confirmed by H. H. COSTA, *in litt.*, 1967).

3. *B. tetrazona tetrazona.*

According to STERBA (1963, p. 293) this species occurs in Sumatra, Borneo, (Thailand?). LADIGES (1951, p. 77) found *B. tetrazona partipentazona* in both turbid and clear waters in Sumatra. ALFRED (*in litt.*, 1967) states that it inhabits "clear or turbid shallow (*ca* 2 feet) moderately flowing streams with sandy (and rocky) substrata. Its distribution in Malaya suggests that it prefers hard waters with calcium carbonate exceeding 40 parts/million. Submerged vegetation may occasionally be present."

4. *B. stoliczkanus.*

According to STERBA (1963, p. 298) this species occurs in the basin of the lower R. Irrawaddy in Burma. According to our importers our specimens came from Thailand. This may explain the differences between our specimens and those present at the British Museum (see p. XI). No literature could be found referring to their type of habitat in Thailand.

5. *B. cumingi.*

According to STERBA (1963, p. 284) this species occurs in Ceylon, in mountain forest streams. Other authors agree that it inhabits mountain streams in the hill country of Ceylon (DERANYAGALA, 1930, p. 20-21: *Puntius cumingi*; MUNRO, 1955; LADIGES, 1951, p. 70; MENDIS & FERNANDO, 1962,

p. 116: *Puntius cumingi*; confirmed by H. H. Costa, *in litt.*, 1967). According to Costa the water in which it lives "is very clear except during heavy rains, when the water tends to be turbid due to the washed down silt from the hills. The depth of the water at normal times is about 1-2 feet. The bottom is usually sandy with a few stones here and there and is devoid of any plants. Neither are there any floating plants but there is a luxurient growth of grass and other macrophytes on either side of the banks."

In the laboratory all five species breed the whole year round (constant light and temperature conditions, see p. xi). It is possible that at least some of the species, under natural conditions, have a reproductive season forced upon them by a monsoon climate. According to Costa, most fishes in the hills of Ceylon spawn with the arrival of the monsoon rains, but it is not certain whether *cumingi* does the same.

SECTION III. COMPARISON OF THE ADULT COLOUR PATTERNS OF THE FIVE SPECIES

1. The black body markings.

The species differ from each other in the number and size of the black markings on the body. However, there are some similarities as well. All five species have a black marking on the tail above the anal fin. In *conchonius* and *stoliczkanus* it is a rounded spot, in *nigrofasciatus* and *tetrazona* it is a cross-bar and in *cumingi* it is more or less intermediate between a spot and a bar. In *nigrofasciatus* and in *tetrazona* the dorsal and ventral ends of the bar sometimes fade relative to the central area, thus leaving a rounded dark spot in the middle reminiscent of the tail spot in *conchonius* and *stoliczkanus.* More or less the same phenomenon may be seen in *cumingi* when the dorsal connection between the tail spots at both sides fades while the spots themselves remain dark. In all five species the tail mark has a deep velvet-like black colour. This is due to melanophores lying relatively deep in the skin in addition to more superficial ones. These deeply situated melanophores may be easily seen in live specimens under a binocular microscope at low magnification, especially if the more superficial melanophores have contracted a little. (see also ontogeny, section V).

All species, except *conchonius,* have a black marking just behind the insertion of the pectoral fin. In *conchonius* a few scales in this area are slightly more strongly pigmented than the others in some individuals (p. 2). In *stoliczkanus, nigrofasciatus* and *cumingi* the pectoral marking lacks the velvet-like hue of the tail marking. Only superficial pigmentation on the scales is present here. In *tetrazona* also deeper melanophores are present and the pectoral bar looks like the tail bar. (see also ontogeny, section V).

A black bar below the dorsal fin is present in *nigrofasciatus* and, in a rudimentary form, in *cumingi*. This bar is absent in the other three species.

A distinct bar on the head is only present in *tetrazona* and, less well developed, in *nigrofasciatus*. In the other three species there is some black pigmentation on the head above the eyes, but it isn't shaped into a distinct bar. (see also ontogeny, section V).

In all five species there is a more or less well developed rim of melanophores on the bases of the caudal fin rays. This is mostly more conspicuous in sub-adults than in full-grown adults (see also ontogeny, section V). Only in *tetrazona* there is a distinct black bar on the tail itself, at the base of the caudal fin.

2. Other colour patterns.

There are a number of similarities between the colour patterns of *conchonius* and *nigrofasciatus*. In both species the females have no red colours at all, whereas the males are coloured brightly red during reproduction. At the same time the males of both species have pitch black dorsal, anal and ventral fins. In both the females have a rather greyish back-ground colour compared to the more silvery colour of *tetrazona* and *stoliczkanus*. The males never completely loose their red colour on the body. In both species at least the males have a green iridescence on the back. When the black of the body markings fades, little or none green colour is shown below. The contrast between the black body markings and the back-ground is not very sharp.

Tetrazona and *stoliczkanus*, on the other hand, resemble one another in having a much less spectacular sexual dichromatism than the fore-going two species, in having a more silvery back-ground colour and in lacking the green iridescence on the back. In both species the emphasis of fin colouration is on the dorsal fin and in both this fin is coloured vividly black and red at least in the males and most brightly so during reproduction. In *tetrazona* the red colour of the ventral fins of the males, which is continuously present, becomes brighter during reproduction; in *stoliczkanus* the normally transparent ventral fins of the males become reddish during reproduction. In both species there are some males which in addition have a few small black marks in the ventrals. When the black of the body markings fades a bright green colour becomes visible in both species and both have a very sharp line of contrast between the black body markings and the back-ground.

In both pairs of species there is one species with fewer and smaller black body markings and one with more and larger ones. In both pairs the species with the least black markings has a golden area in front of the tail spot. When the species with comparable amounts of black markings on the body are compared, it is clear that *stoliczkanus* has a more rhomboid shape and is more strongly laterally compressed than *conchonius*; the same differences exist between *tetrazona* and *nigrofasciatus* respectively.

Cumingi differs from either of the two pairs in the combination of characters. It is like *conchonius* and *nigrofasciatus* in the greyish back-ground colour and in the lack of a sharp line of contrast between the black body markings and the back-ground. It is like *tetrazona* and *stoliczkanus* in the presence of a layer of green iridocytes under the black body markings and in the absence of a green iridescence on the back. Its sexual dichromatism is even less pronounced than in *tetrazona* and *stoliczkanus*. As far as fin colouration is present, the emphasis is on the dorsal fin, like it is in *tetrazona* and *stoliczkanus*, but there are little or no red colours and little black in the dorsal (see also section VIII).

It will be shown in chapter III that a comparison of the behaviour of the five species leads to a similar classification.

SECTION IV. INTERSPECIFIC HYBRIDS

The following interspecific hybrids were bred:

	number
B. conchonius x *B. stoliczkanus* 1)	17
B. cumingi x *B. stoliczkanus*	>140
B. stoliczkanus x *B. cumingi*	7
B. nigrofasciatus x *B. stoliczkanus*	±20
B. stoliczkanus x *B. nigrofasciatus*	±20
B. cumingi x *B. nigrofasciatus*	11
B. nigrofasciatus x *B. conchonius*	>60

All hybrids are roughly intermediate in general shape between the parental species. As all hybrids were males, at least phenotypically (see p. 39), only male colour patterns are known. The colour patterns on the body are also intermediate between those of the parental species. Reciprocal hybrids were practically identical in shape and colour patterns. The black body markings of some hybrids at sub-adult to adult stages are represented in figg. 20-23, p. 32. The black body markings of the *B. nigrofasciatus* x *B. conchonius* hybrids, which are not represented in the figures, are almost identical to those of *B. stoliczkanus*.

The *B. nigrofasciatus* x *B. conchonius* hybrids are brightly red on the entire bodies during reproductive behaviour; the *B. conchonius* x *B. sto-*

1) According to a common use the name of the female parent is given first.

liczkanus, *B. nigrofasciatus* x *B. stoliczkanus*, *B. stoliczkanus* x *B. nigrofasciatus* and *B. cumingi* x *B. nigrofasciatus* hybrids are all more or less reddish on the bodies; the *B. cumingi* x *B. stoliczkanus* and *B. stoliczkanus* x *B. cumingi* hybrids were not red at all on the bodies.

It is difficult to judge the intensity of the green iridescence on the back, but green is absent in the *B. stoliczkanus* x *B. cumingi* and the *B. cumingi* x *B. stoliczkanus* hybrids. At least some green is present in all other hybrids.

A bright golden area in front of the tail marking is present in the *B. cumingi* x *B. stoliczkanus* and the *B. stoliczkanus* x *B. cumingi* hybrids. Some golden colour is present in all other hybrids (also in *B. cumingi* x *B. nigrofasciatus*, though it is present in sub-adult *cumingi* only).

It must be emphasized that in the hybrids considerable changes in the intensities of the colour patterns may occur as they grow older, even long after they have reached adulthood. The colour patterns of the fins, which do not always seem to follow the rule of intermediacy, won't be discussed here.

SECTION V. THE ONTOGENY OF THE BLACK BODY MARKINGS IN THE FIVE SPECIES AND IN THE HYBRIDS

Introduction.

All five species were bred in the laboratory. A great number of young fishes of various stadia — from just free-swimming larva to sub-adult — were observed under a binocular microscope and detailed drawings were made. Special attention was given to the development of black pigmentations. The occurrence of red, green and golden colours was not consistently studied. No study was made of the formation and migration of the black pigment cells themselves.

Young *nigrofasciatus* of several stages of development were cut serially in the transverse plane at 7μ. Fixation was in 4% formalin; the sections were stained with hemalum-eosin.

Successive stadia of all five species are represented in figg. 1-5. The first figure of each series represents a larva during the first days of free-swimming life. The last figure represents a sub-adult. In this last stage the species-specific pattern of black markings on the body is complete.

Since the younger stages are transparent, also internal structures and their pigmentation may be easily observed. The internal black pigmentations are represented in figg. 6-10. The successive stages in these figures are the same as those in figg. 1-5 (sub-adults are omitted). Each drawing represents a specific individual and not an idealized abstraction of a number of individuals of the same stage. Individual differences within each species are very small.

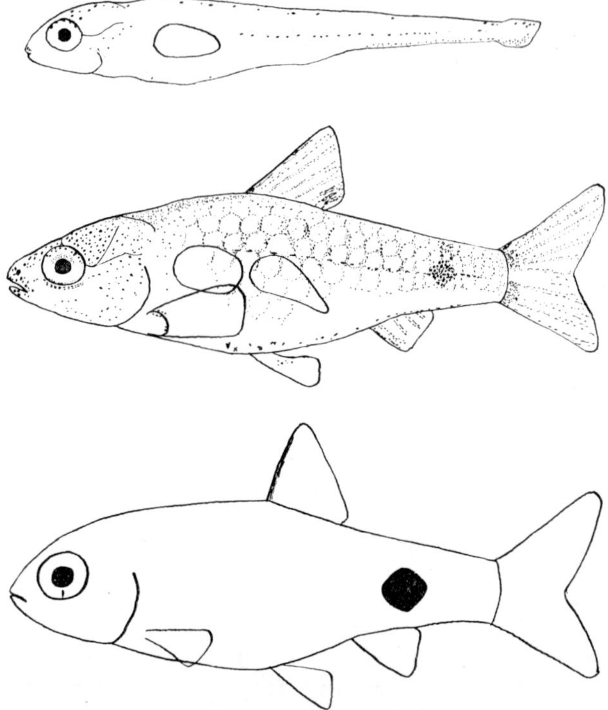

Fig. 1. *B. conchonius*; young of (standard length) 4.2/10.0 mm., and sub-adult; superficial markings.

In the drawings of the more superficial structures (figg. 1-5) the rough contours of the skull and the swimming-bladder are drawn as points of reference.

Measures in this section are given in standard length only.

1. *B. conchonius*.

Larvae become free-swimming when about 4 mm long. The pectoral fins are well-developed by then; the tail fin is rudimentary and the other fins have not yet differentiated. The swimming-bladder is undivided and filled with air.

At this stage the eyes are pigmented, especially at the dorsal side. There are some melanophores on the snout and on the skull. A distinct longitudinal line of melanophores has developed on the side, about where the lateral line system is going to be situated. When the lateral line develops some of the melanophores get associated with it. Further there is a less distinct row of

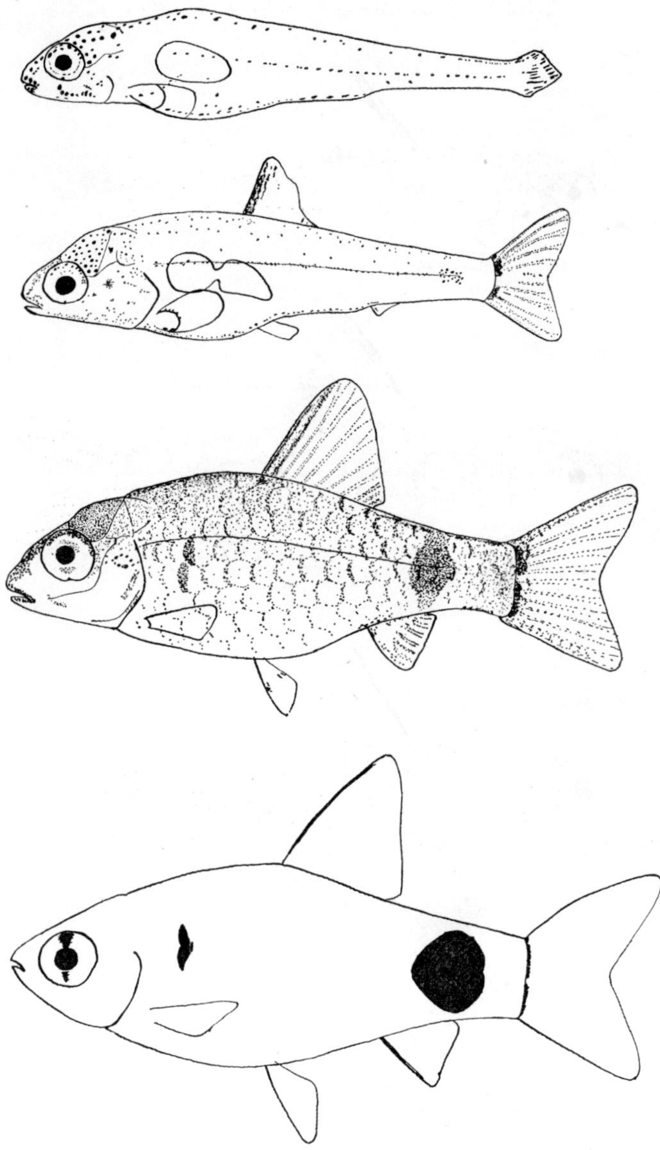

Fig. 2. *B. stoliczkanus*; young of 5.1/6.8/15.2 mm, and sub-adult; superficial markings.

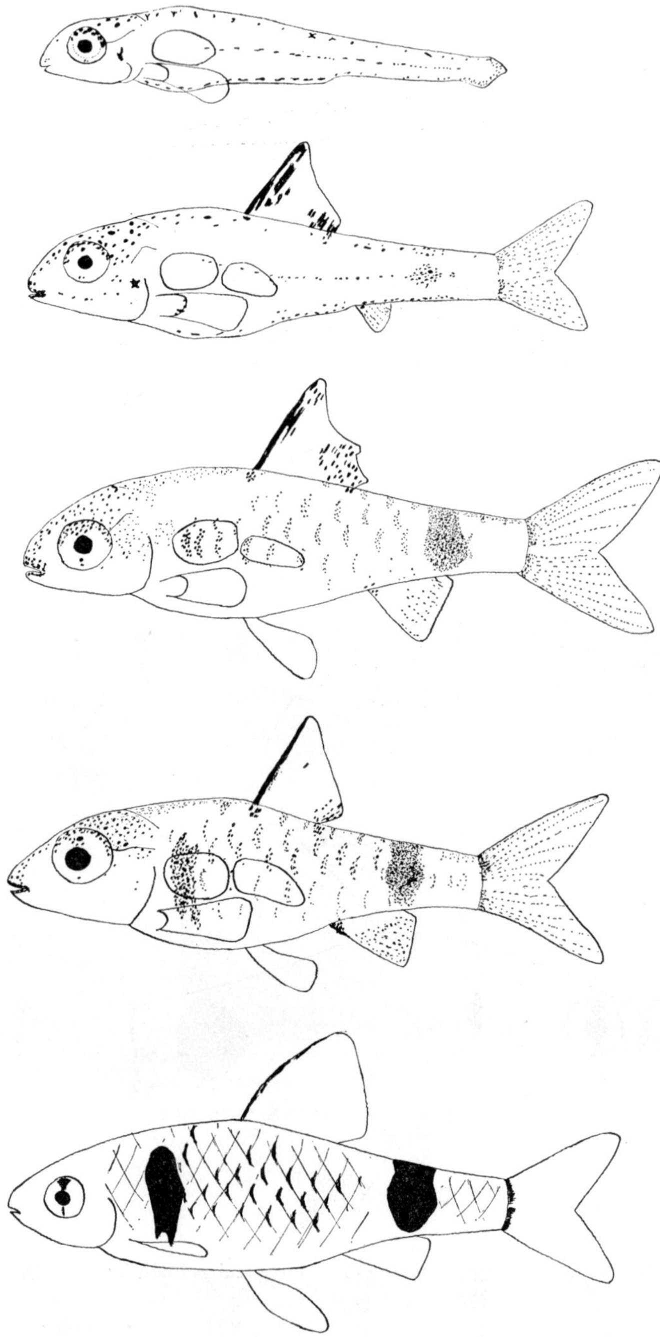

Fig. 3. *B. cumingi;* young of 4.2/6.5/9.1/10.0 mm, and sub-adult; superficial markings.

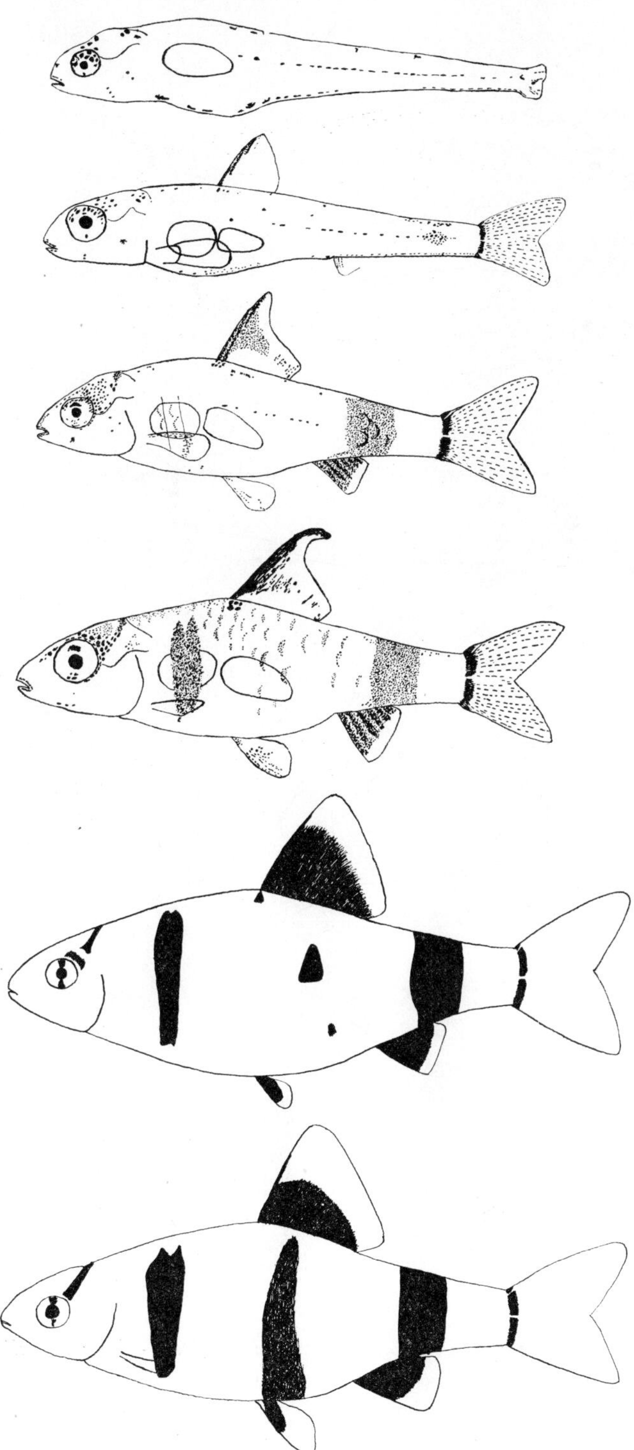

Fig. 4. *B. nigrofasciatus*; young of 4.1/6.8/9.7/11.1/19.3 mm, and sub-adult; superficial markings.

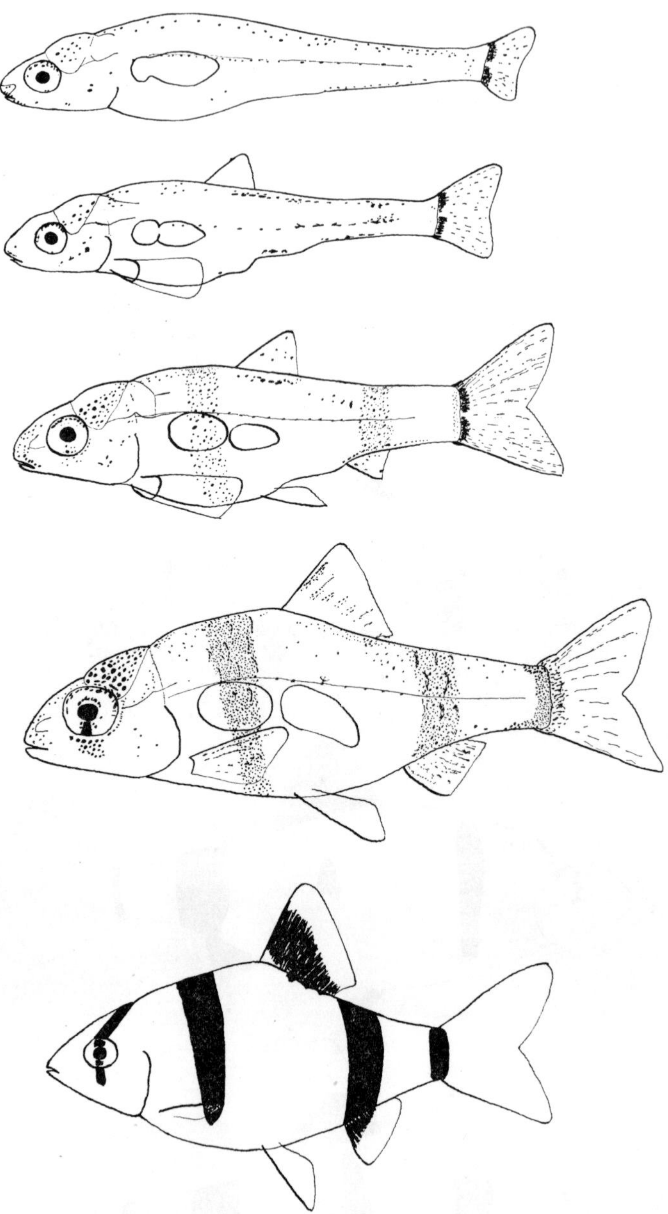

Fig. 5. *B. tetrazona*; young of 5.2/6.0/7.3/10.1 mm, and sub-adult; superficial markings.

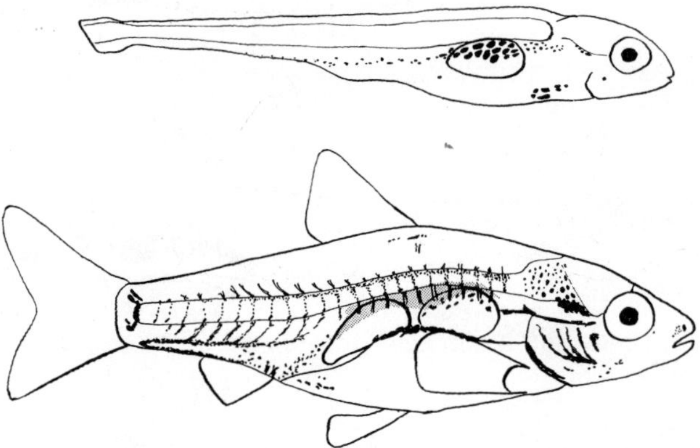

Fig. 6. *B. conchonius*; same young as fig. 1, no sub-adult; internal pigments.

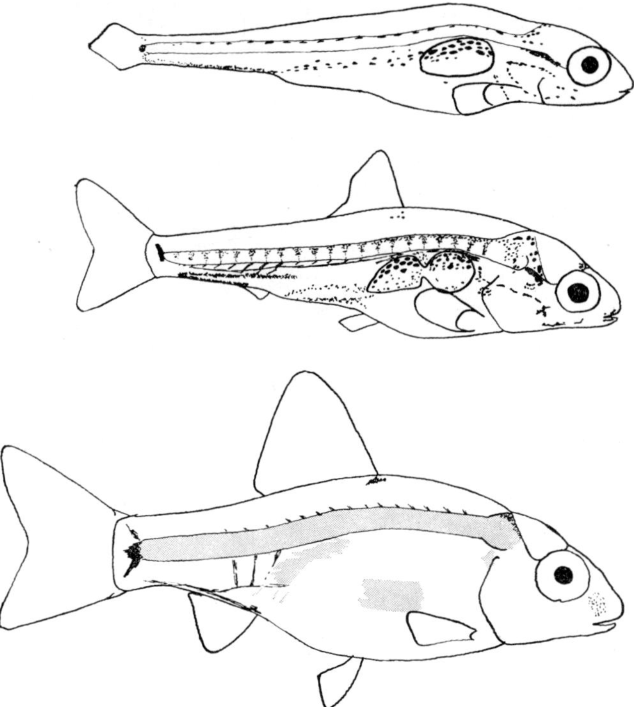

Fig. 7. *B. stoliczkanus*; same young as fig. 2, no sub-adult; internal pigments.

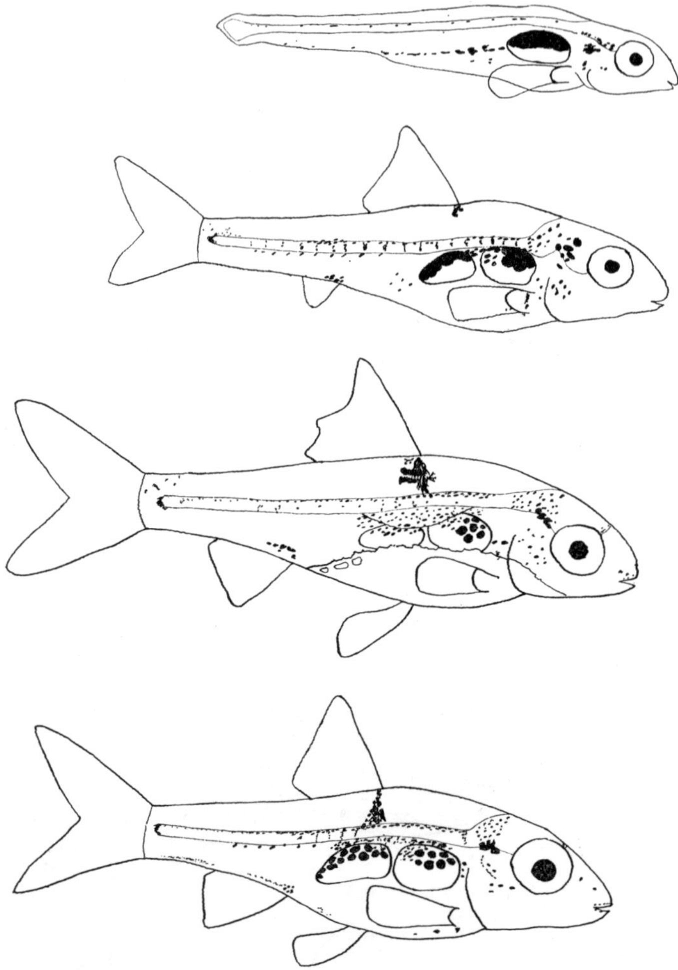

Fig. 8. *B. cumingi*; same young as fig. 3, no sub-adult; internal pigments.

melanophores on the back on both sides of the dorsal median, a string of melanophores along the ventral side of the tail and a few dispersed melanophores on the belly.

Internally there is a very distinct cap of melanophores on top of the swimming-bladder, some more melanophores on the skull, a concentration of large melanophores around the heart and a string of melanophores running from the head to the swimming-bladder and from the swimming-bladder to the anus. This string merges with the superficial melanophores on the ventral side of the tail to form one continuous strand.

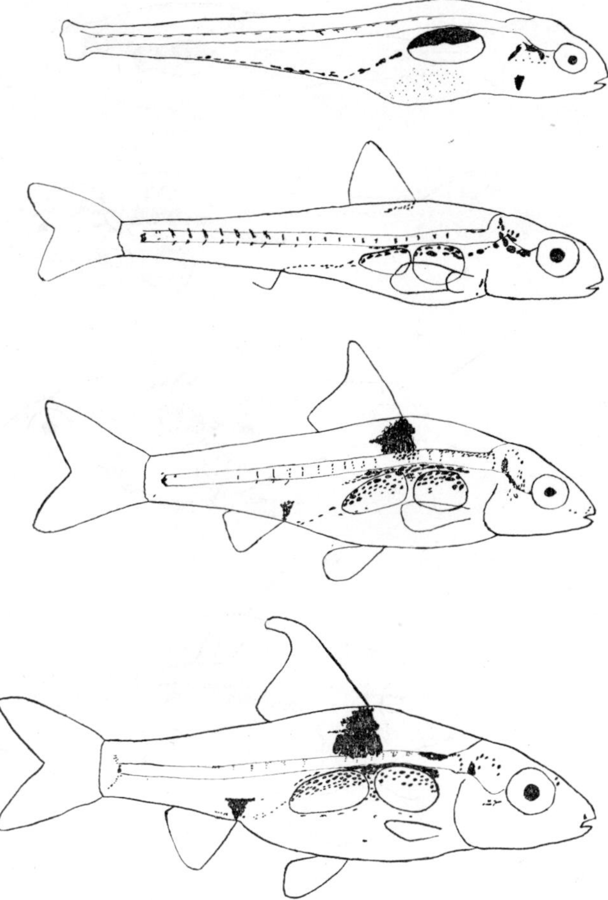

Fig. 9. *B. nigrofasciatus*; same young as fig. 4, last stage and subadult omitted; internal pigments.

As the larva grows the anterior chamber of the swimming-bladder becomes filled with air, the fins develop and become pigmented along the rays and a weak though distinct concentration of melanophores appears at the bases of the caudal fin rays. This concentration is divided in a dorsal and a ventral half. All pigmentations grow gradually. Internally the gills, the spine with its dorsal and ventral processes and the peritoneum also become pigmented. Near to the bases of the dorsal and anal fins a number of large irregular melanophores appear around the median plane.

At a length of about 7 mm the tail spot starts to develop, melanophores first appear both dorsally and ventrally of the lateral line and very close to it.

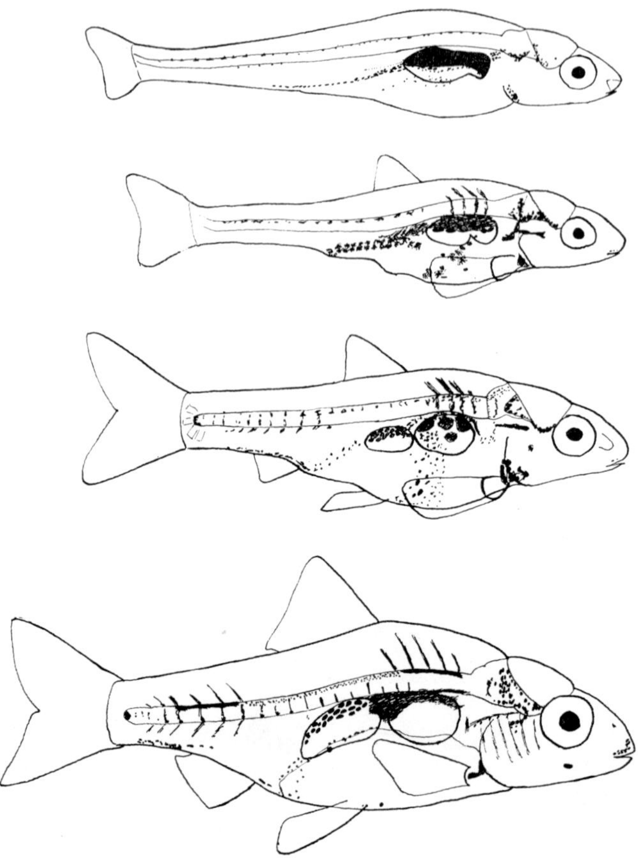

Fig. 10. *B. tetrazona*; same young as fig. 5, no sub-adult; internal pigments.

As more melanophores are added a rhomboid spot gradually develops around the lateral line. At about the same time the scale pigmentation begins to develop, starting in the tail region and rapidly spreading over the entire body and tail except the ventral side. Melanophores first appear near to the bases of the scales and then spread around the rims. The scale melanophores have a relatively superficial location in the skin. The melanophores of the tail spot, on the other hand, are situated much deeper in the skin, probably below the level of the scales; the structure of the scales is in no way reflected in the structure of the spot. Later also the pigmentation of the scales on top of the tail spot becomes stronger than that of the surrounding scales, thus intensifying the black colour of the spot, but the deeper melanophores remain always visible giving the spot its velvet like hue (*cf.* p. 8).

Soon after the tail spot and the scale pigmentation have developed, the fishes become opaque and any further development of the internal pigmentations can no longer be observed directly. At the outside there are no essential changes in pigmentation until the fishes are several cmm. long and the males begin to develop nuptial colour patterns.

As stated above the development of the golden and green colours was not studied in detail, but the green iridescence on the back and the golden area in front of the tail spot are at least present well before the adult stage is attained.

2. *B. stoliczkanus, B. cumingi* and *B. nigrofasciatus.*

B. stoliczkanus, B. cumingi and *B. nigrofasciatus* are very similar to each other and to *B. conchonius* in the development of the black marking patterns. Young free-swimming larvae of all four species look very much like each other in the distribution of both superficial and internal black pigmentations. In *stoliczkanus, cumingi* and *nigrofasciatus* the first of the external black body markings to develop is a tail spot. This spot is formed in exactly the same way as in *conchonius*. However, whereas in *conchonius* the development of body markings comes to an end when the tail spot has been formed, the other three species develop a black marking in the pectoral region as well and *nigrofasciatus* (and *cumingi*, though in a rudimentary form) develop a central bar below the dorsal fin. In addition, the tail spots of *nigrofasciatus* and, less completely, of *cumingi* grow into a vertical bar.

The pectoral mark of *stoliczkanus, cumingi* and *nigrofasciatus* and the central marking of *cumingi* and *nigrofasciatus* are all formed through an intensification of the normal scale pigmentation. The melanophores, like those of the normal scale pigmentations, are located in relatively superficial layers of the skin. At no stage of development the pectoral and central markings contain more deeply located melanophores, which are typical for the tail spot. The difference in location of the melanophores of the pectoral marking on the one hand and the tail marking on the other may be observed very clearly in the cross-sections of young *nigrofasciatus* of 9.0 mm or more. Whereas in the pectoral mark there is only one layer of melanophores, situated to the outside of the scales, in the tail marking two distinct pigment layers are present, one at the outer surface of the scales and one at the inner surface. In all three species the formation of the pectoral marking begins in a few scales which are situated just below the lateral line (the same location where some individual *conchonius* show a slight intensification of scale pigmentation when adult). In *stoliczkanus* the pectoral mark remains confined to about three scales; in the other two species it becomes extended in both ventral and dorsal direction to form a vertical bar.

The internal pigmentations, which are almost identical in all four species during the youngest stages, grow progressively stronger during the later stages in the order: *conchonius, stoliczkanus, cumingi, nigrofasciatus.* This is especially clear in the internal pigmentations near to the bases of the dorsal and anal fins (see figg. 6-10). (In judging the intensity of the other internal pigmentations in these figures it must be emphasized that during the later stages the increasing opaqueness of the fishes interferes more and more with the visibility of the internal structures). The same differences in intensity of black pigmentation between the species are apparent in the dorsal fin and in the pigment concentrations on the bases of the caudal fin rays.

The rates of development of the black body markings and of the scale pigmentation, relative to body length, were studied in some more detail. To this purpose a large number of young of *conchonius, stoliczkanus, cumingi* and *nigrofasciatus* were examined under a binocular microscope; the stage of development of their black pigmentations was assessed and their length was measured.

Body length was measured as follows. The fishes were put in a watch-glass containing a few drops of water. This was placed upon a piece of paper on which two slowly diverging pencil lines had been drawn. The glass was shifted until the position was found in which the young fitted exactly between the two lines. This point was marked on the paper. The distance between the lines was measured later by means of a mm-scale. Care was taken that the measurement was not influenced by the refraction of the water in the watch-glass. Though this may seem a rather crude way of measuring, the fact that the successive stages show very little overlap in length indicates that it was very accurate, at least as a relative measure.

The results are represented in figg. 11-14. The length of the fish is plotted on the ordinate. On the abscissa a number of successive developmental stages are plotted at (arbitrarily) equal distances. Two stages in the development of the tail spot have been distinguished on a rather arbitrary criterium: one in which at least a few melanophores of the spot are present and one in which there are at least a sufficiently large number of melanophores to give the impression of a complete spot. One further stage has been added before the development of the tail spot: where the tail fin becomes distinctly forked as opposed to its trunked shape before. The data of the graphs are from many different individuals, though some individuals have probably been used several times at different stages of development.

In general there is very little overlap in length between the successive stages. This indicates a close correlation between length and stages of

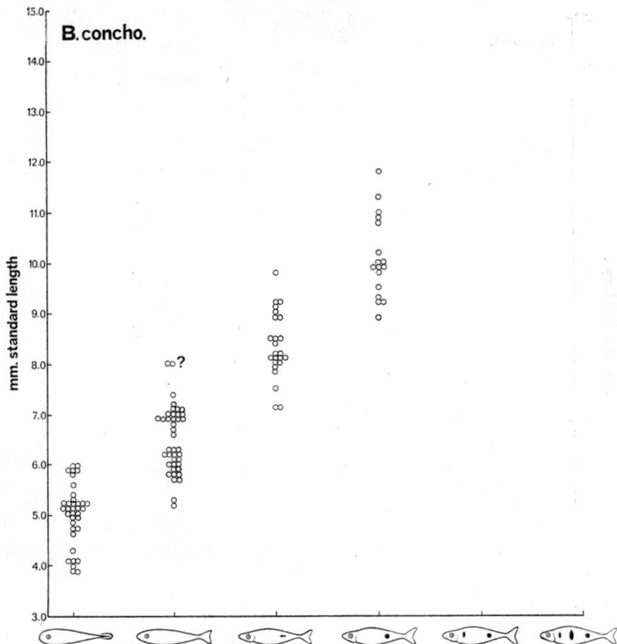

Fig. 11. *B. conchonius*; relation between development of black markings and size.

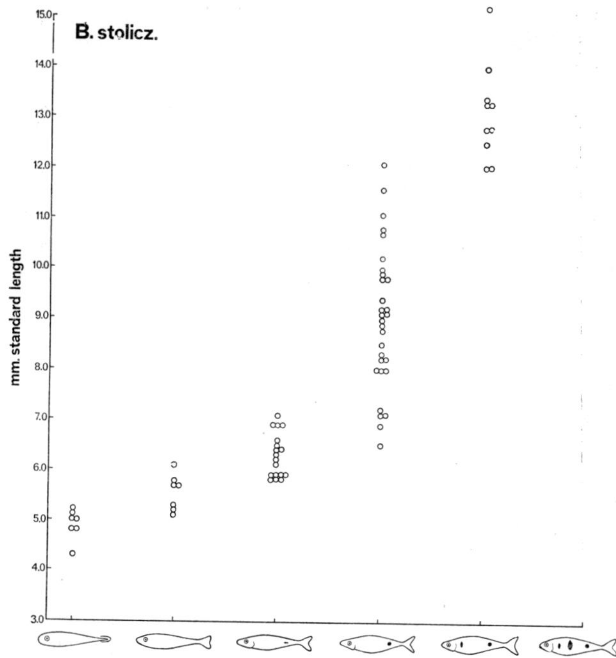

Fig. 12. *B. stoliczkanus*; relation between development of black markings and size.

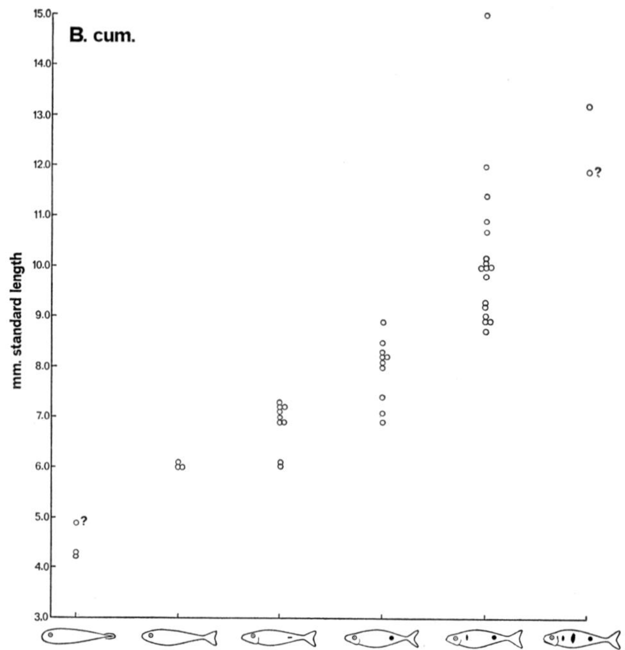

Fig. 13. *B. cumingi*; relation between development of black markings and size.

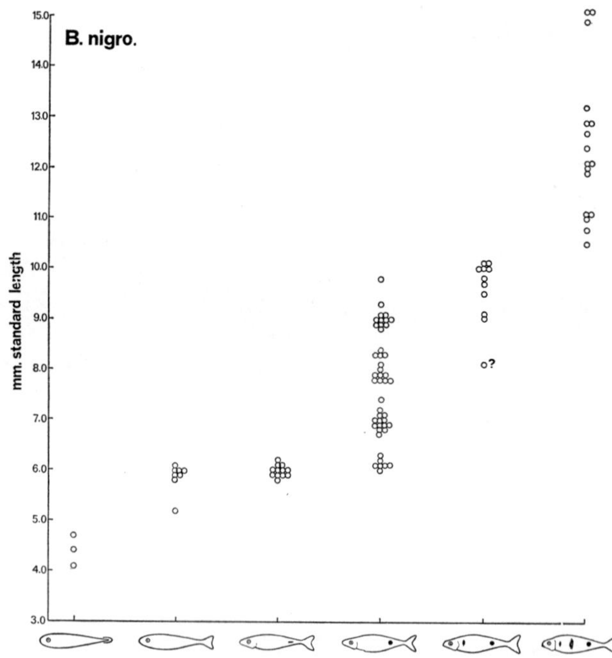

Fig. 14. *B. nigrofasciatus*; relation between development of black markings and size.

development. The ages of the young fishes used are not exactly known. Usually young of a number of different ages were present in the same tank. Yet I am sure that in some tanks young grew much faster than in others. This means that young with the same body length may have different ages. The fact that, in spite of the differences of growth rates of the individuals used, a close correlation between length and developmental stages is found suggests that it is length and not age which is the decisive factor in the development of the black markings. It must be added that if growth is *very* much retarded, the black markings may develop even though the typical body length has not been reached. In these cases, however, there usually are considerable deformations, especially in the head region, which make the measurement of length uncertain. Such individuals have not been included in the graphs.

In fig. 15 the data from the four species are combined. Each point in this graph represents the minimum length found for each stage. We may assume that this is the length at which the stage begins in all individuals. From the graph an interesting phenomenon becomes apparent. The tail spot develops later (at a greater body length) in *conchonius* than in the other three species. The pectoral spot develops later in *stoliczkanus* than in *cumingi* and in

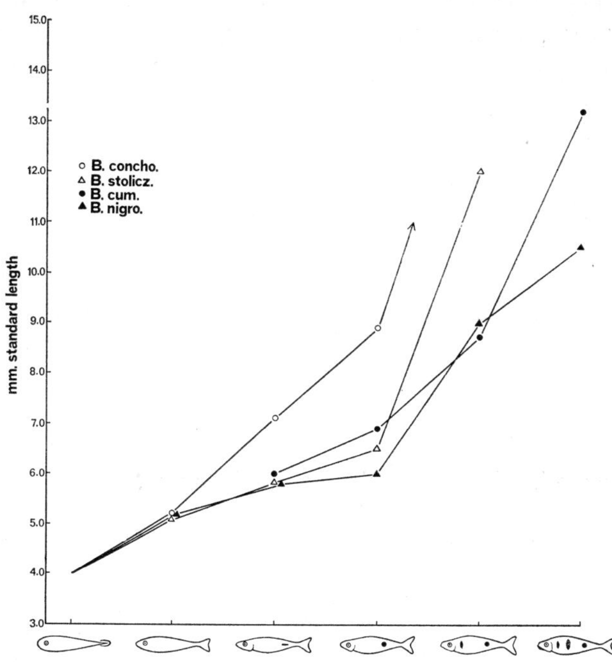

Fig. 15. Relation between development of black markings and size; 4 species.

nigrofasciatus [1]) and the bar-like pigmentation below the dorsal fin comes later in *cumingi* than in *nigrofasciatus*. It is not clear exactly when this last structure starts its development in *cumingi* (because of its dispersed nature) but it is certainly not yet there at the time that the middle bar of *nigrofasciatus* starts to develop. Apparently the development of the marking pattern is slower the fewer markings a species has when adult. This phenomenon is clear only for the last marking that develops in each of the species. To this we may add that in those individual *conchonius* in which there is rudimentary pectoral mark it develops very late, probably not before adulthood is attained.

This phenomenon, in combination with the fact that also internal pigmentations and fin pigmentations are stronger in the order *conchonius, stoliczkanus, cumingi, nigrofasciatus,* and in combination with the fact that there are almost no pigmented areas in *nigrofasciatus* (the most heavily pigmented species) which are not at least slightly pigmented even in *conchonius* (the least pigmented species) suggests that the differences in black pigment patterns between the species are of a quantitative rather than of a qualitative nature. It seems possible that the differences are due to one species just producing more melanin and/or starting the production earlier than another species. This hypothesis might be directly tested through the administration of extra Melanin Stimulating Hormone to for instance *B. conchonius* from an early age and see whether a *cumingi*- or *nigrofasciatus*-like pattern of black body markings ensues.

There seems to be at least one notable exception to the general rule. The pigmentation of the scales develops at approximately the same length in all four species. This is demonstrated in figg. 16-19. In these graphs, for each species, a number of rather arbitrarily chosen phases in the formation of scale pigmentation are plotted along the abscissa at (arbitrarily) equal distances. The phases are distinguished from each other as to the proportion of the body surface that is covered by pigmented scales. The lengths of the fishes is plotted along the ordinate. Two things are apparent from the graphs: (a) once scale pigmentation has begun, it very rapidly spreads over the entire body; (b) the length at which scale pigmentation develops is roughly the same in all four species (around 8 mm).

The synchrony of the formation of scale pigmentation in all four species seems to plead against the hypothesis offered above. However, in evaluating

1) It is not always easy at first sight to fix the startingpoint of a marking which arises from an intensification of the normal scale pigmentation. The criterium used was whether the pigment of at least 2 or 3 scales was distinctly stronger than that of the surrounding scales.

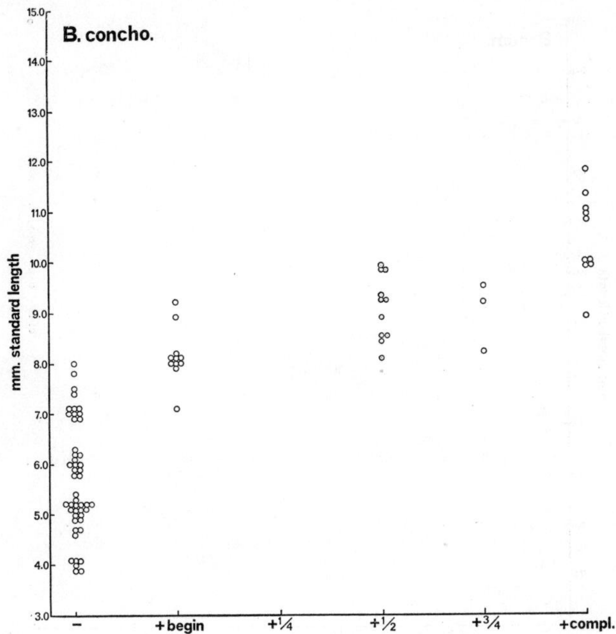

Fig. 16. *B. conchonius*; relation between development of scale pigmentation and size.

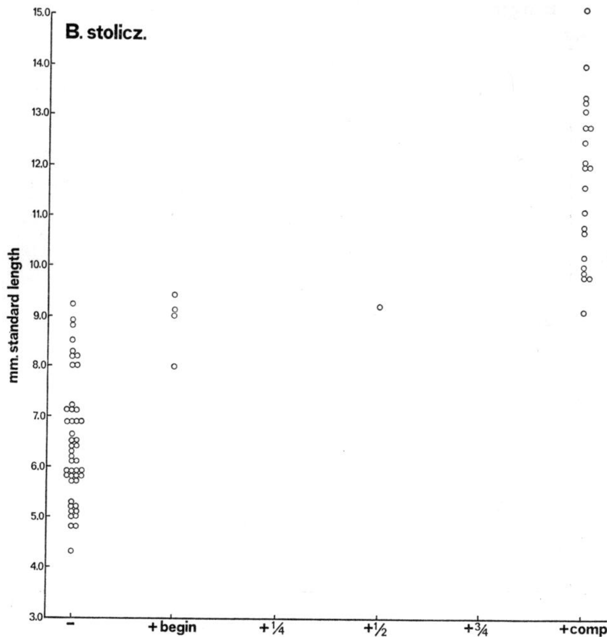

Fig. 17. *B. stoliczkanus*; relation between development of scale pigmentation and size.

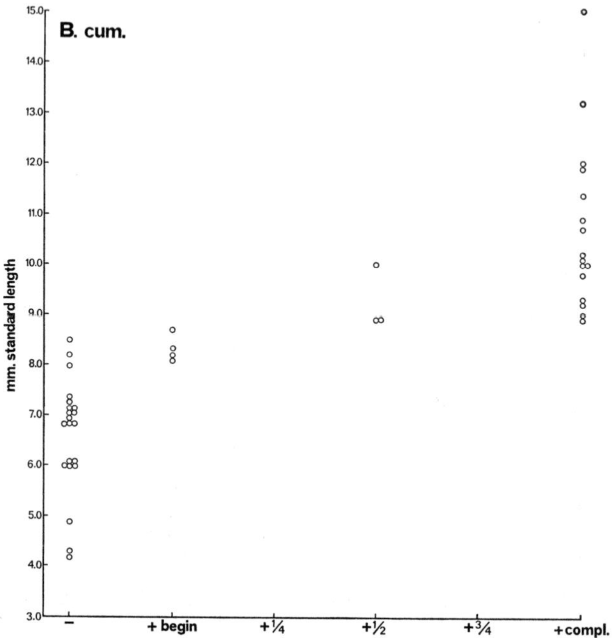

Fig. 18. *B. cumingi*; relation between development of scale pigmentation and size.

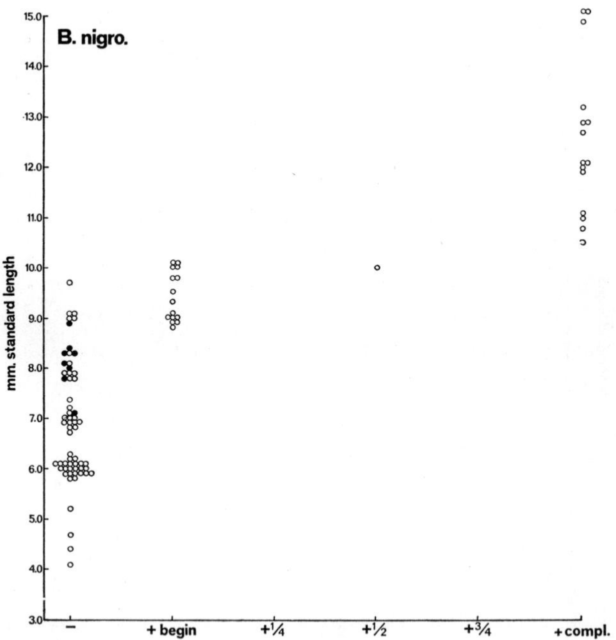

Fig. 19. *B. nigrofasciatus*; relation between development of scale pigmentation and size.

this point the development of the scales themselves will have to be taken into account. The development of the scales was studied in the serial cross-sections of young *nigrofasciatus* mentioned earlier in this chapter (p. 11). Completely developed bony scales at all sides of the body and the tail were found in a fish of 9.0 mm long (+ scale pigmentation ½) and in a larger one (+ scale pigmentation complete). Bony scales were also found on the lateral sides of body and tail in a fish of 8.4 mm (— scale pigmentation), but here the scales were apparently still in process of being formed from the corium, their outer surfaces in very close contact with the latter. Very thin but distinct bony plates were even found along the lateral sides of body and tail in a fish of 7.0 mm but here they were still very rudimentary. Since the melanophores forming the scale pigmentation are apparently situated on the outer surface of the scales, it seems plausible that they cannot appear before the formation of the scale has been completed. Therefore I conclude that the synchrony in the formation of scale pigmentation in all four species is due to a synchrony in the rates of development of the scales themselves, rather than to pigment production.

One disturbing phenomenon has still to be discussed. Scale pigmentation is often present right on top of the tail spot well before the normal scale pigmentation on the rest of the body and tail starts to appear. Such pigmented scales are clearly visible in the third stage of *nigrofasciatus* in fig. 4. In fig. 19 black dots in the column marked "— scale pigmentation" indicate some individuals in which such pigmented scales on top of the tail spot were clearly visible. Scale pigmentation in the tail spot could easily be recognized in the cross sections of the 8.4 mm young *nigrofasciatus*. In some of the cross-sections in the tail-spot area of this individual there was a distinct layer of melanin on the outer surface of the scales in addition to the (more continuous) layer to the inside of the scales.

This phenomenon seems to contradict the assumption made above that melanophores cannot intrude between the bony scales and the inner corium layers before the scales have attained some state of completion; the more so because, as judged from the cross-sections, the scales in the tail-spot region do not differ in differentiation from those on the rest of the body. However, there is an important difference between the pigmentation of the scales in the tail spot on the one hand and the normal scale pigmentation on the other. Whereas in normal scale pigmentation melanophores start to appear near the bases of the scales and only later spread along the rims, in the tail-spot region the caudal edges of the scales are pigmented and the bases are not. (The same phenomenon may be seen in the pectoral and tail bar of young *tetrazona*; see fig. 5). It is conceivable that the thin caudal edge of each scale is completed earlier than the rest of the scale and that, in regions where much pigment is being formed below the scales, some pigment cells migrate around the hind edge of the scale and intrude between the outer surface of the scale and the corium (see also *tetrazona* p. 31).

It must be emphasized that data in this section are on larval up to sub-adult stages only. It is obvious that the conclusions do not apply to the formation of reproductive black patterns in the adults.

3. *B. tetrazona tetrazona.*

The ontogeny of this species is discussed separately because the development of its colour patterns differs in some respects from those of the other four species. Only a small number of young of this species were reared and too few measurements are available to assess precise correlations between successive stages in the development of the colour patterns on the one hand and body length on the other.

Tetrazona larvae become free-swimming at about the same size as do the other four species (± 4 mm). At this stage their patterns of black pigmentation are very similar to those of the other species. One obvious difference is that the pigment concentrations on the bases of the tail fin rays are already relatively strong at an early age. A tail spot (later developing into the first tail bar) develops in the same way as it does in the other species and at about the same body length (± 6 mm). Like in the other species it consists of melanophores situated in the deeper layers of the skin. It gradually grows into a cross-bar consisting entirely of deeply situated melanophores (like the tail bar of *nigrofasciatus* and *cumingi*).

So far the development of the patterns of black pigment doesn't differ essentially from those of the other species. However, the pectoral bar in *tetrazona* develops in a quite different manner. The first melanophores of the pectoral bar appear at the dorsal side of the body as early as the 6.0 mm stage (fig. 5 stage 2). From there they gradually spread in ventral direction. At a body length of 7.2 mm (that is to say well before the pectoral marking appears in any of the other species) a great number of melanophores lie scattered over the entire area that is going to be covered by the pectoral bar (fig. 5 stage 3). All of these melanophores, like those of the first tail bar, are situated in the deeper layers of the skin. In contrast to the tail bar, there is no tendency at all for the melanophores to concentrate around the lateral line. At the same stage (7.2 mm) melanophores have appeared at the end of the tail near to the base of the tail fin: the beginnings of the second tail bar. Also these melanophores lie deep in the skin and also here there is no concentration around the lateral line.

At a length of *ca* 10 mm (fig. 5 stage 4) when there is no sign yet of scale pigmentation on the rest of the body, the hind edges of the scales on the pectoral and the first tail bar are heavily pigmented. The same kind of pigmentation appears on the second tail bar later on. The black pigment on the bases of the tail fin rays is then gradually covered by the pigmented scales growing in caudal direction. The pigmentation of the hind edges of the scales in the regions of the cross-bars (at a stage when normal scale pigmentation is not yet there) is strongly reminiscent of similar scale pigmentation on the tail marking of *nigrofasciatus* and the other species (p. 30). Its presence supports the hypothesis proposed to account for its origin in those species.

At the 10 mm stage also the eye-bar is present in a rudimentary form, the dorsal part being "shaped" out of the normal pigmentation on the skull, the part ventrally of the eye consisting of deeply situated melanophores.

The internal pigmentations of *tetrazona* are essentially the same as in the other species except for two points:

(a) In the zones of the pectoral and the first tail bar all internal pigmentations are much more strongly developed than in the surrounding areas (fig. 10 stage 3 and 4). This phenomenon is not present in any of the other four species.

(b) The internal clots of black pigment near to the bases of the dorsal and anal fins are absent.

The special features of the ontogeny of the more superficial black patterns in *tetrazona*, as compared with those of the other species, may be summarized as follows:

(c) The early and strong development of black pigment on the bases of the caudal fin rays;

(d) The early and different type of development of the pectoral marking;

(e) The presence of a sub-orbital part of the eye-bar;

(f) The presence of a second tail bar which, in the course of its development, covers the original pigmentation on the bases of the caudal fin rays.

It seems possible that some of the special features of marking development in *tetrazona* (especially (c), (d) and perhaps (a)) may be related to a functional demand for a clear-cut contrast pattern at an early age, perhaps in connection with the species' relatively strongly developed schooling tendencies (see chapter III, p. 147). It is not clear, however, how feature (b) could be explained in this way. Features (e) and (f) do not seem to be specially related to young stages (see further chapter IV, section II B).

4. *Hybrids.*

Detailed drawings of ontogenetic stages have been made of the hybrids: *B. conchonius* x *B. stoliczkanus*, *B. cumingi* x *B. stoliczkanus*, *B. cumingi* x *B. nigrofasciatus* and *B. stoliczkanus* x *B. nigrofasciatus*. These are represented in figg. 20-23 and figg. 24-27. which are set up in the same way as those of the pure species.

It may be seen from these figures that the development of black pigment patterns in the hybrids is qualitatively similar to those of the parental species. The intensity of the pigmentations seems to be roughly intermediate to those of the parental species at most stages. Too few individuals have been measured to be sure of the exact timing of the development of the black

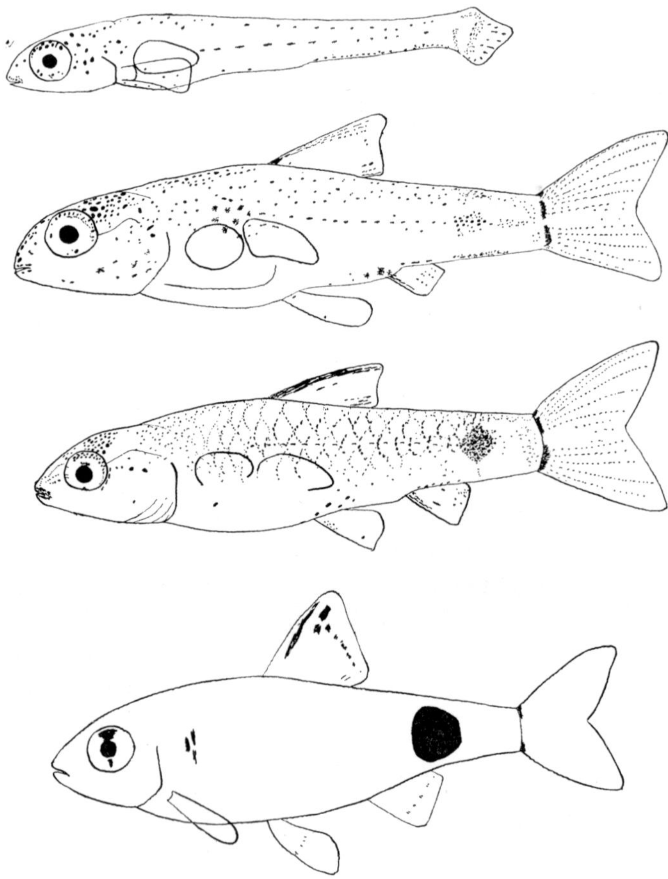

Fig. 20. *B. conchonius* × *B. stoliczkanus*; young of (standard length) 5.3/8.3/10.0 mm, and sub-adult; superficial markings.

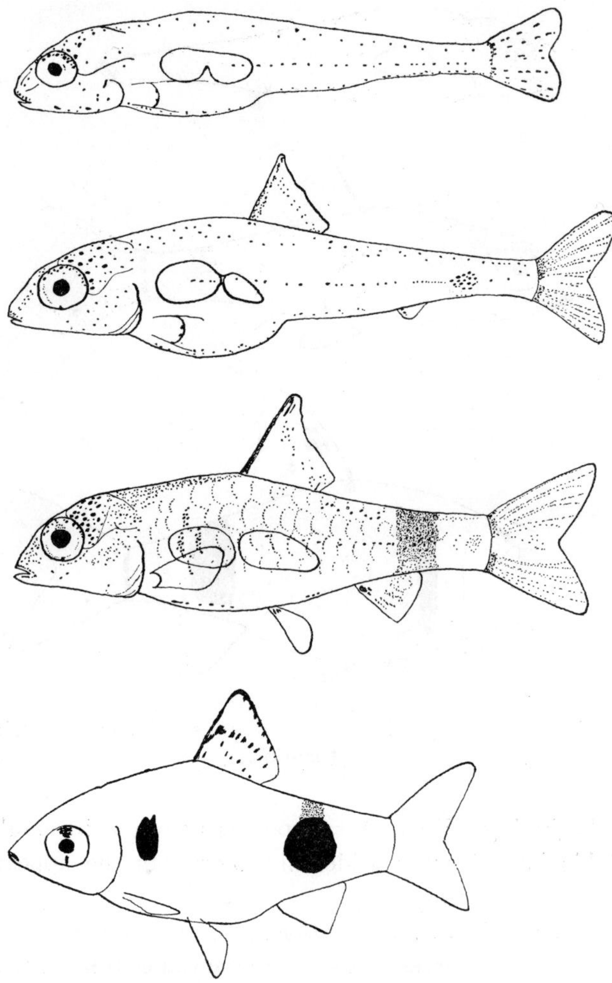

Fig. 21. *B. cumingi* × *B. stoliczkanus*; young of 6.2/7.0/11.0 mm., and sub-adult; superficial markings.

patterns in relation to body length. The following scattered observations may be given here.

In one *B. conchonius* x *B. stoliczkanus* measuring 6.2 mm the tail spot hadn't appeared yet. The beginnings of a tail spot were present in four individuals measuring 7.1, 7.5, 7.8 and 7.9 mm. More complete tail spots were present in four individuals of 8.2, 8.7, 10.0 and 10.2 mm. These data seem to suggest intermediacy between the parental species. Three individuals of 13.0, 15.5 and 17.3 mm didn't show any sign of a pectoral mark yet,

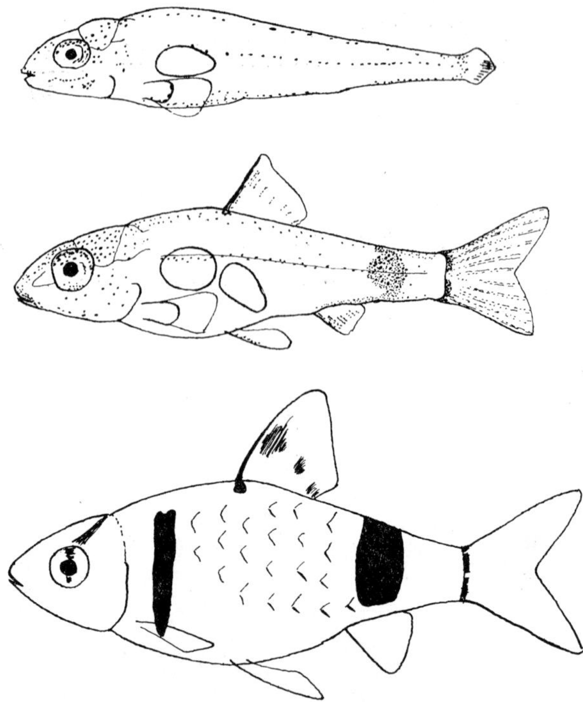

Fig. 22. *B. cumingi* × *B. nigrofasciatus*; young of 5.5/7.2 mm, and sub-adult; superficial markings.

whereas the beginning of a pectoral spot was present in another 17.3 mm individual. This is at least considerably later than the appearance of the pectoral mark in *stoliczkanus*.

One *B. stoliczkanus* x *B. nigrofasciatus* of 6.0 mm had just started to develop a tail spot. This corresponds well to the values found for both parental species. The first signs of a pectoral spot were found in three individuals of 9.4, 9.8 and 10.4 mm respectively. Two individuals of 11.0 hadn't got beyond the very beginnings of the pectoral spot. On the other hand, four individuals ranging from 9.5 to 10.1 mm didn't show any signs of a pectoral spot yet. Though these data are nearer to those found for *nigrofasciatus* than to those of *stoliczkanus* there is some suggestion of intermediacy here too.

In the reciprocal hybrid *B. nigrofasciatus* x *B. stoliczkanus* four individuals of 9.3, 10.1, 10.2 and 10.9 mm had no pectoral spot yet. A pectoral spot was present in five individuals of 11.1 mm and in another five individuals of 11.2, 11.3, 11.3, 11.4 and 11.5 mm. This fixes the timing of the first appearance of the pectoral spot pretty accurately on 11.0 mm. This is clearly

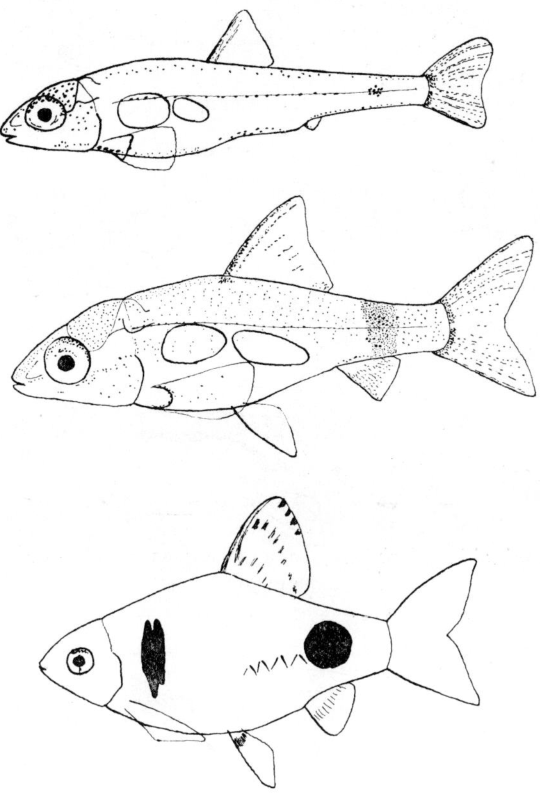

Fig. 23. *B. stoliczkanus* × *B. nigrofasciatus*; young of 6.0/9.8 mm, and sub-adult; superficial markings.

intermediate between the values found for the parents, though perhaps somewhat nearer to *stoliczkanus* than to *nigrofasciatus*. Comparison of the reciprocal hybrids may suggest a tendency towards patrocliny.

In the *B. cumingi* x *B. stoliczkanus* hybrids one individual of 6.4 mm didn't show the beginnings of the tail spot yet. This is rather late for both parental species. Further, two individuals of 8.1 and 8.6 mm hadn't yet developed a pectoral spot but five individuals measuring 9.0, 9.1, 9.3, 9.4 and 9.5 mm clearly showed the first beginnings of the pectoral spot. This corresponds rather well to the values found for *cumingi* and differs greatly from those of *stoliczkanus*.

On the other hand, in *B. stoliczkanus* x *B. cumingi* hybrids in three individuals of 9.1, 9.8 and 10.3 mm the pectoral marking was still absent. It was present in two individuals of 12.5 and 13.0 mm. This at least suggests

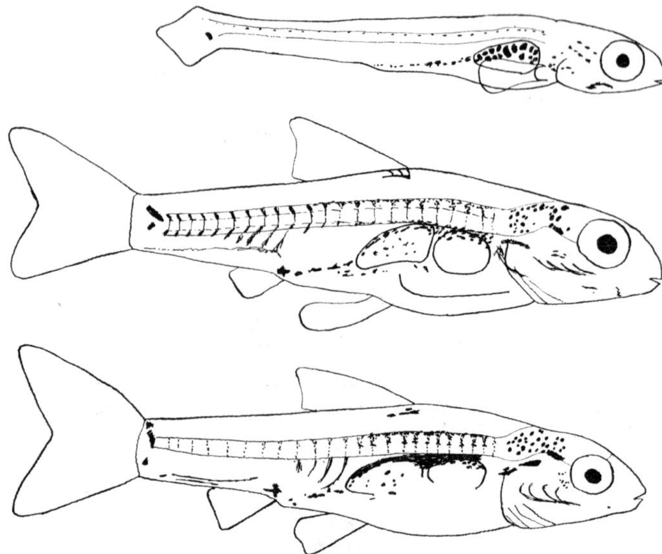

Fig. 24. *B. conchonius* × *stoliczkanus*; same young as fig. 20, no sub-adult; internal pigments.

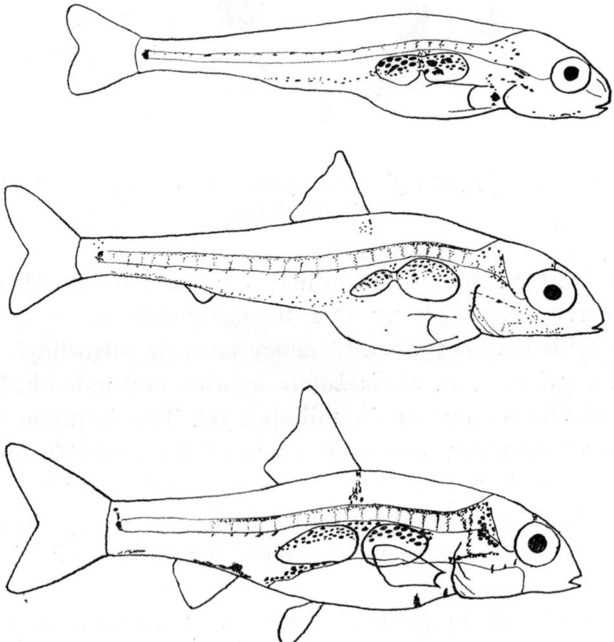

Fig. 25. *B. cumingi* × *B. stoliczkanus*; same young as fig. 21, no sub-adult; internal pigments.

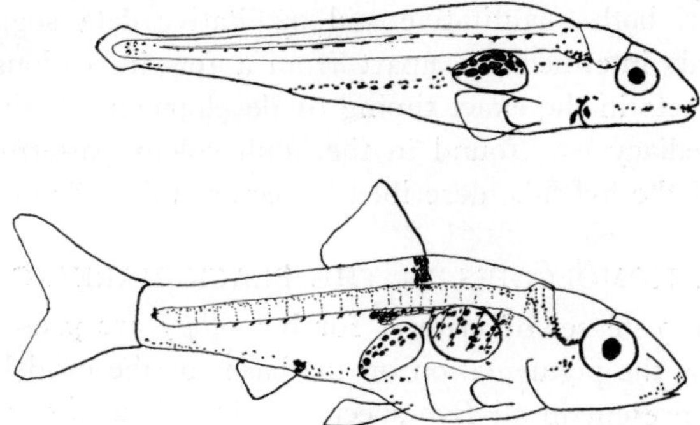

Fig. 26. *B. cumingi* × *B. nigrofasciatus*; same young as fig. 22, no sub-adult; internal pigments.

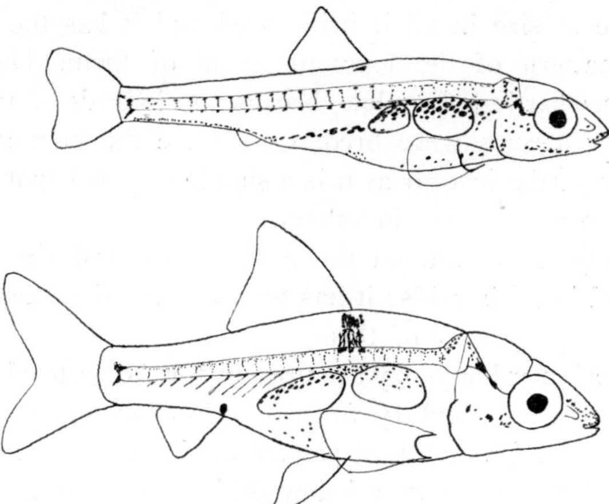

Fig. 27. *B. stoliczkanus* × *B. nigrofasciatus*; same young as fig. 23, no sub-adult; internal pigments.

that in this hybrid the pectoral marking appears later than in *cumingi* (either at the same time as in *stoliczkanus* or at an intermediate stage). So the data of the last two hybrids suggests a matroclinic effect for the same character which in the hybrids between *stoliczkanus* and *nigrofasciatus* appears to show a patroclinic tendency. No comparable data are available of hybrids between *nigrofasciatus* and *cumingi*. However, these couldn't give evidence about matrocliny or patrocliny anyway, because in these two species the development of the pectoral marking starts at practically the same stage.

It must be emphasized that quantitative data presented in this sub-section are few. Many more data are needed, for instance to investigate the amount of individual variation in the hybrids as compared with the parental species.

So far however, both quantitative and qualitative data suggest a general tendency towards intermediacy, apart from a few indications of patro- or matro-clinic effects in the exact timing of development. A similar tendency towards intermediacy was found in the adult colour patterns on the head, body and tail of the hybrids, described in section IV of this chapter.

SECTION VI. HOMOLOGIES IN THE BLACK MARKING PATTERNS

The most complete set of criteria for homology are present for the tail marking and for the pigmentation on the bases of the caudal fin rays. The tail marking is present in all five species. It has about the same location in all of them. It has the same structure in the adults of all five species in that it consists of two layers of pigment cells, one inside and one outside the bony scales. It has the same pattern of development in all five species. It is present and intermediate in size in all hybrids bred and it has the same structure and the same pattern of development in all of them. The evidence for *tetrazona* is somewhat less complete, because no hybrids of this species with any of the other species were bred. Of course the homology of the tail markings is only partial insofar as it is a simple rounded spot in some species and develops into a cross-bar in others.

Likewise, the pigmentation on the bases of the tail fin rays is present in all species and in all hybrids; it has the same location, the same ontogeny and the same structure in all of them.

The same criteria for homology are present for the pectoral marking except that it is present in rudimentary form only in *conchonius* and that it has a different structure and a different pattern of development in *tetrazona.*

The pigmentation in the region below the dorsal fin in *nigrofasciatus* and *cumingi* also show similar location, structure and development. It is present in more or less intermediate form in the hybrid between the two species and in some of the hybrids with other species. A comparable structure is absent in *conchonius, stoliczkanus* and *tetrazona.*

Pigmentation on the head is again present in all five species and in the hybrids. It shows the same general location and development in all of them but only in *nigrofasciatus* and *tetrazona* it is "shaped" into an eye-bar. An infra-orbital part of an eye-bar is present in *tetrazona* only. The second tail bar such as is found in *tetrazona* is confined to that species too. Its location, structure and development differ from anything found in the other species.

Most of the other superficial and internal pigmentations may be considered to be homologous in all five species according to the same criteria.

It is difficult to assess whether markings with a comparable location are situated on the same segment or segments in different species. In general, comparison of numbers of segments between species is difficult. However, the fact that all hybrids show the same markings as the parental species in an intermediate form, and not for instance double markings on adjoining segments, may suggest that the identity of location between the species is indeed complete. The same effect might result if one of the parents were always dominant in this respect. However, the general trend seems to be for intermediacy.

SECTION VII. SYSTEMATIC STATUS OF THE FIVE SPECIES

A. Species status.

As already mentioned in section IV of this chapter, all hybrids were males phenotypically. All of them, when adult, displayed full reproductive colours. Most of the hybrids were seen to display complete or nearly complete reproductive behaviour at some time or another. (More detailed descriptions of the behaviour of the hybrids will be given in a future paper).

Several attempts to produce F2 back-crosses of *B. conchonius* x *B. stoliczkanus* hybrids with *conchonius* females failed, although normal matings were frequent and eggs were seen to be produced on several occasions. The following hybrids were dissected and their gonads studied by Mrs BREDEROO at the histology department of our laboratory:

B. conchonius x *B. stoliczkanus*	2	individuals
B. cumingi x *B. nigrofasciatus*	2	,,
B. stoliczkanus x *B. nigrofasciatus*	2	,,
B. nigrofasciatus x *B. stoliczkanus*	2	,,
B. cumingi x *B. stoliczkanus*	10	,,

The gonads were fixated in formaline 4% for three days, transferred step-wise to alcohol 70%, embedded in paraffine and cut serially in a plane perpendicular to their long axes. Every 400 μ 6 sections of 5 or 7 μ were mounted and stained with hemalum-eosin.

The two *B. nigrofasciatus* x *B. stoliczkanus* hybrids (both only about 7 months old and probably not yet in full reproductive condition) had unripe testes. In one of them only spermatogonia were found, in the other spermatogonia and spermatocytes. The testes of all other hybrids were ripe and contained normal spermatogonia and spermatocytes I and II. However, the spermatids were in general abnormal and spermatozoa were absent in all of them. The testes of an adult male *B. conchonius,* which was taken as a control, were found to be full of spermatozoa.

The fact that all hybrids were males and the fact that all of them were probably completely sterile suggest that at least *conchonius, stoliczkanus, cumingi* and *nigrofasciatus* are true species, at least relative to each other.

Theoretically it is possible that some of them represent extremes of a continuous cline only. This point cannot be decisively assessed before more information on their geographical distributions is available. However, since most of the species do not seem to be sympatric or even contiguous (see section II), they are best considered as separate species. The argument of the hybrids cannot be used for *tetrazona* but, since *tetrazona* is in some respects the most deviating of the group of five and since it also is geographically the most remote, it seems reasonable to consider it a separate species too.

B. Sub-group status.

Sections I and V of this chapter have revealed some striking similarities in the location, structure and development of the black pigment patterns of the five species. These similarities become even more striking if the five species are compared with some other Asiatic *Barbus* species. Fig. 28-31 and 32-35 represent relevant stages in the ontogeny of the black pigment patterns in four species: *B. oligolepis, B. semifasciolatus, B. filamentosus* and *B. arulius.* All four species were bred in the laboratory and drawings of ontogenetic stages were made in the same way as those of the first five species. The figures are arranged in the same way as those of the first five species.

Like in the first five species the basic patterns of black markings on the head, body and tail are the same in both sexes. No details on sexual dimorphism will be given here.

1. *B. oligolepis* and *B. semifasciolatus.*

Oligolepis, when adult, has five small black spots on or in the direct proximity of the lateral line: the first immediately behind the head, the second in the pectoral region, the third below the dorsal fin, the fourth on the tail above the anal fin and the fifth at the tip of the tail near to the base of the tail fin. In adult *semifasciolatus* black markings are present at the same locations; the second, third and fourth marks are narrow, relatively short cross-bars instead of spots. Some more spots or very short cross bars may be intercalated between these markings. Very small intercalated spots may also occur in *oligolepis.* In both species all of these markings consist of melanophores in the deeper layers of the skin, below the level of the scales.

The development of the markings in the two species is remarkably uniform [1]) and very much different from those of the first five species. All markings, except the second tail spot start their development as small

1) Data on the more advanced stages of *semifasciolatus* are based on one individual only: fig. 29, 33, stage 3 and 4.

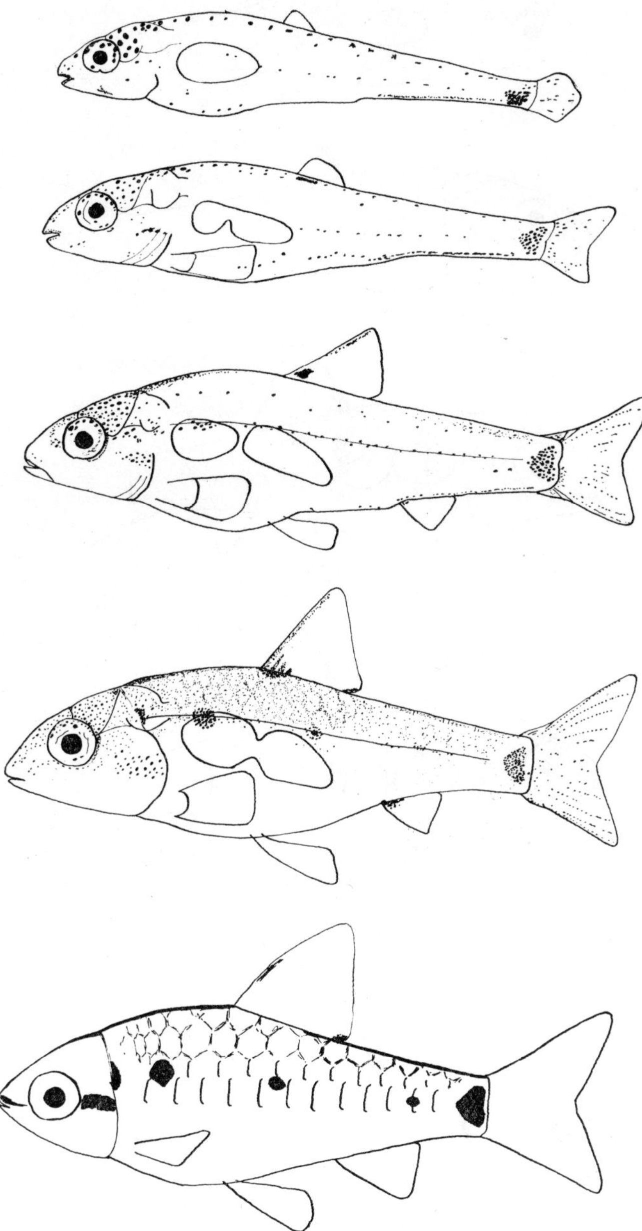

Fig. 28. *B. oligolepis*; young 5.2/5.9/8.1/10.3 mm, and sub-adult; superficial markings.

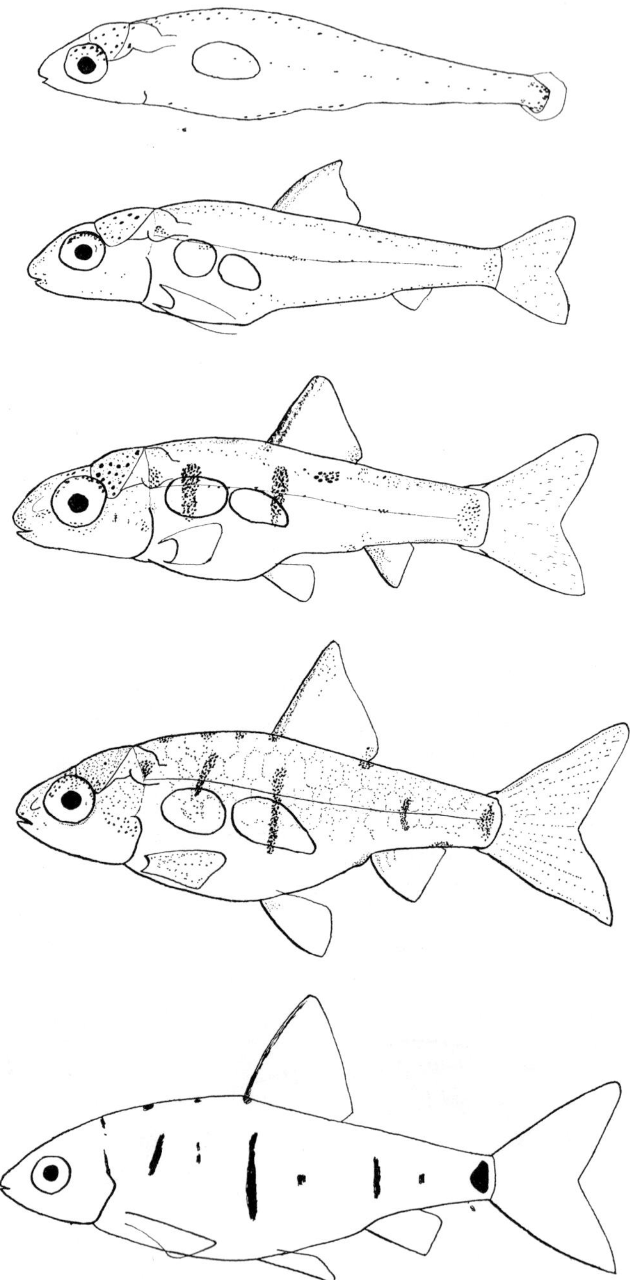

Fig. 29. *B. semifasciolatus*; young of 4.9/5.9/8.9/13.1 mm, and sub-adult; superficial markings.

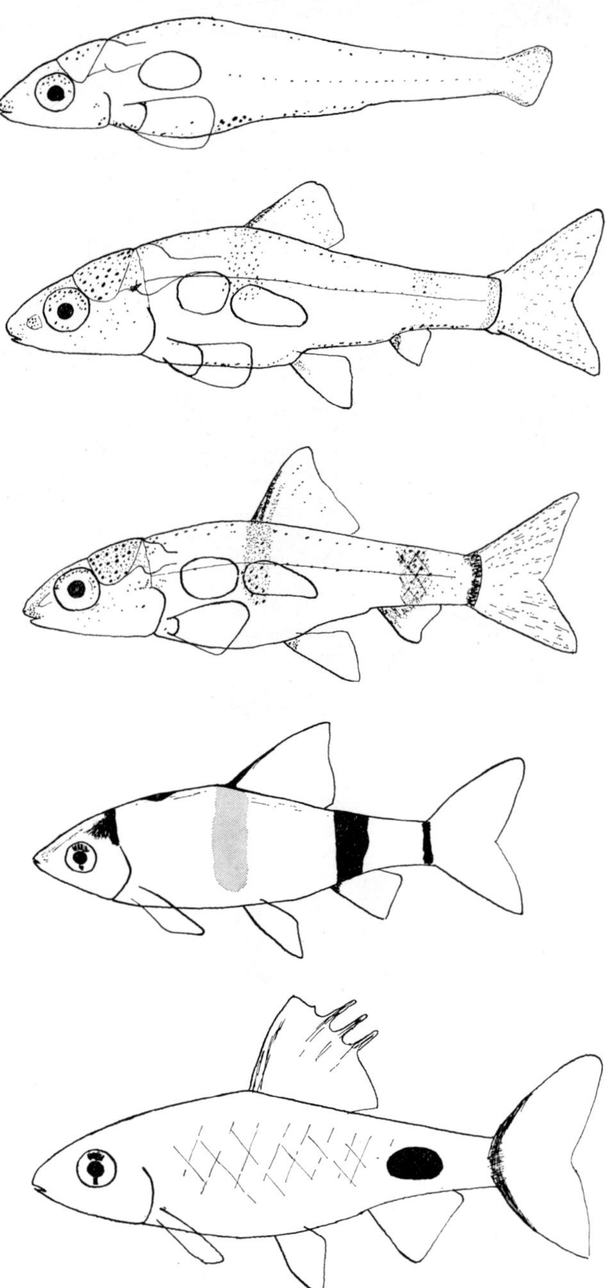

Fig. 30. *B. filamentosus*; young of 5.4/9.0/11.1 mm., sub-adult, and adult ♂; superficial markings.

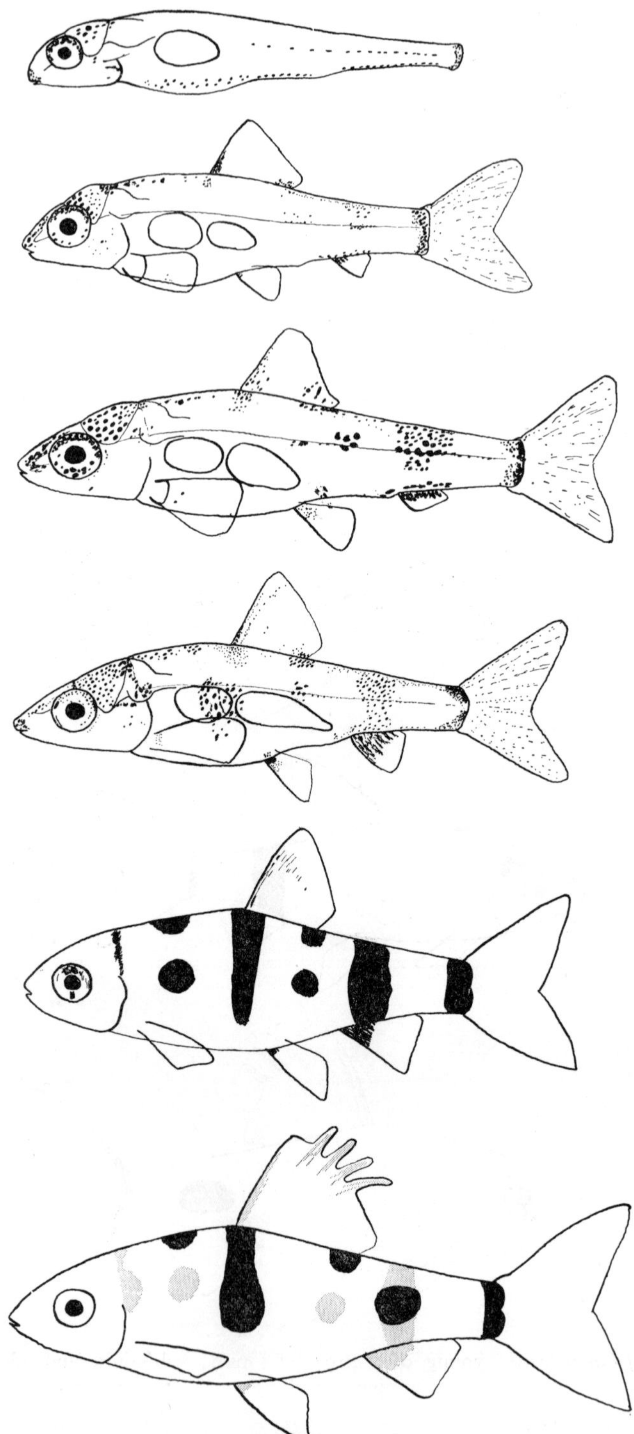

Fig. 31. *B. arulius*; young of 4.8/8.7/9.1/11.5 mm, sub-adult, and adult ♂; superficial markings.

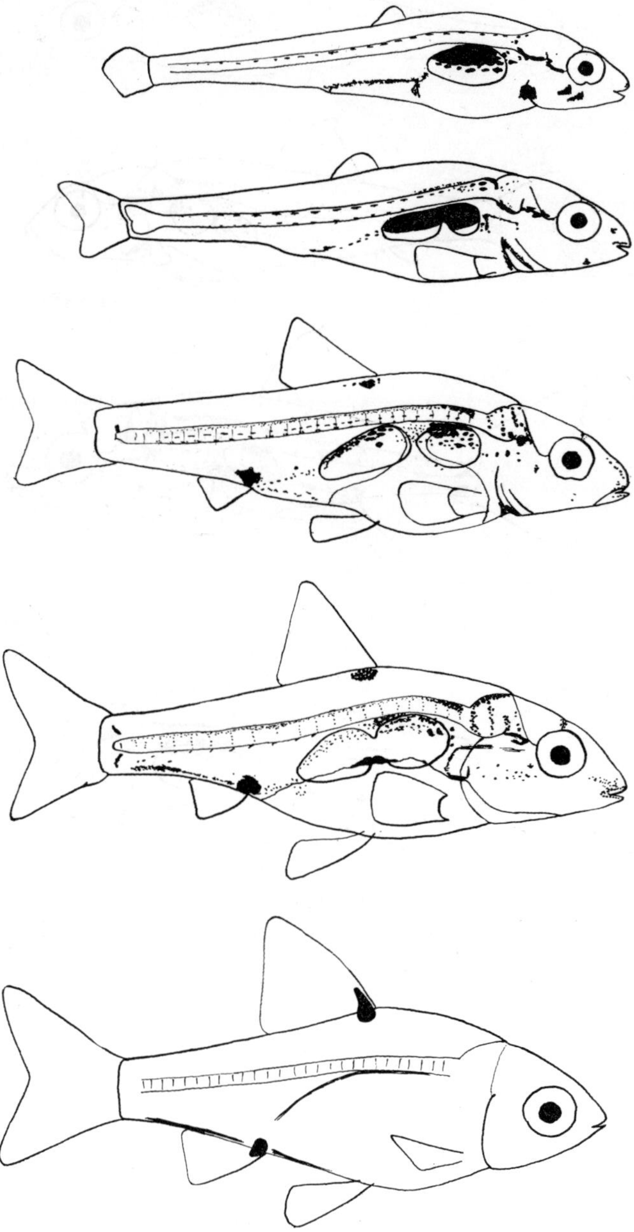

Fig. 32. *B. oligolepis*; same young as fig. 28, and sub-adult; internal pigments.

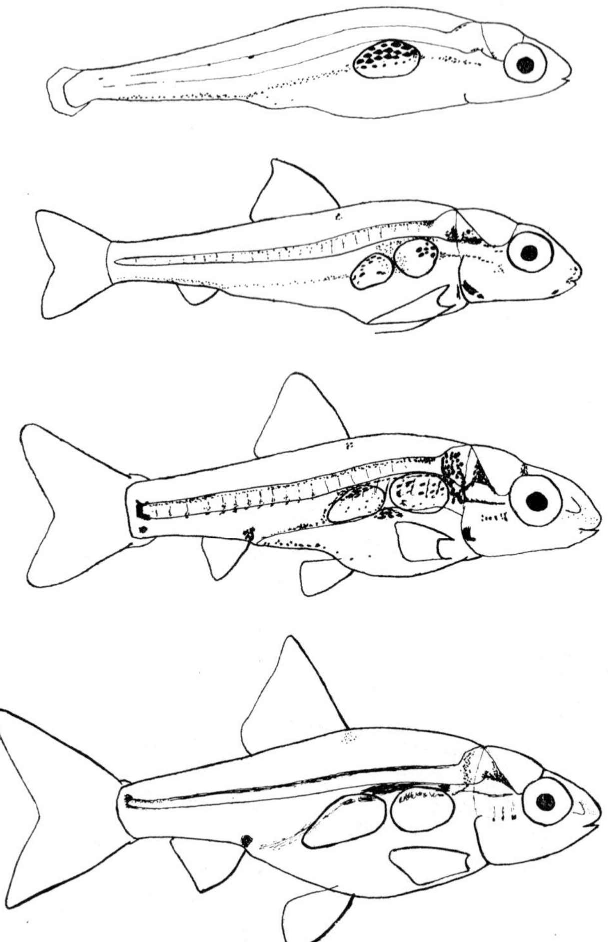

Fig. 33. *B. semifasciolatus*; same young as fig. 29, no sub-adult; internal pigments.

rounded concentrations of a few melanophores around or very near to the lateral line, much in the same way as the first development of the tail spot in the first five species. In *oligolepis* the ventral half of the second tail spot is already strongly developed soon after the larva has become free-swimming. In *semifasciolatus* it comes later, about at the time that the first two markings start to develop. Of the other markings the pectoral spot is one of the first to appear. The first spot, the one immediately behind the head, often appears about simultaneously but, at least in *oligolepis*, it may be somewhat retarded. The pectoral spot is followed soon by the third and fourth spot in that order.

Though the order in which the markings appear is rather consistent between

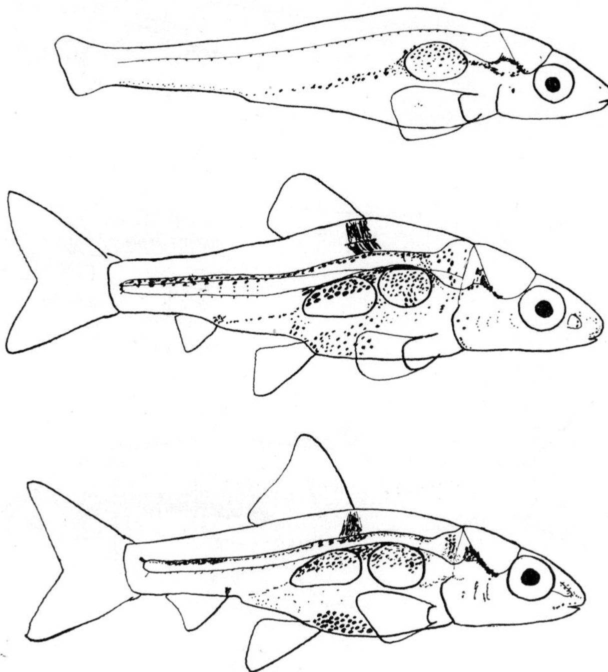

Fig. 34. *B. filamentosus*; same young as fig. 30, no sub-adult, no adult; internal pigments.

individuals, the timing of the development of the markings relative to body growth seems to be much less exact than in the first five species. In both *oligolepis* and *semifasciolatus* the markings on the two sides of the body of the same individual may often differ considerably in stage of development.

The black pigmentation of the fins and the internal pigmentations are in many ways similar to those of the first five species. Internal concentrations of black pigment are present at the bases of the dorsal and anal fins. However, one marked difference with tht first five species is that both in *oligolepis* and in *semifasciolatus* melanophores on the bases of the caudal fin rays are completely absent.

The arrangement and the structure of the black markings on the body and tail in *oligolepis* and *semifasciolatus* are probably representative of a whole group of species to which belong *B. binotatus, B. pentazona, B. everetti* and *B. fasciatus*. (The last species, which has a number of longitudinal stripes when adult, bears cross-bars as a sub-adult very much like those of *pentazona*). The ontogeny of the colour patterns of these species hasn't been studied so far.

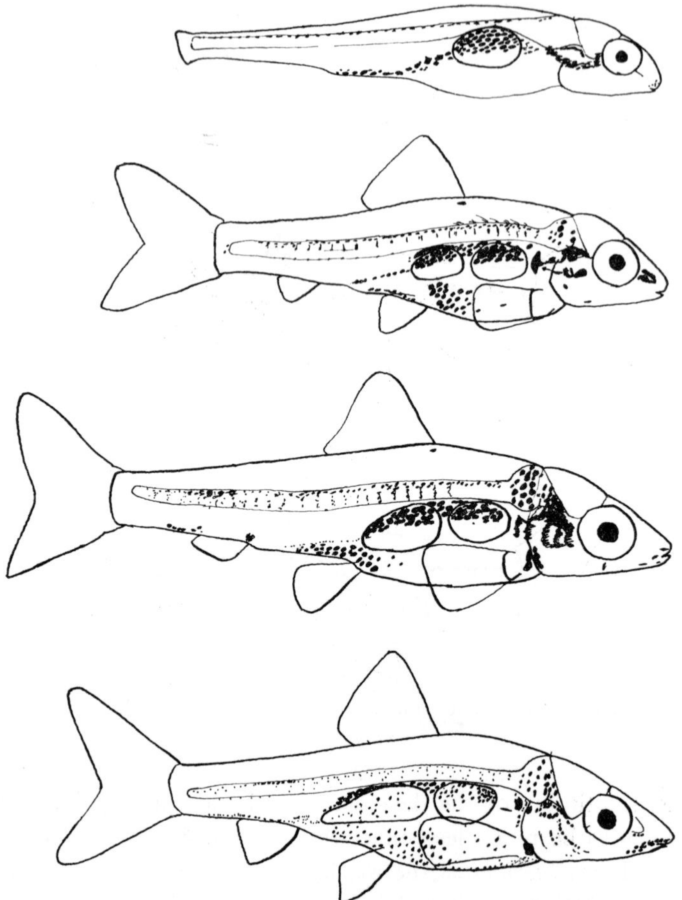

Fig. 35. *B. arulius*; same young as fig. 31, no sub-adult, no adult; internal pigments.

2. *B. filamentosus* and *B. arulius*.

Again these two species are similar to each other in the arrangement, structure and development of the black body markings. They clearly differ in these characters from both the first five species and the *oligolepis-semifasciolatus* group. The sub-adult *filamentosus* has the following markings: a rather amorphic black area at the dorso-lateral side just behind the head, a broad black cross-bar below the base of the dorsal fin, a broad black bar on the tail above the anal fin and narrower black bar on the tip of the tail. In many, but not in all, individuals there is a black spot on the dorsal side about mid-way between the nape and the base of the dorsal fin. Sub-adult *arulius* have the same markings plus a few more: all individuals have the black spot on the dorsum between the nape and the dorsal fin, a similar spot near to the end of the dorsal fin and a rounded lateral spot below each of these dorsal spots. The second tail bar is broader than in *filamentosus.* In both *filamentosus* and *arulius* all of these markings consist of deeply situated melanophores, in addition to more superficial pigment on the scales.

In both species part of the markings are gradually lost again in the adult. In *filamentosus* only a large tail spot remains (the central area of the original first tail bar); in *arulius* the black bar below the base of the dorsal fin, the dorsal black spot near to the caudal end of the dorsal fin, the first tail bar above the anal fin (especially the central area of it) and the second tail bar remain. In the fig. 30 and 31 a picture of an adult male is added to show the elongated dorsal fin rays which are characteristic of the males of both species. These two pictures represent males in process of loosing part of the body markings.

In both species all markings begin their development as rather scattered groups of melanophores in the deeper layers of the skin without any relation to scale structure. Only later pigmentations on the scales are added. In all lateral markings, in the beginning, some tendency is present for melanophores to concentrate around the lateral line but nowhere is this tendency as clear as it is in the developing tail spot of the first five species or in the lateral markings of the *oligolepis-semifasciolatus* type.

The internal pigmentations in many ways resemble those of all fore-going species. Neither of the two species has any black pigmentation on the bases of the caudal fin rays.

No other species have been found so far which might be considered as closely related to these two.

When compared to the *oligolepis* and the *filamentosus* types, the arrangement, structure and development of the black marking patterns in the first five species look remarkably uniform. This may be considered an important argument to put them together in one closely-related sub-group of the genus, Another sub-group might be erected which contains *B. oligolepis, B. semifasciolatus, B. binotatus, B. pentazona, B. everetti* and *B. fasciatus.* A third sub-group would contain *filamentosus* and *arulius.*

It must be conceded that, so far, there is little further evidence to support this classification. The following points may be brought forward, each in partial support of the classification proposed.

(a) Young free-swimming larvae of *oligolepis* and *semifasciolatus* are very similar to each other in general shape. Likewise, young larvae of *filamentosus* and *arulius* are very similar to one another and very different from those of the former two species. It is very difficult to define precisely what details constitute these general impressions. For instance the larvae of *filamentosus* and *arulius* are very long and slender compared to those of all other species studied. However, detailed morphological study of the young larvae will be necessary to substantiate the general impressions. Young larvae of the first five species are certainly different in shape from those of both the *oligolepis* and the *filamentosus* type. However, I think I am too familiar with them and know too well how to tell them apart to be able to say how uniform they are among each other compared to the larvae of the other two groups.

(b) All five species of the first group, except *tetrazona*, have relatively Western to North-eastern distributions (Ceylon, India, Burma, Thailand; see section II of this chapter). Even *tetrazona* is linked to this general area *via* its subspecies *B. tetrazona partipentazona* which, according to STERBA (1963, p. 293) occurs in South-east Thailand [1]). On the other hand, most of the species of the *oligolepis*-group have much more Eastern distributions: *oligolepis* in Sumatra, *semifasciolatus* in South-east China, *binotatus* in Further India, Great Sunda Islands, Banka, Biliton, Bali, *pentazona pentazona* in Malay Peninsula, Singapore, Borneo, Sumatra, *pentazona hexazona* in Central Sumatra, *everetti* in Singapore, Borneo, *fasciatus* in Sumatra, Borneo, Banka, Malay Peninsula (Johore) (STERBA, 1963, pp. 282-297).

The geographical argument cannot be used to support the division between the first five species and the *filamentosus* group. According to STERBA (1963, p. 287, 280) *filamentosus* occurs in South-west India (Further India, Thailand?) and *arulius* in South-east India, Travancore, Cauvery.

(c) The results of hybridizations between *conchonius, stoliczkanus, cumingi* and *nigrofasciatus* are very uniform: sterile phenotypic males only in all cases. (It must be mentioned here that, though only 7 out of the 12 possible crossings between these four species have been produced, the absence of the remaining 5 does not mean that those hybrids were not viable; I just did not try to breed them yet. The same is true for crossings between *tetrazona* and the other species). This uniformity pleads for a rather constant degree of "genetic compatibility" (DUYVENÉ DE WIT, 1964) between the four species involved. However, the uniformity doesn't set these species apart from the *oligolepis* and *filamentosus* groups. Intergroup crossings should be attempted. Only one such inter-group crossing was carried out: *conchonius* females were paired with *filamentosus* males. Complete courtship ensued and fertilized eggs were produced. A large number of larvae hatched but all of them failed to get the swim-bladder filled with air at the water surface in spite of their very

---

1) See also *B. phutunio* and *B. gelius*: section VIII of this chapter.

persistent attempts to do so. They all died soon afterwards. This single experiment seems to indicate a lower degree of genetic compatibility between *conchonius* and *filamentosus* than between the first four species. However, the experiment should be repeated to make sure that the failure was not due to specific conditions and experiments with other species should be conducted.

TABLE I

Some numerical characters of Barbus species

species	D	A	LL	barbels
conchonius	3/7–8	2/5–6	24–28	0 pairs
nigrofasciatus	3/8	2–3/5	20–22	0 „
cumingi	3/8	3/5	21	0 „
phutunio	2–3/8	3/5	20–23	0 „
gelius	2–3/8	3/5	23–24	0 „
stoliczkanus	2–3/8	2/5	23–25	0 „
tetra. tetra.	4/8–9	3/5–6	21	0 „ *)
tetra. parti.	3/8	3/5	19–20+1–2	1 „
oligolepis	4/8	3/5	17+2–4	1(–2) „
semifasciolatus	3/8	2/5–6	22+2	1 „
binotatus	4/8	3/5	23–27	2 „
penta. penta.	3/8	3/5	22–25	2 „
penta. hexa.	3/8	3/5	22–25	2 „
everetti	4/8	3/5	22–25	2 „
fasciatus	3–4/8	3/5–6	27–30	2 „
arulius	3/8	2/5	21-23	1 „
filamentosus	3/8	2/5	21	0 „

*) 1 pair according to SCHULTZ (1957) and a figure in ALFRED (1963).

No adult morphological characters could be found as yet which might support the classification. On the contrary, if the classification is real, species of such different shapes and sizes as *oligolepis* and *binotatus*, or species with such strikingly different colour patterns as for instance *nigrofasciatus* and *stoliczkanus* would be combined in one group. Only for *filamentosus* and *arulius* also the adult body shape and the presence of elongated fin rays in the dorsal fins of adult males strongly support their supposed close relationship. Table 1 presents a few numerical values of adult fishes (taken mainly from STERBA, 1963, p. 280-298) which are valuable in the identification of the species. These counts do not seem to support the classification either (except perhaps, to some extent, for these species, the presence or absence of barbels). Further, it must be emphasized that only a small fraction of the numerous species of Asiatic barbs have been examined in this study. It is quite possible that further investigations would reveal all kinds of intermediate patterns of development of black marking patterns.

Nevertheless, the criteria on which the classification is based (especially the development of the colour patterns) seem to me to be sufficiently important to maintain the classification awaiting further information.

The classification of *tetrazona* into one group with *conchonius, stoliczkanus, cumingi* and *nigrofasciatus* may remain more or less problematic: it has a more or less aberrant distribution and the ontogeny of its black marking patterns differs on some points from those of the other four species. However, the way of development of the tail spot and the presence of black pigment on the bases of the caudal fin rays and further its general resemblance in colour patterns and in behaviour (see chapter III, section II) to *stoliczkanus* seem to justify its (tentative) classification with the other four.

SECTION VIII. *B. PHUTUNIO* AND *B. GELIUS*

Two more species may be briefly described: *B. phutunio* and *B. gelius*. Both probably belong to the same sub-group with the first five species (section VII of this chapter) and both show some peculiar morphological and behavioural characteristics as compared with the first five species (see also chapter III p. 164ff.). Both species are characterized by small size and by very little sexual dichromatism. Their portraits are given in figs 36 and 37.

Imported wild adult *phutunio* never grew larger in the laboratory than 2½-3 cm standard length. The body shape is rather rhomboid; the mouth is rather pointed; the lateral compression of the body is relatively strong; there is often an obtuse angle between the dorsal surface of the head and the dorsum.

The most conspicuous markings on the body and tail are two black spots, one in the pectoral region and one on the tail above the anal fin. These spots have roughly the same size and shape and the same location as the pectoral and tail marks of *cumingi*. As in *stoliczkanus, cumingi* and *nigrofasciatus* the tail spot consists of melanophores situated deeply in the skin in addition to more superficial melanophores situated on the scales, whereas the pectoral mark consists of superficial melanophores only. Like in the first five species there is a distinct vertical line formed by melanophores situated on the bases of the caudal fin rays. Further there is a tiny black spot on the body at the base of the dorsal fin. (Such a spot is also present in juvenile *nigrofasciatus* and *cumingi* (see fig. 4 stage 3-5 and fig. 3 stage 3 and 4) but it is easily overlooked in those species because of its small size relative to the large internal pigment concentration at the base of the dorsal fin).

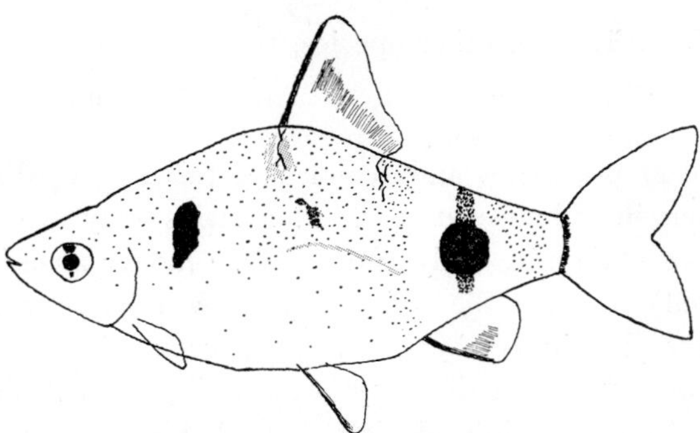

Fig. 36. *B. phutunio*; superficial markings (black) and externally visible internal pigments (grey).

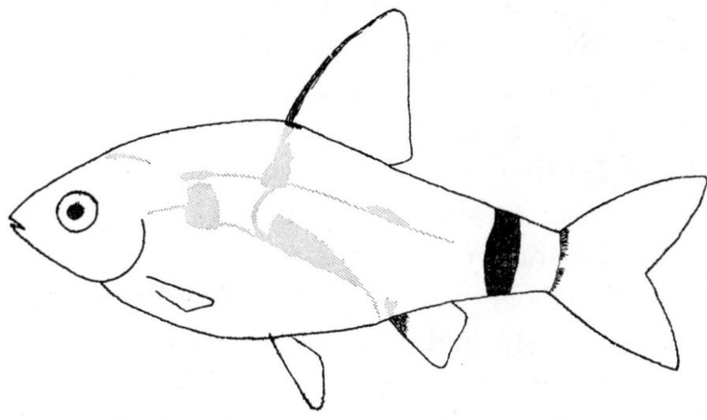

Fig. 37. *B. gelius*; superficial markings (black) and externally visible internal pigments.

The whole body and tail are slightly greyish as a result of melanophores lying along the rims of the scales. Like in *cumingi* there is a lighter zone in front of the tail spot. Also like in *cumingi,* the area below the dorsal fin is slightly darker than the rest of the body and tail as a result of the normal scale pigmentation being slightly more intense than in the surrounding areas. In *phutunio* the intensity of this pigmentation increases in caudal direction until it ends abruptly at the transition to the lighter zone in front of the tail spot

The fishes tend to be slightly transparent and at the base of the dorsal fin an internal string of black pigment, running from the base of the dorsal fin to the vertebral column, is weakly but unmistakably visible. In some individuals also the black cap on top of the swimming-bladder is faintly visible from the outside.

Like in *conchonius, stoliczkanus* and juvenile *cumingi,* there is a small golden area right in front of the tail spot. A similar though smaller golden area is present in front of the pectoral marking. (A similar structure was not found in any of the other species studied so far).

Like in the first five and other species, the hard rays of the dorsal, anal and ventral fins are blackish.

As mentioned above, males and females of this species differ very little from each other in colour patterns. The males may become somewhat brownish-black on the entire body and tail (the lighter zone in front of the tail spot excepted). Further the males may show a transverse greyish band in the dorsal fin and a similar, but much more weakly developed, band in the anal fin; the anal and ventral fins may become slightly reddish. These colour differences are shown mainly during reproductive activity (for further details see chapter III, p. 168).

The species' geographical distribution is in Eastern India, North-eastern Bengal and Ceylon (STERBA 1963, p. 294). The location and the structure of the black marking patterns seems to be sufficient argument to place *phutunio* in the same sub-group with the first five species.

Gelius is even smaller in size than *phutunio*. Adult wild imports never grew larger in the laboratory than 2-2½ cm standard length. The body shape is definitely rhomboid; the mouth is pointed and lateral compression is rather strong. The body and tail are quite transparent; many of the internal organs and pigmentations are clearly visible from the outside. Only most of the intestine and swimming-bladder is rendered invisible by an opaque whitish layer, full of guanin, covering the greater part of the belly.

A great number of black markings are present in the body and tail. A cross-bar is found on the tail, behind the insertion of the anal fin. This bar is similar in structure to the tail markings of the first five species. There is a strongly developed vertical line consisting of melanophores situated on the bases of the caudal fin rays. Like in *phutunio* there is a tiny black spot on the body at the base of the dorsal fin. All other markings are internal pigmentations. Most of these closely resemble the normal internal pigmentations found in juveniles of for instance *nigrofasciatus* or *cumingi*. A large roughly triangular internal concentration of pigment is present between the base of the dorsal fin and the vertebral column. It is often even larger than in the specimen represented in fig. 37. It closely resembles the large black triangle typically found in juvenile *nigrofasciatus* (fig. 9 stages 3 and 4). A smaller internal concentration of pigment is present at the base of the anal fin. There are two distinct concentrations of pigment on top of the vertebral column, one just behind the head and a second approximately below the hind edge of the dorsal fin. The first concentration is also typically found in juvenile *nigrofasciatus* (see fig. 9 stages 3 and 4) and to a lesser extent also in *cumingi* (fig. 8 stages 2-4). The second concentration is also found in many, though not all, juvenile *nigrofasciatus* and *cumingi*, but it is never as strongly developed in these species as it is in *gelius* (see *cumingi* fig. 8 stages 2 and 4). A third concentration of pigment on the vertebral column is found inside the ring formed by the tail bar. A similar concentration has never been found in young *nigrofasciatus* or *cumingi* but it does recall the intensification of internal pigmentation on the spine in the region of the tail bar found in *tetrazona* (fig. 10 stage 4) Internal pigmentation is also visible on top of the posterior parts of the skull, similar to those found in juveniles of all species studied so far. Also the black cap on top of the swimming-bladder is visible from the outside, just above the rim of the opaque ventral membrane; it is augmented by peritoneal pigment. Finally, a large concentration of peritonial pigment is visible in the pectoral region, partly above the rim of the opaque membrane, partly "breaking" through it. Peritoneal and intestinal pigment is found in other species as well — it is clearly visible in the cross-sections of young *nigrofasciatus* discussed earlier in this chapter — and the opaque membrane on the belly is also present in other species. However, a pectoral marking with a structure similar to that found in *gelius* was not found in any of the other species studied so far (see also chapter IV, p. 309).

The front edge of the dorsal fin is marked by a thick black line reaching from the base to the tip; towards the base it widens into a small black triangle. A similar blackish triangle is present on the base of the anal fin. The hard rays of the anal and ventral fins are not specially pigmented.

There are very little colour patterns other than the black pigment patterns described above. Some patches of golden colour, both internal and more superficial ones, lie scattered over the body and tail. There is extremely little sexual dichromatism. Males typically have a copper-coloured longitudinal stripe at the lateral side of the tail and part of the body. Even this pattern is often difficult to discern when the fishes are not reproductively active (for further details see chapter III, p. 167).

The structure of the tail bar and the presence of black pigment on the bases of the caudal fin rays would plead for the species' close relationship to the first five species and *phutunio*, and so does the general resemblance of the internal black pigment patterns to those of juvenile *nigrofasciatus* and *cumingi*.

The species also has a Western distribution: Bengal and Central India, in still waters (STERBA, 1963, p. 287).

The location of the tail bar differs from those of the first five species in that it is situated more caudad relative to the anal fin. However, this is similar to the location of the tail bar of juvenile *conchonius, stoliczkanus, cumingi, nigrofasciatus* and *tetrazona* (see figg. 1-5).

The transparency of the body and the general resemblance of the black colour patterns of adult *gelius* to those of juvenile *nigrofasciatus* and *cumingi* suggest that *gelius* is a neotenous form derived from a *nigrofasciatus*-like ancestor. This is also in accordance with the relatively strong development of pigment on the bases of the caudal fin rays and on the leading edge of the dorsal fin. Both characters are typical of juvenile and/or sub-adult *nigrofasciatus* and *cumingi* rather than of adults of these species (see fig. 3 and 4 and section I of this chapter). It is also in accordance with the location of the tail bar (see above).

This hypothesis might also explain the species' small size and its extreme paucity of sexual dichromatism (though, of course, neither of these characters is a necessary concomitant of neoteny).

Some of the characters suggesting neoteny in *gelius* are present, though in a much less pregnant form, in *phutunio* too. Also in this species the body is more or less transparent; the internal pigmentation below the dorsal fin and in some individuals also the black cap of the swimming-bladder are visible from the outside.

Phutunio is also characterized by small size (only slightly larger than *gelius*) and by relatively little sexual dichromatism (perhaps slightly more than *gelius* but definitely less than even *cumingi*).

This would suggest that also *phutunio* has neotenous characters though to a lesser degree than *gelius*.

These hypotheses on neoteny in *gelius* and *phutunio* might even throw some light on some of the characteristics of *cumingi*. It is the smallest in size of the first five species (though definitely larger than *phutunio* or *gelius*) and it shows the least sexual dichromatism. Though its body is not by far as transparent as in *gelius* or even *phutunio*, some tendency towards transparency is certainly present, at least in some individuals.

I would submit, therefore, that *cumingi*, *phutunio* and *gelius* all are more or less neotenous descendants from a *nigrofasciatus*-like ancestor. Neotenous characters are most clearly present in *gelius*, less clearly so in *phutunio* and least clearly in *cumingi* (see also the behavioural characteristics of the three species: chapter III p. 169). It might be possible to test the hypothesis by the administration of Thyroxin to individuals of the three species from an early age and see whether they could develop more "adult" colour patterns.

CHAPTER II. BEHAVIOUR I

'The elements of the pattern'

INTRODUCTION

This chapter contains two sections. In the first the behaviour elements used in this study are listed and briefly described. The description is mainly based upon the behaviour of the species *B. conchonius* and *nigrofasciatus*, but it may be considered as largely representative of the other species as well. Behaviour elements which are found in the other species but absent in *conchonius* and *nigrofasciatus* are also described. Specific features in which each of the five species: *conchonius, nigrofasciatus, tetrazona, stoliczkanus* and *cumingi,* deviates from the more or less average picture presented in this chapter are described in the first section of chapter III (p. 103). Specific features of the species *B. phutunio* and *gelius,* which both have largely the same repertoire of behaviour elements, are described briefly in the second section of chapter III (p. 164).

The second section of this chapter presents a preliminary motivational analysis — based mainly on form analysis only — of several of the reproductive behaviour elements listed in the first section. A more comprehensive analysis, involving quantitative data on *conchonius* and *nigrofasciatus,* will be given in a separate paper (KORTMULDER, in prep.).

SECTION I. BEHAVIOUR ELEMENTS

A. Non-reproductive behaviour.

1. Locomotion.

Propulsion. When undisturbed, barbs swim around slowly for most of the time, following a rather erratic course. Single propulsory beats with the tail, rather irregularly spaced in time, are alternated with braking movements, usually followed by a turn in the horizontal plane. When speeding up the tail movements become faster and rhythmic. All fins, except the tail fin, are more or less loosely folded during forward movement. They are more strongly depressed when locomotion is fast.

Braking. In braking all fins are spread. The main braking force seems to be exerted by the pectoral fins which are flicked forward and downward a few times in rapid succession. In addition the tail fin, the caudal part of the dorsal and the anal fin are bent, all three to the same side. The ventrals are raised and spread, the more caudal rays moving out laterally thus increasing the resistance of the fin surface. The outward movement and raising is stronger in the ventral fin of the same side to which the un-paired fins are bent. The combined actions of the un-paired fins and the ventrals alone would tend to make the fish turn to one side when braking (judging impressionistically). As this doesn't happen a compensatory force must be exerted, probably by the pectoral fin at the contra-lateral side making a stronger forward movement than the other. This could not be confirmed by direct observation. Braking in the barbs is a much more complicated movement than in such fishes as *Gasterosteus* or *Macropodus* which often brake by movements of the pectorals only. The complexity of the braking movement in barbs is probably functionally related to their comparatively high degree of hydrodynamic instability. This would require stabilizing movements to accompany braking.

Horizontal turns. Turning in the horizontal plane is effected by bending the front part of the body sharply into the new direction, striking forward of the pectoral at the inner side while the other remains adducted and abduction of the outer ventral fin. Subsequently the rest of the body is brought in line with the front part through a powerful sideward beat of the tail. It may at first sight seem surprising that it is the outer ventral which is abducted and not the inner. However, the extended ventral probably functions as a pivoting point around which the fish turns. This would enable the fish to turn more sharply if the outer ventral is extended instead of the inner.

Vertical turns. Shallow vertical turns during forward movement may be made through a slight lifting or lowering of the pectorals which lie backwards against the body. Sharp turns in a vertical or inclined plane are made by rolling on one side and using the same technique as used in horizontal turning.

2. Feeding.

Much of the time not spent in reproductive behaviour is spent feeding. Feeding barbs mostly swim slowly along the bottom or plants. Propulsion is again by the tail; all other fins are used in manoeuvring. When feeding from the bottom a fish usually swims with the body tipped forward to about 30°. To recover a horizontal position both pectorals may be struck forward and downward simultaneously. From time to time a feeding fish *fixates* a point at the substrate with both eyes. This may be followed by *snapping* a particle or *tearing off* a piece of algae. The object may be *spit out* again, sometimes followed by repeated snapping and spitting. A snapped particle may be subsequently *chewed* and *swallowed.* When feeding from the bottom often a mouthful of sand is snapped up and chewed. The sand is then spit out again, partly through the mouth, partly through the gill covers. Edible particles, present in the sand, are probably sifted out and swallowed.

Besides these more normal ways of feeding, barbs may snap food from the water surface (for instance scattered dry food) or from mid-water (for instance live *Daphnia* or other small animals). At times they may give chase to more mobile animals such as *Notonecta* or free-swimming fish larvae (own species included). Such attempts are only rarely met with success.

3. Comfort behaviour.

Yawn-stretch. The mouth is opened and closed comparatively slowly. Simultaneously the body becomes rigid through contraction of body musculature and the fins are raised. In some species the dorsal part of the body musculature is more strongly contracted than the ventral part for the spine, which is normally arched, becomes slightly more straight. Typically there is a close concordance between the degree to which the mouth is opened and the degree of fin-raising.

Stretch. The body is quickly flexed laterally into a U-shape and relaxed. This movement may occur as an isolated element but is also often seen superimposed on a horizontal turn.

Headshake. The body is flexed laterally just behind the head a few times in rapid succession. There is a gradation of intermediate movements between stretch and headshake.

Finflicker. The dorsal fin or the ventral(s) or both simultaneously are very rapidly flicked up and down a few times. The movement is so rapid that individual flicks in a series can hardly be distinguished. The pectoral fins may be flicked rapidly against the body a few times in succession and with small amplitude. The caudal fin may be rapidly opened and partly closed a few times with small amplitude. Though especially flickering of the pectorals is unconspicuous and is easily over-looked, I am sure that both pectoral and caudal fin flickering are much rarer than dorsal and ventral fin flickering. Anal fin flickering has never been observed.

Chafe. The fish performs a rapid dash forwards or forwards and downwards, often rolls over more or less to one side, and brushes the ventral or latero-ventral surface of the body against the bottom or a plant.

Mouthflicker. The mouth is opened and closed a few times in rapid succession with small amplitude. Often some particle is seen to be expelled from the mouth with this movement.

4. Schooling.

All species studied so far are definitely schooling fish. Schooling behaviour consists of two tendencies: keeping together and parallel orientation. The latter tendency is relatively weakly developed in most species; it is strong, however, in *B. tetrazona* (see also chapter III, section II). Schooling is most often displayed when the fishes are not reproductively active. (For interactions between schooling and reproductive behaviour in different species see chapter III, section II). Intensive feeding behaviour tends to disrupt schools. Strong schooling is commonly elicited by changes in the environment such as are brought about by adding new individuals to a group, by cleaning a tank or by moving a whole group of fishes to another tank. Also a drop in temperature often provokes strong schooling.

5. Sleeping.

During the night, and during any longer period that the room is dark at day-time, the fishes stand still with the body tipped forward up to 45 or 60°, usually between plants or near to the bottom. When lights are lit it takes them a few minutes to start moving around.

B. Reproductive behaviour.

1. Male agonistic behaviour.

Attack. The fish swims towards another fish with the snout pointing at him. Locomotion is of the normal type (see above). The fins are more or less depressed dependent on the speed of swimming.
Distance attack. The same as the former but the attacker doesn't approach to within biting distance.
Chase. Prolonged attack on a fleeing fish, mostly accompanied by frequent biting or butting (see below).
Distance chase. The same as the former but the chasing fish doesn't approach to within biting distance.
Inhibited attack. An attack may be inhibited in several ways: (a) by being combined with threat (see below) which brakes the forward movement (b) by being followed by braking movements before the opponent is reached (c) by the locomotion itself being of an inefficient and inhibited-looking type. Of the three types the first and the last are most common. Inhibited attacks of the last type are usually very short, rarely exceeding 1 or 2 seconds in duration; they are also often followed or interrupted by braking movements; they are often intermingled with stretch and headshake movements; they are

often poorly aimed. The term 'inhibited attack' when used without further specification, will hereafter designate type (c) only.

Bite. When near to the opponent the fish suddenly darts forward and bites the opponent.

Butt. The same as the former but the opponent is touched with the closed mouth only.

Intention butt. The same movement as the former but the fish stops just short of hitting the opponent.

Tap. A sudden sideways movement, comparable to the forward dash in biting or butting, and touching the opponent with the mouth from the side.

Intention tap. The same movement as the former but without actually hitting the opponent.

Flight. The fish swims away from the opponent, typically as a response to attack. The locomotion is of the normal type. Fleeing may follow a horizontal course but often it is directed obliquely upwards. Prolonged fleeing is often directed towards vegetation or to other places (such as the corners of the tank, behind a stone or just near to the bottom) where an inferior fish often hides from the attacks of a dominant fish. Pressing against the bottom is typically seen in a fish which is in process of becoming inferior to another. It is obvious that during this phase fleeing may be directed slightly downwards. This is assumed to be equivalent to fleeing to any other hiding place. As inferiority becomes more clearly established, all attempts to hide near to the bottom are abolished and fleeing becomes more consistently directed upwards. In a very inferior fish practically all activities are confined to the surface layers irrespective of the presence or absence of floating vegetation.

Lateral threat. At highest intensity all fins, including the pectorals and the caudal fin, are maximally erected and spread. Unlike in braking, no parts of fins are bent sideways. At lower intensities often the dorsal fin seems to be erected first. However, at least once in *nigrofasciatus* the ventral and anal fins have been seen to be spread while the dorsal remained comparatively lax. It is possible that species differ in the order in which fins are spread with growing intensity of threat. The pectorals are spread only at highest intensities. The entire body becomes rigid, probably as a result of simultaneous contraction of body musculature at both sides of the body and tail. The body is often bent into a S-curve with two relatively sharp kinks, one in the region between the ventral and anal fins and one just behind the head, or into a C-curve (see fig. 39 p. 81). During threat there is a strong tendency to orient with the side towards the opponent, though all other orientations in the horizontal plane do also occur (see also frontal threat, below). If the opponent is below or above the threatening fish, the latter rolls over so far as to maintain lateral orientation. If an S-curve is present the head is always curved away from the opponent. With C-curve the head is always curved towards the opponent. Threat is often accompanied by very strong and rhythmic propulsory movements of the tail. In spite of the strength of these movements progress is typically slow, probably because of the friction exerted by the erected fins and the rigid body on the surrounding water.

Waggle. The threatening fish may perform a series of relatively slow pivoting movements while staying at the same place. Waggling is only performed when the orientation of the threatening fish is lateral and especially when it is standing in front of the threatened fish. It is most often performed during S-curve threat. Apparently waggling produces a strong water current directed at the opponent, originating from the concave lateral surface of the tail. It is functionally comparable to tail-beating in other fish. Tail-beating was not found as a typical element in any of the 5 species which are the main object of this study; it was found to be present as a common behaviour element in *Barbus titteya.*

Frontal threat. Frontal threat differs from lateral threat in orientation only; instead of lateral orientation, frontal orientation is actively sought. Frontal threat is relatively rare in all 5 species except *tetrazona*; mutual frontal threat (see mouth-fight) is common in that species, occurs regularly in *stoliczkanus* though always of short duration and is virtually absent in the other three species.

Head-down threat. The posture is essentially the same as in threat but the body, the long axis of which is kept horizontal during normal threat, is tipped forward by an angle varying from a few to 90°. As the angle increases, especially when surpassing some 45°, the body musculature becomes progressively more lax and the fins are kept less stiffly extended. Like in normal threat there is a strong tendency for lateral orientation relative to the opponent (and perhaps for frontal orientation in *tetrazona*). The downward inclination doesn't seem to bear any direct relation to the position of the opponent. If in a certain situation the total amount of head-down threat increases relative to the total amount of horizontal threat (this may for instance happen if a third male is introduced in a tank where two males are fighting a duel) the same shift occurs in threat of all orientations. On the other hand head-down threat, when compared to horizontal threat, seems to be more often accompanied by flight (usually not straight away from the opponent then, see DUNHAM *et al.*, 1968 p. 17 — 'Plough') and less often by attack.

Ventral roll. The fish rolls around an axis running through the bases of the pectoral fins and parallel to the longitudinal axis, directing its ventral surface towards the opponent. At the same time the fins are actively adducted and the body musculature seems to be relaxed. With its low muscle tone, active adduction of the fins and display of the narrow ventral surface ventral roll is the opposite of threat. Whereas threat normally may prevent attack from a conspecific by intimidation (*i.e.* by eliciting flight tendencies which in their turn inhibit the opponent's aggressive motivation) ventral roll withdraws all flight- and attack eliciting patterns from the view of the opponent and thus makes him loose interest. Ventral roll may thus be considered an appeasement activity. (DUNHAM *et al.*, 1968).

Dorsal roll. The same as the former except that, instead of the ventral surface, the dorsal surface is shown to the opponent. The effect upon the opponent is much the same as that of ventral roll. Whether ventral or dorsal roll occurs is at least in part determined by the relative positions of the attacker and the rolling fish. If the attacker comes from below and/or behind practically always ventral roll is performed; if the attacker comes from above and/or from the front often dorsal roll is shown.

In both ventral and dorsal roll orientation relative to the opponent is adjusted continuously if the relative positions of the 2 fishes shift.

At least in *tetrazona* frontal threat may sometimes merge gradually into dorsal roll through a series of intermediate postures, for instance: frontal head-down threat — frontal head-down threat with the snout pointing below the opponent instead of directly at him — concurrent relaxation of body musculature and of fin erection — dorsal roll. This may in its turn be followed by ventral roll and/or flight.

Besides the individual behaviour elements described above, a number of stereotyped actions involving two — and occasionally more — individuals may be distinguished. These actions make up the repertoire of duels (most often between two males, sometimes involving two females or a male and a female).

Parallel threat swim. The two fishes swim forward, parallel to one another, both threatening. Often the threat is of the S-curve type. The distance between the two fishes is mostly more than one cm. While progressing they often slowly diverge from each other. Forward movement may be in a horizontal plane but often a slight upward component is present (see also chapter III, section I). When head-down threat is performed instead of normal threat, the action takes the form of mutual *plough*; the divergation of the courses of the two fishes is much stronger than in normal parallel threat swimming (see DUNHAM *et al.*, 1968, p. 17). The course is always horizontal or slightly downwards in this case. *Parallel threat* may be performed without forward movement; *Parallel swim* may occur without threat. Both actions are much more rare than complete parallel threat swimming and are probably representative of relatively low fighting motivation.

Caroussel. The two fishes swim in a circle, antiparallel to each other, both threatening. The distance between the two fishes may be quite large (up to some 20 cm) but in the most intensive form of caroussel they are very close to each other, both touching the tail of the other with the mouth. When carousselling at greater distances threat is often of the S-curve type; when carousselling at close quarters, especially if circling is fast, the C-curve is more common. Close carousselling is invariably performed in a horizontal plane; carousselling at a distance may also be performed in an inclined plane. Inclinations above about 60° are virtually absent in the species discussed in this study. During horizontal caroussel both fishes may stay at the same level, but often they either rise or drop slowly during the performance (see also chapter III, section I). Caroussel may be performed with head-down threat; circling is always very slow in that case. *Antiparallel threat* may be performed without circling but this is relatively rare; *Circling without threat* in one or both fishes is very rare and always of short duration. Like the comparable actions of *parallel threat* and *parallel swim*, the last two actions are probably representative of low fighting motivation. Species differ from each other in the maximal speed of circling during caroussel (see chapter III p. 103 ff.). This probably has no simple relation to differences in strengths of fighting motivations.

Exchange of bites. The two fishes are close together and take turns in biting while the other turns broadside. Like in the other actions during a duel, typically both fishes are threatening during the performance.

Mouth-fight. This action is common and of long duration in *tetrazona* only; it often occurs in *stoliczkanus* too but it is always of short duration in the latter species (rarely more than one second). It is absent in *conchonius, nigrofasciatus* and virtually absent in *cumingi*. It is probably absent too in *phutunio* and *gelius*. 1)

In mouth-fighting the two fishes are oriented face to face, bite vigorously at each other's mouth and both are threatening. Often jaws get interlocked and they then may tug vigorously at each other's mouth. One or both may display head-down threat. Both threat and head-down threat are usually performed with the body quite straight. Often the fishes drop slowly during the performance (see also chapter III, section I). At rare occasions threat may fail in one or both fishes.

1) Mouth-fighting of a slightly different type is also common in some other species, *e.g. B. arulius* and *filamentosus*.

In a duel the actions described often alternate with each other. However, there is a tendency for parallel threat swim to occur preponderantly at the beginning of the duel; it is gradually replaced by carousselling at more and more close quarters; in *tetrazona* carousselling is in its turn more and more replaced by mouth-fighting. Exchange of bites tends to occur more often in later phases of a duel.

2. Female agonistic behaviour.

Females are usually rather placid, especially as long as active males are present in the same tank. In an all-female tank, however, or if the males present are very inactive, females may fight furiously among each other showing all elements of male fighting including parallel threat swim, caroussel, exchange of bites and — in *tetrazona* — mouth-fighting. Females do also attack males but, as a male as a rule wins quickly in a fight with a female, intensive rounds of duelling are most often seen between males or between females. Females commonly flee and/or roll as a reaction to a male's attacks or courtship.

For differences between the species in the occurrence of female fighting see chapter III, sections II, III.

3. Male courtship.

Courting posture. The male approaches the female, usually from behind, and takes a position with his long axis parallel to hers. Typically the male stands just below the female and a little to the side. The body is bent in a slight C-curve, concave to the side of the female, and is inclined towards the female (*leaning*). The body musculature is comparatively lax; body and tail make slight lateral quivering movements. The median and ventral fins are actively adduced; the pectorals are often flicked forward repeatedly with high frequency (about the same as in normal braking) and small amplitude. If the female moves forward the male moves with her; pectoral fin movements are stopped and the pectorals adduced during forward movement.

The various components of the posture may vary in intensity or in kind. The C-curve, leaning and quivering may be more or less pronounced. Instead of below the female the posture may also be performed beside the female or even somewhat above and to the side of her. If leaning is present in that position it is to the same side as in courting posture below the female. This means that, unlike in ventral and dorsal roll, not always the same side of the body is turned towards the female; if courting posture is performed below and to the side of the female, the female is more likely to see the dorsal surface of the male whereas she is more likely to see his lateral side if he courts beside or slightly above her. Further the variation in degree of leaning during courting posture below the female doesn't seem to be directly related to the relative positions of male and female and to a need to direct the male's back accurately towards the female (see also below: leading *s.l.*). Rather, there is a tendency for the male to lean more strongly and to stay more consistently below the female when courtship takes place in deeper water layers near to the bottom of the tank; when courting near to the surface both the position below the female and leaning tend to be abolished (see also chapter III, section II, p. 111). In addition to the above described variations in courting position the male may also stand perpendicularly under the female. Both leaning and C-curve are abolished then and the body of the male is directed more or less upwards, often touching the female gently from below with the mouth. In this position the male is relatively immobile; if the female moves on, the male immediately resumes normal courting position.

From the courting posture, which takes a central place in male courtship and which usually is the first element in a courting sequence, the male may proceed to either of two groups of elements: *leading s.l.* or *mating attempts*. In the first category the following elements may be distinguished:
Leading s.s. (referred to hereafter as leading). Starting from the courting posture the male suddenly swims forward, away from the female. He may follow a horizontal or a more or less upward course. (In *tetrazona* leading is also often directed more or less downward). As the male proceeds the C-curve disappears, the adducted fins are more or less relaxed and leaning is gradually abolished. If leading is obliquely upwards, leaning is abolished at once. Movements of the pectorals, if present, are stopped and the pectoral fins adducted. Locomotion is rhythmic and displays more or less the same quality as the quivering movements made during courting posture. Locomotion differs from normal fast locomotion in that the body musculature is more lax and the amplitude of the tail-movements seems to be smaller. (Perhaps also the frequency is higher.) Leading, both horizontal and upward, is often directed towards a clump of vegetation. This is particularly conspicuous in leading in *tetrazona* (see also chapter III, p. 105).
Circling. Instead of leading in a straight line the male, after swimming away in front of the female, may turn and circle below the female. The type of locomotion is like in leading, the C-curve is distinct and leaning is very strong. The distance of the circling male to the female may vary greatly; at one extreme the male turns almost on one place below the female, at the other

extreme he may describe circles with 20 cm or more diameter. After leading upwards the male may also circle above the female in the same manner as normal circling. Leaning in this case is particularly strong. After having completed one or more circles the male may go on leading in a straight line — usually again in front of the female — or he may return to courting posture.

Circle front and *half eight.* Instead of circling the female completely the male may turn towards the female either in front of her, which results in a brief *vis à vis* position, or, after having completed a half circle, obliquely from behind. Both movements may end with the male butting the female gently on the mouth or in the anal region. During the circling parts of both movements locomotion is again similar to that in leading, a distinct C-curve is present, leaning is strong and the fins are adducted. When approaching the female at the end of both movements C-curve and leaning are abolished, like it is in the courting posture perpendicularly below the female (see above). Both movements are always performed closely below the female and the distance to the female in the horizontal plane is usually relatively small during the circling parts.

Circle-leading. All kinds of combinations between circling and leading may occur, the male describing more or less elliptic courses, the long axes of which usually extend in front of the female. The plane may be horizontal or slant upward in front of the female. In the latter case leaning is usually strong at the point farthest away from the female as the male starts to return to her. As a general rule, leaning is always very strong whenever the male is courting above the female and starts to return to lower levels.

Starting from courting posture the male may, instead of leading *s.l.*, proceed to mating attempts. Whereas leading *s.l.* occurs more frequently in a male courting in open water, and apparently serves to induce the female to enter clumps of vegetation, mating attempts are more frequent within, or at least in contact with, vegetation clumps.

Mating attempt. From courting posture the male moves upward or forward and upward and slides alongside the female. Quivering becomes more intense, the C-curve remains but leaning is abolished. The fins are kept folded. Two intensities were arbitrarily distinguished: *mating attempt 1* in which the male covers about half of the female's side or less, and *mating attempt 2* in which the male covers more than half of the female's side. Often the male's orientation is not precisely parallel to the female but his head is slightly more close to her body. From this position the male often taps or intention taps the female's side. The male's orientation is less precisely parallel and tapping is more frequent during mating attempt 1 than during mating attempt 2. Table 2 presents frequencies with which mating attempt 1, mating attempts 2 without subsequent mating and mating attempt 2 with subsequent mating were accompanied by at least one tap or intention tap; data are from *nigrofasciatus* males.

TABLE 2

Percent of mating attempts accompanied by tapping movements (nigrofasciatus)

		% + t.m.			n		
		m.a.1	m.a.2	m.a.-mate	m.a.1	m.a.2	m.a.-mate
♂ 1	♀ 1	94	58	18	296	64	161
	♀ 2	92	76	26	411	238	112
	♀ 3	94	71	35	85	35	48
♂ 2	♀ 4	92	76	20	681	240	290
	♀ 5	94	88	43	149	331	243

It follows from the data presented in the table that the nearer a male is to completed mating, the lower the probability of tapping during mating attempts. At rare occasions a male may, from mating attempt, switch to a completely frontal orientation; this may be followed by a gentle butting on the side of the female. This resembles the butting which may follow circle front or half eight (see above). It probably only occurs if courting motivation is low.

Mating. When the male has come completely alongside the female, he tips head-down a little relative to the female and bends his tail over the back of the female and around her. Often the pair rolls over (maximum 90°) with the male on top. The male's fins remain adducted and quivering reaches a maximum culminating in an intensive brief jerk. At this moment a few eggs are released by the female and the male probably ejects some sperm simultaneously. At the same moment male and female part. The whole act lasts a fraction of a second only. Only in *tetrazona* the mating position may occasionally be held for a second or more before the jerk ensues. The male usually resumes courting immediately after mating by courting posture, leading or a renewed mating attempt.

Keeping-off. When a courting male is approached by another male or female he swiftly takes a position between the intruder and the courted female and tries to stay between them. In doing so he compromises between keeping parallel to the courted female and turning broadside towards the intruder. Keeping-off is most often accompanied by threat or head-down threat, more often so when the intruder is a male than when it is a female.

Female intruders are usually easily discouraged. If the intruder is a male he may flee as a reaction to keeping-off or he may be more persistent. He may either keep swimming straight towards the female as if he wants to swim right through the keeping-off male or he may try to manoeuvre around the keeping-off male to keep him off in his turn. He may also keep-off a third male from the female plus the first male. Sometimes neither of two male manages to keep-off the other. They then either follow the female, both threatening and pressed very closely together, or they may loose the female and start parallel threat swim and/or caroussel.

4. Female courtship.

The female plays a much less active role in courtship than the male. Still, by performing or not performing the appropriate response to the male's courtship, the female may decisively influence the course of courtship. The following general rules are valid for all species (for species differences see chapter III, section II). A male often stops courting a female if she (a) attacks or threatens him (b) flees too fast or rolls too intensively in response to his approaches (c) stands still completely (d) displays too much other activities such as intensive feeding (e) doesn't respond positively to his advances for periods exceeding a few minutes. The female's positive responses may be: (a) swimming actively to a place where mating can take place; she may do so by following a leading male or by selecting the place and the moment herself during the male's courtship. (b) allowing the male to slide alongside and to mate. It is not quite clear what the active role of the female in the mating act is but if she remains completely passive the male doesn't succeed in mating. If a female is willing and if a courting male is beside her but doesn't come near enough to mate, she may actively sidle up to him and mating may ensue. This response was only seen in *tetrazona*. If a female is very willing and no courting male is present, she may perform mating movements *in vacuo*: pressing against a plant, rolling to one side and quivering. This behaviour may also be observed in willing females when courted by a male. She then leans away from the male. The male usually responds by instant mating. The female's leaning resembles male courtship leaning in that she doesn't seem to show any specific side of the body to the male (such as in ventral and dorsal roll). It differs from male leaning in that it is with the back away from the partner. The occurrence of female leaning in normal courtship just before mating suggests that the rolling over of the pair during mating is initiated by the female. This suggestion fits the fact that female *vacuum* leaning is only displayed near to the water surface and that rolling during mating is also stronger in higher water levels (see table 3 and also chapter III p.143, table 5).

TABLE 3

Relation between degree of rolling-over during mating and depth of pair in water (proportion of total water height above bottom). B. nigrofasciatus

height	mean angle *)	n
0–1/10	87°	33
1/6–1/2	82°	59
3/5–5/6	79°	17
9/10–1	63°	28

*) 90° = vertical.

5. Male-male courtship.

Males of all species are often seen to court conspecific males if no females are present. For male-male courting in mixed groups see chapter III, section II, III.

When two males are closely pressed together during mutual attempts at keeping-off (see p. 69) one of them may suddenly perform complete mating with the other. No other elements of courtship are seen in this situation. See also chapter III, p. 124: homosexuality).

6. Female-female courtship.

If no active males are present, females may perform male courtship directed at other females. This may include courting posture, leading *s.l.* and mating attempts. Complete mating between females was never observed.

SECTION II. MOTIVATION OF SOME REPRODUCTIVE BEHAVIOUR ELEMENTS AND COMPONENTS

A. Agonistic behaviour.

Animal fighting behaviour is often considered as a product of two antagonistic tendencies: one to attack and one to flee (for a review see HINDE, 1966, chapter 16). The two tendencies are assumed to be irreducible to one another. The strengths of the motivations under-lying the two tendencies are thought to be mutually more or less independent.

It follows from this hypothesis that agonistic behaviour may take two basic forms: (1) one of the two tendencies may be present alone and thus give rise to pure attacking or pure fleeing behaviour. Differences in strength of the dominant motivation may result in different intensities and/or kinds of aggressive or fleeing behaviour. For instance, a low aggressive motivation may, in the absence of flight tendencies, lead to low intensity attack; a higher aggressive motivation may lead to stronger attack and/or to biting. (2) Both tendencies may be present simultaneously (or in rapid alternation) and give rise to various forms of behaviour which may be commonly referred to as conflict behaviour.

Of the various kinds of conflict behaviour described and discussed in the literature, the following seem to be relevant to the present study: (a) ambivalent behaviour, *i.e.* behaviour patterns in which elements of both tendencies are recognizably present; such patterns may or may not be entirely built up out of such elements. (b) Compromise behaviour (ANDREW, 1956) *i.e.* behaviour patterns in which components which are common to both tendencies are present, may be in an exaggerated form. This category is a rather mixed one including such different examples as tail-flicking in a bird during an approach-avoidance conflict (tail-flicking being an intention movement of both approach and avoidance behaviour) and the assumption of the female solliciting posture by a male *Emberiza* as a compromise between the tendencies to copulate and to flee from the female (ANDREW, *l.c.*, p. 42-43). Compromise behaviour may occur in the absence of certain incompatible expressions of the two conflicting tendencies (as in the first example) or elements of one or both tendencies may be also present; in the second example the assumption by the male of the female solliciting posture instead of copulating may be interpreted as an expression of a fleeing tendency. Apart from the fact that ambivalent and compromise behaviour may occur in one and the same pattern, the delineation between compromise and ambivalent behaviour may often be vague. The latter point is also demonstrated in the second example. (c) autonomous responses, *i.e.* changes which result from activation of the autonomous nervous system and the adrenal hormone systems under influence of the state of conflict [1]). Examples of such responses in mammals and birds are pilo-erection and vasodilatation (MORRIS, 1956); in fishes colour changes (see CAMPBELL, 1970 for reference) changes in swimbladder volume (CAMPBELL, 1970; STEEN, 1970; FÄNGE, 1943) and changes in general muscle tone are among the most obvious examples.

The omission from this listing of other categories of conflict behaviour — such as displacement or redirection — does not imply that these do not play a role in *Barbus* behaviour; they are relatively irrelevant to the present discussion however.

In addition to the two tendencies and their interactions described above, several other, equally independent, tendencies have been postulated as necessary for the causal explanation of fighting behaviour. For instance, BARLOW (1963, p. 49) suggests that ventral rolling in *Badis*, a posture occurring in

1) This is not meant to imply that no other than adrenal hormones may be brought into action by a state of conflict or even as a result of the activity of one behaviour system only.

duels between males, is derived from female sexual behaviour. (This does not necessary imply sexual *motivation* in the case of a duel). BARLOW (1961, p. 186; 1962, p. 41-45) also suggests the existence of *defense* as a more or less independent sub-system of the fright system to account for fin-spreading (especially the spiny rays) and dorsal roll in *Badis*. (The fright system in his terminology may be held more or less equivalent to the fleeing system in the present discussion).

LORENZ (*e.g.* 1964, p. 40-43) and others have suggested that threat postures in cichlid fishes, though originally of purely ambivalent nature, have become ritualized in the course of phylogeny — both by addition of 'new' motor patterns to the original posture and by changes in threshold of threat behaviour relative to pure attack. Both processes necessarily imply the assumption of at least some degree of emancipation of threat postures from their original dual motivation. Some authors have claimed the existence of 'symbolic' and therefore completely emancipated occurrence of threat and inferiority postures (*e.g.* SEITZ, 1943).

For the purpose of the present analysis I will take the hypothesis of the dual motivation of fighting behaviour as a working hypothesis, taking into account the various possible forms of conflict behaviour listed above. I will not challenge the dual nature itself. For reasons of parsimony, however, I will attempt to interprete the whole of *Barbus* fighting behaviour in terms of the two tendencies and their interactions alone; additional independent tendencies will be postulated only if interpretations in terms of the working hypothesis fail.

(a) attacking and butting movements.

For the purpose of this discussion the term 'butting movements' will be used to denote the whole category of bite, butt, intention butt, tap and intention tap. Of the various forms of butting movements, *biting* seems to be the most intensive and complete movement. *Butting* may be considered as a less intensive form, because the final forward dash is mostly less intensive than in biting and the jaws are not clearly opened and shut as in biting.

In contrast to biting and butting, *tapping* is typically given from a more or less lateral orientation. This may indicate the presence of a conflict between straightforward butting and a tendency to turn away from (avoid) the other fish. Tapping and intention tapping often accompany lateral threat and (in male courtship, see below) mating attempt.

Both *intention butting* and *intention tapping* are characterized by the fact that the other fish is not actually touched. This might conceivably be due to the other fish speeding up or moving away at the moment that the aggressor starts the butting or tapping movement. However, close observation suggests that the failure to hit is rarely if ever due to the behaviour of the other fish but instead to the aggressor stopping short of hitting. At least in intention butting the initial forward dash does not seem to be less strong than in actual butting. This indicates that intention butts and intention taps are inhibited butts and taps rather than low intensities. (Because of the speed of the movements I was unable to see whether braking movements were involved). It is reasonable to assume that in intention taps the same conflicting tendency which causes lateral deviation is also effective in preventing actual hitting of the other fish.

It may be concluded that the form of butting movements is determined (1) by the strength of the aggressive motivation (2) by conflicting tendencies which may act in two ways: (a) by stopping the butting movement just before hitting (b) by forcing the fish into a lateral orientation thus preventing straight butting. For reasons of parsimony it may be assumed that in both cases the inhibition of butting movements is due to the fleeing tendency. (For tapping movements occurring during threat and during mating attempts see p. 86, 99). The presence of two different kinds of conflict (inhibition of the butting movement and lateral orientation) may be explained by assuming that in the first case the fleeing tendency conflicts with the butting (biting) component of aggressive behaviour and in the other case with the attacking component (see below).

As far as fighting behaviour is concerned [1]), butting movements occur mostly during *attack* (which may or may not be accompanied by threat). Tapping and intention tapping may also occur during stationary lateral threat. As may be seen from figg. 55, 63, 71 (chapter III section IV, p. 205, 235, 268) the number of butting movements given in a certain period of time is strictly correlated with the time spent attacking in the same period. Attacks are not always accompanied by butting movements, but butting movements without attack are extremely rare except during stationary lateral threat [2]) (see below). These facts suggest that butting movements

1) For butting occurring during courtship see sub-section B.

2) It might be argued that this asymmetry arises from the fact that butting can only occur after the distance to the other fish has been sufficiently reduced. However, close proximity between two fishes also often occurs during schooling and incidentally between fishes moving about in a random fashion. In these situations butting movements are rare; rather, head-on orientation seems to be avoided in the absence of aggressive tendencies.

and attacking are expressions of the same major aggressive tendency and that butting movements reflect a relatively high aggressive motivation.

Distance attack differs from attack only in that the distance to the attacked fish is greater. Like in intention butting the greater distance between the two fishes might in principle be due to stronger avoidance in the attacked fish or to a less strong approach tendency in the attacker. However, close observation suggests that the occurrence of distance attack is practically always due to the latter cause. In *distance chase* the chased fish never moves so fast that the aggressor could not catch up if he intended to do so. Unless distance attack is accompanied by threat there are no outward signs of conflict such as might be expressed by the simultaneous occurrence of strong locomotion and braking. In this respect distance attack differs from intention butting. These facts suggest that distance attack is a low intensity form of attack, reflecting a relatively low aggressive motivation.

As will be shown elsewhere (KORTMULDER, in prep.) there is no direct relation between the time spent attacking in a certain length of period and the ratio of distance attack to attack. Thus it seems that the strength of the aggressive motivation on the one hand and the time during which aggression is the dominant motivation on the other are mutually independent variables. A similar phenomenon may be found in the lack of correlation between the time spent attacking on the one hand and the number of butting movements per time spent attacking on the other (see chapter III, section IV, figg. 55, 63, 71, p. 205, 235, 268).

Inhibited attack differs from distance attack in the following points: (1) the distance to the other fish, though often remaining relatively large, may be as small as in attack; (2) whereas distance attacks are variable in duration (durations of 10-20″ are not rare), inhibited attacks are almost always short (1-2″ at most); (3) in contrast to distance attack, there are obvious outward signs of conflict. These may be of three kinds (see description p. 60): (a) braking movements, (b) poor aiming, *i.e.* slight to gross lateral deviations, (c) inefficient locomotion. (a) and (b) are similar to the types of inhibition found respectively in intention butting and in tapping; (c) may reflect an internal conflict resulting in peripheral disorder (see also below: p. 96).

For reasons of parsimony it may again be assumed that all three types of inhibition arise from the fleeing tendency. This interpretation is in accordance with the fact that inhibited attacks are typical of inferior fishes in small groups (2 ♂♂ or 3 ♂♂ for instance) whereas both attack and distance attack are typical of dominant fishes in the same kinds of group and of fishes of roughly equal strength. Inhibited attacks may also be frequent in dominant fishes with low aggressive motivation (as judged by the time spent in full attack). In this case there are no reasons to suppose that the fleeing motivation is less strong than in periods in which the aggressive motivation is high (see KORTMULDER, in prep.). Thus inhibited attacks in this situation probably also arise from a conflict between attacking and fleeing, though perhaps at a lower level of intensity of the two motivations than in the inferior fishes.

It may be concluded that there are no reasons so far to postulate more than one single aggressive motivation — interacting with a fleeing tendency — to account for the various forms of aggressive behaviour discussed. On the contrary, similarities in form and orientation of the various elements and their close temporal association with each other (except for inhibited attacks) suggest that they all are expressions of a common aggressive tendency. One reservation must be made: apparently the fleeing tendency may interact differentially with different components of the aggressive behaviour. One of these components is the butting movement which involves movement of the whole of the head, body and tail; the other is the attacking component which seems to involve action of body and tail only. It is further suggested that conflicts between the two tendencies may be relatively more peripheral (as in intention butting, tapping and intention tapping) or more central (as in inhibited attack).

Apart from differences in speed and in duration, all *fleeing* behaviour is so uniform that there is no difficulty in ascribing it all to a single fleeing tendency. Since fleeing is often directed upwards and/or towards cover, it is reasonable to assume that hiding behaviour — such as may be shown by inferior males in small groups [1]) — and seeking high water levels are products of the same fleeing tendency.

(b) threat behaviour.

Threat postures are often interpreted as typical examples of ambivalent behaviour, containing both attacking and fleeing intention movements. As mentioned above, various degrees of emancipation from an originally ambivalent motivation have been claimed for some species of fishes. In the following analysis it will be shown that both interpretations are largely unsatisfactory in the case of *Barbus* threat postures. For the sake of the argument I will first attempt to interprete the various components of threat (as described on p. 61 in terms of the ambivalence hypothesis. Only by doing so it may become clear in how far the hypothesis is unsatisfactory in this case.

1) See chapter III, sections III and IV.

The preponderance of *lateral orientation* during threat may be interpreted as a result of a balance between the tendencies to turn towards (attack) and to turn away (flee) from the other fish. A similar interpretation may fit the *simultaneous contraction of the body and tail musculature* of both sides. Turning in a horizontal plane during normal locomotion was described on p. 58. The initial stage of such turns involves a bending of the front part into the new direction. This bending must be effected through contraction of the rump musculature of the ipsilateral side. It is quite conceivable that a fish, which has its side turned towards an opponent and which is motivated simultaneously to turn towards and away from the opponent, will contract rump muscles at both sides. The contraction of the rump musculature during threat may thus be interpreted as a combination of attack and fleeing intention movements. The same interpretation may account for the *rolling movements* by which a threatening fish manages to maintain broadside upon an opponent even when it is above or below it. As was described on p. 58, sharp turns in a vertical or inclined plane are made by rolling to one side and then bending the body. Intention movements to turn in a vertical or inclined plane is exactly what a threatening fish is expected to show on the basis of our hypothesis, when it is above or below the opponent.

In the light of the analysis presented so far, the common presence of C and S curves during threat gets a special meaning. C curve may indicate a relative preponderance of the aggressive tendency. Accordingly, C curve typically obtains during close carousselling and is then accompanied by frequent butting or prolonged mouth-to-tail contact. S curve involves a double curving, one in the anal region and one just behind the head. The anal curvature is topographically more similar to the curvature seen in turning during normal locomotion. It might be interpreted as a (prepondering) intention movement to turn towards the other fish. The pectoral curvature is topographically different but — assuming the possibility of behavioural homonomy of consecutive body segments — it might be interpreted as a (prepondering) intention movement to turn away [1]). This interpretation of the S curve would imply that the balance between attack and flight is slightly different in different parts of the body. Analogous phenomena are known from display postures in birds. For instance, in the aggressive upright threat posture of the herring gull the arms are slightly lifted, which results in a raising of the carpal joints, but the hand parts of the wings remain

---

1) I couldn't say for sure that bending of the body in this region does not occur during normal locomotory turning, in addition to and to the same side as the curvature in the anal region. VON HOLST (1935, p. 660) describes lateral bending of the front part in spinal *Cobitis* attempting to change their direction of movement.

adducted (TINBERGEN, 1959). The opposite occurs in two-sided wing-lowering and in waltzing of cocks: the hand feathers of the wings (or of one wing as in waltzing) are fully spread whereas the arms remain adducted or are lifted a little way only (KRUIJT, 1954, p. 61-63). In certain human intimidation postures the upper arms are lifted and moved slightly forward, while the lower arms and hands are depressed (LORENZ, 1964; PRECHTL, 1954).

In some cases differences in balance between two tendencies in different parts of the body may be understood by assuming that the part of the animal which is nearer to the opponent is relatively more afraid, while the other part is relatively more aggressive. KRUIJT (1964) is quite successful in interpreting in this way the complex posture of a waltzing cock. In the case of S curve in a threatening barb, however, this kind of interpretation cannot explain why the body is always turned towards and the head away from the opponent and never the other way round 1). (S curving during frontal threat might be understood as a result of relative retreat of the head part, because it is nearest to the opponent, but in barbs S curving is typically absent during frontal threat). S curving during lateral threat might be explained by assuming, by way of a secondary hypothesis, that the head is *subjectively* nearer to the opponent because the fish has some 'expectancy' that, if it is going to hit the other fish, it is going to do so with its head. A similar interpretation would fit the depression of the lower arms and the hands in the human posture mentioned above.

The pivoting movements made in *waggling* during lateral threat might be interpreted as an alternation of attack and flight intention movements (successive ambivalence) superimposed on the simultaneous ambivalence of threat itself. Waggling may culminate in either butting the opponent (mostly on the side) or fleeing from the opponent (mostly combined with prolonged threat). There are several reasons to believe that the occurrence of waggling reflects a state of balanced conflict between two extremely strong tendencies: first, it is only seen in lateral orientation; second, it often coïncides with S curving; third, it is most pronounced when the performer stands in front of the opponents mouth. To my general impression it is the front end of a conspecific which may induce a maximum of flight *and* attack behaviour.

It may be emphasized that the turning movements made in waggling cannot be effected in the normal way because of the rigidity of the body; strong lateral beats with the tail are probably responsible for the lateral displacement. It is only when waggling merges into overt butting or fleeing that the normal bending of the body may resume its function.

---

1) The concept of 'typical intensity' or 'typical compromise' MORRIS, 1957a) may serve as a label here but raises questions rather than answering them.

The *erection of the fins* during threat may, according to our hypothesis, be derived from the raising of fins during turning and braking. The absence of lateral deviations in the unpaired fins and the absence of asymmetry in the extent of raising of the paired fins may be explained by the simultaneous presence of the tendencies to turn to either side [1]).

The presence of *strong locomotory movements* of the tail combined with braking may be interpreted as another expression of strong conflict.

It would follow from the above analysis that the conflict producing threat is mainly between the fleeing tendency and the attacking component of the aggressive tendency; *i.e.* the same kind of conflict which causes tapping movements (see p. 74: inhibition type (b)).

However well the ambivalence hypothesis may seem to fit the *Barbus* threat posture — even in as detailed an analysis as presented above — there are some features in *Barbus* threat behaviour which cannot be accounted for by the hypothesis and even raise serious doubts as to the validity of the hypothesis as a whole.

One of the basic implications of the ambivalence hypothesis is that, wherever two antagonistic sets of muscles are simultaneously active, it must be possible to ascribe the activity of either set to one of the postulated tendencies. This is not always possible in the case of *Barbus* threat. For instance, if a threatening fish is approaching an opponent, the strong locomotory activity of the tail may be ascribed to the attacking tendency, the braking movements to the fleeing tendency; on the other hand, if the same fish is swimming away from the opponent, the flight tendency may be held responsible for locomotion and the attacking tendency for braking. This implies that, if a threatening fish first approaches and subsequently swims away from the opponent, the two tendencies must switch roles in command of muscle groups at the point of nearest proximity to the opponent. Fig. 38a gives a schematic representation of a course that may be followed by a threatening fish.

One might expect, on the basis of the above suppositions, that the locomotory movements would stop and the threat posture (at least the erection of the fins) be interrupted or at least relaxed at the point of nearest proximity. This is often not the case. As a rule the threat posture is maintained continuously throughout the whole manoeuvre. If anything, the threat posture

---

1) Photographs show that not always both pectorals are extended to the same degree. The fin to the side of the opponent may be spread while the contralateral one is adducted. No sufficient material is available to know whether this is typical or whether the opposite may occur too.

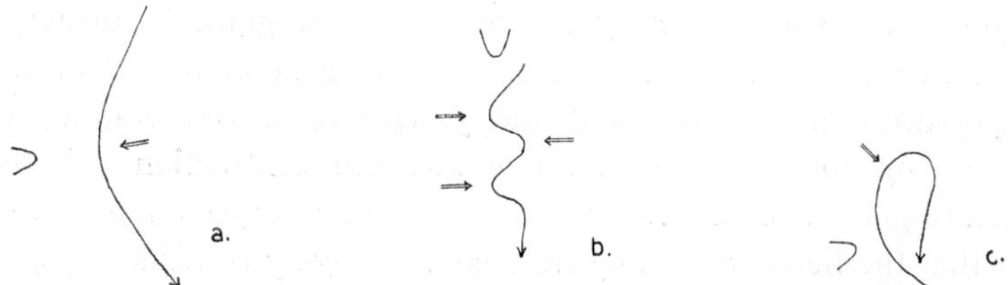

Fig. 38a-c. Courses, relative to opponent, followed by a threatening fish; top view; front end of opponent indicated; explanation see text.

is strongest at the point of nearest proximity. This observation casts doubts upon the ambivalent nature of the locomotion-braking antagonism.

More or less the same may be demonstrated for the ambilateral contraction of the body musculature. Figures 38 b, c exemplify some typical courses which a threatening fish may follow relative to its opponent. In both cases the attacking and fleeing tendencies must again be expected to exchange command of the lateral muscle systems at the points marked with an arrow. Nevertheless, the threat posture may be maintained uninterrupted throughout the courses shown. To this it may be added that ambilateral contraction also occurs during frontal threat, when neither side is nearer to the opponent. Finally, from close observation, one often gets the impression that — with sufficient excitation of the threatening fish, such as may obtain during prolonged threat — the direction of the locomotion relative to the opponent may become more or less arbitrary. The performance of the fish in the last example is vaguely reminiscent of the unoriented 'compulsory' locomotion described by von Holst (1935, p. 660) in spinal *Cobitis*. These observations cast serious doubts upon the ambivalent nature of ambilateral muscle contraction.

The invalidation of the interpretations of the above features in terms of ambivalence also affects the concurrent interpretations of rolling movements, S and C curves and fin raising during threat. For fin-raising the concept of compromise behaviour might be relevant; fin raising may be assumed to belong to both the motor patterns of turning to either side, irrespective of where the opponent is. The same concept might be adduced to account for the impossibility to ascribe the locomotory movements and braking to either of the two tendencies. However, the concept can hardly account for the simultaneous activation of antagonistic muscle systems.

It may now be emphasized that the ambivalence hypothesis also requires the adduction of several secondary hypotheses: behavioural homonomy of consecutive body segments and 'expectancy' in the case of S-curve; a special technique of horizontal turning in the case of waggling. Further, the ambivalence hypothesis does not account for the general resemblance between threat postures on the one hand and yawn-stretch and braking-stabilizing movements on the other. In both ambilateral contraction of the rump musculature goes together with fin-raising. The interpretation also overlooks the fact that the bends obtaining in S curve roughly coïncide topographically with those found in stretch and head-shake.

The anal curvature of the S curve was interpreted above as bending the body, relative to the tail, in the direction of the opponent. It could, however, equally well be interpreted as bending the tail relative to the body. In a relatively short fish like a *Barbus* it is difficult to decide between the two possibilities by direct observation. However, in other fishes (*e.g. Macropodus* or *Badis*) the tail may — during lateral threat — be bent towards the opponent so strongly that it is manoeuvred right in front of the opponents head. In these cases there can be no doubt as to the 'intention' of the bending movement. It is possible that the same movement is present, to a lesser degree, in S curving in *Barbus* during threat. If this is true, it would also affect the interpretation of S curving in terms of ambivalence.

As was described on p. 62, S curve threat plus waggle may produce a strong water current directed at the opponent. The movement may be used by a dominant fish to jet an inferior fish out of a corner in which it was hiding or to sweep it up while resting on the bottom. In the light of the interpretation of the threat posure, S curving and waggling in terms of ambivalence, this would seem to be a rather intelligent use of the posture. Though this is of course not impossible, any interpretation which accounts for the normal biological significance of lateral 'water jetting' should be preferred. The wide-spread occurrence of tail beating (functionally equivalent to threat + S curve + waggle in *Barbus*) in fighting fishes would suggest the relevance and the necessity of such an interpretation.

Finally — and perhaps most importantly — the ambivalence hypothesis can not account for the observation that lateral orientation is often obviously sought by a threatening fish; this feature is quite different from the, equally undeniable, fact that threat is most easily elicited in a fish which has its side turned towards the opponent.

In view of the considerable success of the interpretation of the threat posture in terms of the ambivalence hypothesis and the subsequent failure of the same hypothesis to account for a variety of other features of threat behaviour, it may be tempting to assume that *Barbus* threat originally was a purely ambivalent posture and to ascribe to phylogenetic ritualization and emancipation all those features which do not fit the ambivalence hypothesis. However, such a solution, apart from being facile, would do paltry justice to the respectable concepts of ritualization and emancipation, reducing them to mere interpretational stop-gaps.

A way out of the impasse may be indicated by a few simple experiments I conducted with a male *nigrofasciatus* which had been specially conditioned to fight models. The model used in the experiments was ball-shaped and measured 2.1 cm in diameter. It was fixed at the end of a stiff rod.

Provided that the fish was in a mood to fight the model, the threat posture could be elicited in the manner of a reflex whenever the model was moved near to the side of the fish. This could be repeated at will. On the other hand, if the model was moved to the front of the fish, the latter bit vigorously and repeatedly *but did not threaten.* These experiments strongly suggested that for threat posturing to occur a specific external stimulus is necessary: tactile stimulation of the lateral surface.

C or S curving often accompanied threat in these experiments. Which of the two obtained could be determined at will and apparently depended entirely upon the relative position of the model to the fish. If the model was presented somewhere between the hind part of the body and the centre of the tail, the fish invariably responded with S curve threat, the concave surface of the tail to the side of the model. If the model was presented near to the pectoral region, C curve threat was given, also with the concave side to the side of the model. Figures 39 a, b give a schematic top-view of the S and C postures and the relative positions of the model which optimally elicited them.

Fig. 39a-b. Body curving by threatening fish, dependent on relative position of model; top view; explanation see text.

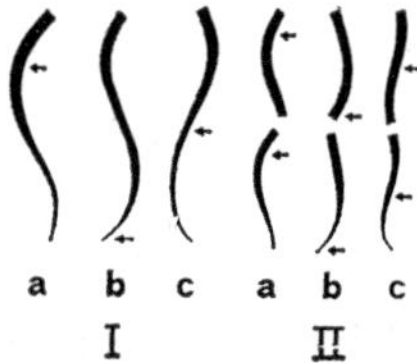

Fig. 40. Body curving by spinal preparations of eels, dependent on relative location of tactile stimulus; top view; (after VON HOLST 1935); explanation see text.

The reactions of the fish to the model bear a remarkable resemblance to some of the responses to lateral stimulation which VON HOLST (1934; 1935) obtained in spinal fish. Figure 40 shows the body postures taken by a spinal eel on lateral stimulation with a fine brush at three different locations.

The first three figures refer to the hind half of an eel which had the spinal chord cut in the middle; the other figures represent the responses of body parts after the hind half of the spinal chord had again been divided by the middle. As may be seen from the figures, the parts react in much the same way as the whole; it is only the *relative* location of the stimulus which determines the posture of the part; the length of the segment and its anatomical position in the whole animal are practically irrelevant. Given sufficient excitability of the preparation, a single short-lasting stimulus first causes the postures as shown in the figures and subsequently the curvature runs backwards along the rump. With prolonged stimulation, however, the postures are maintained (see below).

Most important for the present discussion is the similarity in the location of the stimulus and the resulting response between figures 39 a, b on the one hand and figures 40 I a, c and 40 II a, c on the other. (Situation b of figure 40 could not be produced in the *Barbus* experiments). Since the body of a barb is much shorter than that of an eel (and its pattern of locomotion more of the *carangiform* rather than of the *anguilliform* type; BREDER, 1926) it is perhaps not surprising that the postures shown by the threatening barb are most similar to those of 1/4 eel.

Several other responses were elicited from the same and other *nigrofasciatus* males with the help of models. Some of these responses also were similar to spinal responses reported by VON HOLST for a variety of fishes. If the model was presented at the location indicated in figure 39 a, strong locomotory movements of the tail often accompanied S curving. This response was quite similar to the combination of S curve threat and 'compulsory' locomotion seen in normal fighting between conspecifics (see p. 79). The response was typically shown during the early stages of a presentation of the fish with the model. The locomotory movements then often carried the fish away from or in a circle around the model. While moving away from the model, threat, S curving and locomotion were maintained for a while and subsided together as the animal stayed away from the model or re-attacked it. The combination of body curving and locomotion is also reported by VON HOLST for spinal eels. On one-sided stimulation first the curvature as depicted in figure 40 a obtains; as stimulation is continued the curving becomes progressively stronger until suddenly rhythmic oscillations are started which are similar to normal locomotory oscillations. The oscillations are superimposed on the curvature of the body which is maintained during stimulation. According to VON HOLST, the locomotory movements in this case result from a balance between the exteroceptive stimulation on the one side and proprioceptive stimulation caused by the curvature on the other. (Similar locomotory oscillations may be elicited by simultaneous tactile stimulation at both sides) (VON HOLST, 1935, p. 660). This finding could well account for the 'compulsory' nature of locomotion in barbs during S curve threat.

During the later stages of a presentation of the *Barbus* with the model, weak waggling movements often accompanied S curve threat. Strong waggling together with S curve threat was easily elicited from several *nigrofasciatus* males as a response to lateral stimulation with a blunt-pointed model at the same location where S curve threat was typically elicited with the ball-shaped model. Also in these individuals S curve threat + waggle was most easily elicited during the later stages of a presentation. The waggling response is very similar to the "Rumpfschlagreflex" reported for spinal goldfish and other species in response to lateral stimulation (VON HOLST, 1934, p. 583).

Some of the spinal reflexes described by VON HOLST were derived from fishes as different from barbs as eels and *Cobitis*, and they were derived from spinal preparations rather than from intact fishes as was done in the *Barbus* experiments. Nevertheless, the reflex-like character of the *Barbus* responses and their similarity to those of the spinal fishes suggest that the two groups of responses are homologous.

Whereas in the eel rhythmic locomotory movements could be elicited (in the ways described above) in preparations comprising as little as the posterior 1/5 of the total length, in the goldfish rhythmic movement could only be produced in preparations in which the neural chord was cut as far forward as the posterior end of the medulla or, even better, in the anterior part of the medulla. Apparently, the ability of the neural chord to produce locomotory rhythms, which in the eel extends backwards up to about 4/5 of the total length, is restricted to the anterior part of the spinal chord and the medulla in the goldfish (VON HOLST 1934, p. 591-592; 1935, p. 660). *Cobitis* seems to hold an intermediate position. Assuming that the barbs are more similar to the goldfish than to the eel or to *Cobitis*, it may be surmised that in the *Barbus* experiments it is especially the bend in the pectoral region which — in combination with unilateral exteroceptive stimulation — causes the 'compulsory' locomotory movements. This would be in accordance with the finding that the locomotory movements especially accompany S curve threat. As may be seen from figures 39 a, b, the pectoral region of the body is bent much more sharply in the S than in the C curve [1]).

Two more responses found in the model experiments must be reported: (1) As was described above, the fish which was stimulated with the ball-shaped model responded with C or S curve threat dependent on the location of the stimulus. Whereas it was very easy to approach the fish in the anal region and thus elicit S curve, the fish obviously tried to avoid the pectoral stimulation by moving slightly forward and thus bringing the anal region near to the model; it then promptly responded with S curve. Considerable steermanship of the experimenter was required to keep the model near to the pectoral region and thus provoke C curving. On the other hand, proximity of the model to the anal region was obviously sought by the fish.

(2) Once threat had been elicited in the beginning of a presentation, the fishes obviously sought to repeat and to prolong lateral stimulation by the model. The most striking examples of this response were seen in some of the fishes which were presented with the blunt-pointed models. First threat was elicited once or twice by approaching the fish with the model moved by hand. Subsequently the model was fixed to a little machine which made it move around in a narrow circle (ø *ca* 5 cm). The fish then approached the model and presented broadside to the leading blunt-pointed end of the model. By doing so the fish stimulated itself to repeated threat and moved away again. After a while it approached again and the procedure was repeated. The tendency to present the lateral surface to the model seemed to increase as the time spent threatening increased; at the same time the prolongation of the threat posture seemed to become less dependent on direct stimulation: threat + S curving + locomotion were maintained for some seconds as the fish moved away from the model (see above); eventually the fish threatened continuously, swimming close to the model, for at least several minutes on end. For neither of the two responses (1) or (2) counter-parts could be found in the experiments by von Holst.

Finally, it appears from the experiments that the threat posture itself (ambilateral muscle contraction and fin-raising) obtained by way of a reflex to tactile lateral stimulation. Also for these responses no counter-parts may be found in the experiments by von Holst.

Though the physiological evidence is fragmentary as yet, I hypothesize — on the basis of the model experiments described above — that *all components of the Barbus threat posture are spinal reflexes to tactile stimulation of the lateral body surface.* The fact that fin-raising accompanies ambilateral

1) This point should be checked by means of photographic analysis.

contraction of rump musculature not only in threat but also in yawn-stretch and in braking-stabilizing, suggests that these two responses are in some way reflexively coupled. The contention that it is the tactile stimuli which are relevant is supported by the observation that *Barbus somphongsi*, which had been blinded at both sides through electrical coagulation of the cornea, showed perfectly normal lateral threat when they accidentally met in parallel or anti-parallel orientation.

The spinal reflex hypothesis provides a much more comprehensive explanation of *Barbus* threat behaviour than does the ambivalence hypothesis. Apparently, for threat to be initiated, some kind of sensitizing of the spinal reflexes must take place. This might occur through changes in the degree or kind of control of the higher brain centres over the medullary and spinal centres. One might speculate that such a change could occur by virtue of a disinhibition resulting from a conflict in the higher brain centres. Once sensitizing has occurred, the animal (1) actively seeks the relevant stimulus and (2) the response threshold is lowered as the proportion of time spent threatening increases. Under these conditions threat appears as a separate tendency, more or less independent from the tendencies to attack and to flee. Since lateral orientation to the opponent is actively sought, the threat tendency may even counter-act the occurrence of overt attacking or fleeing. At the same time it is conceivable that the fact that the animal gets more or less 'trapped' in the lateral orientation through its very performance of threat, may also cause an increase in attacking and/or fleeing motivations.

Under most conditions the tendencies to attack, to flee and — in the presence of females — to court are apparently sufficiently strong to prevent complete 'trapping' (continuous threat). It is only in duels and in the model experiments that complete 'trapping' may occur [1]). Complete trapping may be terminated by the fish (or one of the two fishes in the case of a duel) suddenly switching from threat to rolling behaviour (see below).

It may be emphasized that, even though the spinal reflex hypothesis is more comprehensive, the ambivalence hypothesis has not been disproved in its entirety. It is conceivable that ambivalent effects also do contribute in the shaping of threat behaviour insofar as the ambivalence hypothesis fits the facts. It is only for reasons of parsimony that the ambivalence hypothesis is perhaps better discarded for lack of evidence.

It may be clear that the postulation of the independent attacking and fleeing tendencies remains necessary for the interpretation of the whole of

1) The striking resemblance between the behaviour of the fishes in the model experiments on the one hand and duelling on the other will be elaborated elsewhere.

Barbus fighting behaviour and that they cannot be reduced to the threat tendency. Possible influences of threat on these two tendencies have been discussed above. In addition, it may be assumed that threat plus waggling may lead to either attack or fleeing through postural facilitation.

The interpretations in terms of conflict between attacking and fleeing tendencies of intention butt, tap, intention tap and inhibited attack are probably not affected by the spinal reflex hypothesis of threat. An exception may be made for tapping movements which occur simultaneously with threat. In this case tapping movements may result from a conflict between turning towards the opponent (attack) and keeping parallel to the opponent (threat) as well as from a conflict between the tendencies to turn towards and away from the opponent.

Some kind of explanation is still necessary to account for the fact that threat typically occurs in fighting behaviour and most often in situations in which it may be surmised that the tendencies to attack and to flee are more or less balanced; for instance in duels, to some degree in boundary fights between males of the more territorial *Barbus* species (see chapter III) (and very conspicuously in boundary fights of more strictly territorial fishes such as Anabantids and Cichlids). A quantitative elaboration of the dependence of threat on a balance between attacking and fleeing tendencies will be given elsewhere (KORTMULDER, in prep.). First, it may be hypothesized that the occurrence of a balance between attacking and fleeing during fighting may bring the fish into a lateral orientation to the opponent and that this in its turn triggers the initial threat response. Second, as a final completion to the spinal reflex hypothesis, it must be emphasized that the lateral tactile stimuli apparently elicit threat *only if at least some degree of aggressive (and probably fleeing) motivation is present.* It may be that the activation of the aggressive and fleeing 'centres' in the brain is necessary for the 'sensitizing' of the spinal reflexes to occur.

Finally, some remarks are due on the subjects of ritualization and emancipation. It is obvious that the combination of spinal reflexes which is threat is remarkably well adapted to its function of intimidating and fighting conspecifics. The various spinal reflexes may have originated partly as devices for stabilization in turbulent water, partly as fragments of locomotory rhythms (VON HOLST 1934, p. 592). There is little doubt, however, that considerable changes must have occurred, in the course of phylogeny, in their mutual coordination and/or in their absolute and relative thresholds to produce the degree of adaptation mentioned above. These changes may well be put together under the heading of ritualization. On the other hand, the spinal reflex hypothesis implies that threat has never been an ambivalent posture and therefore there is no reason to invoke the evolutionary process of emancipation to account for the relative independence of threat from attacking and fleeing tendencies. Interspecific differences in 'threat or attack specialism' (see chapter III, section IV, p. 206, 261) may be accounted for by relative changes — in the course of evolution — in the thresholds of attacking behaviour on the one hand and the spinal reflexes on the other.

(c) head-down threat.

As was described in p. 62, head-down threat differs from normal threat mainly in that the body axis, instead of being kept horizontal (normal position), is pitched forwards to varying degrees. The pitching does not seem to be an orientation relative to the opponent: it may occur in frontal, lateral, caudal and intermediate orientations.

Close observation of *conchonius* males fighting and courting in a group strongly suggests that often, as soon as a male starts to threaten (or to court, see below, p. 99), it tends to drift upwards. As a rule the fish tries to counter-act this tendency. It may do so in two ways. (a) by pitching forward if it is in normal position (b) by slightly bending the body with the head downwards if it is lying more or less on its side while threatening a fish above or below it. Combined with locomotion both reactions produce a downward force which counter-acts the lifting. The lifting force and concurrently the angle of pitching become stronger as a fish gets to lower water levels. The drifting tendency disappears as soon as the fish stops threatening (for instance by switching to straight attack or flight) or if it goes to higher water levels. At the same time it resumes normal position.

If a *stoliczkanus* or *tetrazona* in head-down threat swims forwards, it typically progresses along a horizontal or near-horizontal course even though the body axis may be inclined forward as much as 45° or more. This also suggests the presence of a lifting force which neutralizes the downward component of locomotion in the pitched position. There are no indications that the lifting force results from the positions or movements of the fins (though of course it is difficult to be sure of this from simple direct observation). It rather looks as if the fish becomes lighter and therefore tends to rise, more or less in the same way as when it has swallowed air during feeding at the surface. Such a change in specific gravity during threat would most probably be caused by distension of the swimbladder.

MÜLLER (1843), EVANS (1925), JACOBS (1940) and FÄNGE (1943) have described the presence and the structure of smooth muscle strands in the wall of swimbladders. The investigations covered some 10 different species of Cyprinids including *Barbus barbus* (EVANS, 1925). Quick contraction of these muscles has been experimentally accomplished by electrical stimulation (FÄNGE, 1943) and by the administration of adrenalin (FÄNGE, 1943) or physostigmin (JACOBS, 1940). Quick relaxation followed the administration of sodium nitrate solution (FÄNGE, 1943) or nicotin (JACOBS, 1940). The wall of the swimbladder is richly innervated by branches of the *nervus vagus* and *nervus sympathicus*. These data suggest that changes in swimbladder volume may be brought about through activities of the autonomic nervous system. MORRIS (1956) has claimed that autonomic responses often occur in states of conflict and play an important part in display postures of mammals and birds. It is a plausible assumption that the same may obtain in fishes. The quite dramatic colour changes which may occur during threat and courtship in fishes plead for this assumption (see also WIEPKEMA, 1961, p. 141-142).

Several techniques other than the one indicated above are known by which a fish may change the volume of its swimbladder. The gas content of the bladder may be increased or diminished (a) by swallowing or spitting air (b) by the secretion or the resorption of gas from the bladder (see *e.g.* JACOBS, 1940). No swallowing or spitting of air has been observed in threatening barbs. The processes of gas secretion and resorption are much too slow (several minutes at least) to account for the changes accompanying threat. Finally, the swimbladder may be passively compressed by the hydrostatic pressure if the fish moves to lower water levels and *vice versa*. However, since the rising tendency in threatening barbs seems to become stronger as the fish seeks lower levels and to be relaxed when it goes towards the surface, the drifting force during threat cannot be due to changes in hydrostatic pressure.

Pitching in response to increased buoyancy is also seen in barbs which have swallowed air while feeding at the water surface. Pitching plus normal locomotion then enables them to follow a near-horizontal though skipping course. Pitching is also reported to occur in *Phoxinus* in response to artificial lowering of atmospheric pressure (DYKGRAAF, 1942). Pitching in head-down threat may be homologized with these responses. The absence of skipping during head-down threat plus locomotion is probably due (a) to the rhythmic movements of the tail (in contrast to the irregular movements during normal locomotion) (b) to the combination of strong propulsion and braking which should result in greater stability.

Though both threat and head-down threat may occur in any orientation relative to the opponent, and both may be accompanied by attack as well as by flight, head-down threat is relatively much more often accompanied by flight ('plough', see p. 62) than is normal threat. This would indicate that distension of the swimbladder tends to occur when the balance between attacking and fleeing tendencies is relatively more to the side of fleeing. This would be in accordance with the observation that in duels between *tetrazona* males the male which is more strongly pitched during the final stages of mouth-fighting usually looses the duel.

The tendency of fleeing fishes to move upwards and the tendency of inferior fishes in small groups to stay at the higher water levels (see chapter III, section IV) indicate that fleeing motivation is highest at lower water levels (see also chapter III, section II, p. 112). In combination with the above considerations on the motivational state during head-down threat, this would explain why the rising tendency is stronger at the lower water levels.

The question remains what induces the fishes to counter-act the lifting force. One reason may be to stay near to the opponent. Aggressive motivation is probably responsible for this tendency. Another reason no doubt is to be found in a tendency to stay at low water levels; otherwise the occurrence of head-down threat during fleeing would be incomprehensible. The fact that dominant fishes of small groups tend to stay at lower water levels and often pay little attention to inferior fishes as long as these stay at the higher levels (attacking them as soon as they come downwards) would indicate that also the tendency to stay low is an expression of aggressive motivation. Thus, apart from orientation relative to the opponent, aggression and flight motivations would seem to be each other's antagonists in that the one brings the fish towards the bottom, the other towards the water surface. (For a further elaboration of depth preferences see chapter III, section II).

It may be (tentatively) concluded that the pitching component of head-down threat is an aggressively motivated response to distension of the swimbladder. The distension is an autonomic response resulting from a conflict (either between aggression and flight or between threat and flight) with a slight emphasis on the flight tendency.

As to the role of the swimbladder, WICKLER (1957 a, b) has developed a similar argument to account for head-down threat in *Nannostomus* and head-up threat in *Xiphophorus* — however, with one important difference. According to WICKLER, the deviations from horizontal position are passive movements due to changes in volume of the posterior sac of the swimbladder relative to the anterior sac. WICKLER's theory implies that either the contraction and relaxation of the swimbladder musculature is confined to the posterior sac (1957a, p. 14) or that the anterior sac is less distensible than the posterior sac. In both cases contraction of the musculature would cause a shift of the body weight towards the tail combined with a general increase in specific weight (as would be the case in *Xiphophorus*); relaxation of the muscles would have the opposite effects (as would be the case in *Nannostomus*). Neither of these implications is supported by the evidence. According to the authors mentioned above, smooth muscle strands occur in the anterior as well as in the posterior sac. According to EVANS (1925, p. 567) "the posterior sac is indistensible and inelastic... the anterior sac is elastic and distensible." MÜLLER (1843, p. 155) reports the same [1]). Though these facts would not disprove WICKLER's theory, they would invalidate his explanation of the connection between relative shift of volume and general body weight.

WICKLER's main argument for the assumption that the head-down and head-up postures are the result of passive lifting or lowering of the tail end is that he cannot identify them as intention movements of either attacking the opponent or fleeing from it (1957 b, p. 214-215). However, at least in the case of *Nannostomus*, the pitching movement could be interpreted as an intention movement to swim downwards in order to counteract the lifting force of the distended swimbladder and stay at one level with the opponent. Though *Nannostomus* apparently are not always able to counter-act the lifting force completely during typical 'vertical fights', when the pitching is as much as 90°, they are apparently able to do so during moderately pitched threat (1957 a, p. 13).

As a result of his assumption that pitching is a passive movement, WICKLER has to face the question why the fishes do not correct their postures towards the normal horizontal position according to the 'positional reflexes' described by VON HOLST (1950 a). He answers this question by stating that the fishes cannot do so because the fins necessary for the corrective reflexes are too much 'absorbed' by the movements of fighting the opponent (1957 b, p. 217-218). Though this statement may be true, it is difficult to prove.

In the case of *Barbus* head-down threat the pitching apparently is not a passive movement because it is perfectly tuned to the strength of the lifting force so as to keep the fish on the same level. It may be that pitching is helped by a relative shift of volume towards the posterior sac of the swimbladder, but the perfect tuning of the degree of pitching proves that pitching is essentially an active movement. If it were not, the fish could hardly be expected to show sustained intermediate degrees of pitching; it would in all cases proceed to 90° pitching and perhaps even fall on its back. The same argument is relevant to intermediate degrees of pitching in *Nannostomus*.

---

1) And so does ALEXANDER (1959).

The theory on the causation of head-down threat in *Barbus* as developed above has the following advantages over the version proposed by WICKLER: (a) it explains pitching as an active movement (b) it does not depend on which part of the swimbladder is contracted or expanded; it is only the total volume that counts (c) it suggests a more general explanation why no 'positional reflexes' are brought into action to bring the fish back to normal position. According to VON HOLST (1950 b, p. 206) the 'positional reflexes' are no reflexes in the sense that a specific stimulus elicits a qualitatively and quantitatively fixed specific reaction. All animals can *actively* deviate from the so-called normal position. Such active deviations may be seen in feeding, in broadside rolling during threat, in ventral and dorsal rolling and in leaning during courtship. Therefore, if pitching is an active movement, there *is* no problem why the fish does not return to normal position.

I do not agree with WICKLER that *Gasterosteus* apparently do not change their specific gravity during head-down threat (1957 b, p. 218). Slow rising may obtain also in this species when pitching has reached 90° (for instance when a male is almost beaten by a rival within its own territory; personal observation). The fact that *Gasterosteus,* in contrast to *Barbus,* often do not show any obvious rising tendency during threat as long as pitching is less than 90° may be ascribed to the special hovering technique (rapid rhythmic movements of caudal and pectoral fins or fin rays) by which *Gasterosteus* are able to maintain a perfectly stable position while standing still. The same may apply to moderately pitched threat in *Nannostomus,* which have a similar hovering technique. Similar hovering techniques are absent in the *Barbus* species discussed here.

(d) ventral and dorsal roll.

Ventral and dorsal roll differ from threat postures in that the belly or the back are turned towards the opponent instead of the side, the fins are actively adducted instead of tautly spread and the body musculature is relatively lax instead of strongly contracted. Rolling is similar to threat in that it is oriented relative to the opponent in such a way that neither the head nor the tail is directed at the other fish. This may again be surmised to result from a balance between the tendencies to turn towards and to turn away from the opponent. For reasons of parsimony these two tendencies might be identified with the aggressive and fleeing tendencies.

However, as was argued in sub-section (b), the preferred technique of making turns is by lateral bending of the body, combined with a rolling movement if the turn has to be made in an inclined plane. This would bring a fish that hesitates between turning towards and turning away from an opponent in a lateral orientation relative to the opponent. (This is so, irrespective of whether threat is explained according to the ambivalence hypothesis or the spinal reflex hypothesis). This means that if ventral and dorsal roll are to be interpreted on the basis of an ambivalence hypothesis, a special explanation must be found why not the side but the belly or the back is turned towards the opponent, a position from which both turning towards and turning away from the opponent are rather laborious processes. Such an explanation may be found in terms of (a) active seeking of dorsal or ventral orientation or (b) avoidance of lateral orientation. Further, some kind of evidence as to the nature of the two allegedly conflicting tendencies would be desirable.

The following observations may help to answer the first question. If one walks along the front of a tank with barbs (notably *nigrofasciatus* or *cumingi*) one often sees those fishes which stood with the side parallel to the front pane roll their backs towards the observer. I did a few simple experiments with a (male) *cumingi* x *nigrofasciatus* hybrid which also showed this behaviour. The fish was alone in a 60 × 35 × 35 cm tank with asbestos-cement sides and a glass front. It had previously been inferior to two other hybrids of the same kind which had been removed from the tank some two days before the experiments. The fish was not shy to the observer but its movements were somewhat hesitant and inhibited, probably because of its previous inferiority. The fish often stood near to the back wall of the tank at about 8 cm from the bottom, the head tipped downward for some 10°, facing the front pane. If I showed a finger (somewhat conspicuous because of an off-white bandage) in front of the glass near to the bottom, the fish fixated it by turning both eyes forward. If I moved the finger from side to side in a horizontal line, the fish pivoted slightly around a vertical axis thus following the finger with the eyes and maintaining binocular fixation. If I moved the finger in the same way but at a higher level, the fish did not tip upwards to follow its movements but rolled to and fro keeping its back towards the finger. The rolling movements immediately followed the pivoting as the finger was moved to a higher level and *vice versa*. If the fish stood facing away from the front pane and I moved the finger in front of the glass below the fish, it immediately showed ventral roll. Dorsal roll could also be elicited by presenting the finger above and behind the fish; ventral roll could be elicited by presenting the finger below and in front of the fish. However, the rolling movements were much more easy to elicit from the positions described first.

These experiments suggest that rolling movements may serve to allow binocular vision if the fish, for whatever reason, does not choose to face the object frontally. If this interpretation is right it might be expected that dorsal and ventral roll are accompanied by movements of both eyes towards the fixated object. I am almost but not quite sure that I have seen such movements to occur, both in the experiments described above and in actual ventral and dorsal roll at a conspecific. However, such movements are difficult to see because they are superimposed on the counter-rolling movements of the eyes which obtain whenever the body is either actively or passively rolled out of normal position.

Though it seems likely that dorsal and ventral roll serve to allow binocular vision both in the finger experiments and in agonistic situations, it is by no means certain that the motivation of binocular fixation is the same in the two situations. Binocular fixation is seen in a variety of contexts. It is very obviously present during feeding. The substrate is fixated as the fish swims around in search of food and so is any particle before it is snapped. Binocular fixation of a substrate (the bottom or a leaf) typically precedes chafing. An inferior *tetrazona* or *stoliczkanus* in a 2 ♂♂ or 3 ♂♂ group often hangs in a corner of the tank and continuously fixates the dominant from a distance (see chapter III, section IV). Fixating is maintained as the inferior fish is approached by the dominant and during inhibited attacks at the dominant. This suggests that binocular fixation always accompanies attacking and butting behaviour, though it is usually impossible to see it because of the speed of the movements. Finally, there is no reason to assume that the fixating in the finger experiments was an intention movement of feeding, chafing or attacking; a more general attention to unfamiliar objects seems more likely here.

The context in which it occurs might suggest aggressive motivation to be responsible for binocular fixation in agonistic roll. The following observations also plead for this interpretation: extremely strong and sustained roll is seen (a) in a fish which has just lost a duel (that is: immediately following sustained aggressive behaviour); it is accompanied by very slow fleeing for the chasing winner in this case. (b) in the dominant of two males, very early in the morning, if it is approached by the inferior. In such 2 ♂♂ groups it often takes the dominant some time (*ca* half an hour or an hour) to start attacking the inferior and to re-establish his dominance. In the meantime the inferior often performs inhibited attacks at the dominant who responds in the way described. Roll in this case is typically accompanied by standing still. (c) in a male which has just fled for a rival in a period of intensive reproductive behaviour in a large group in a large tank (see chapter III, section II). It is typically accompanied by standing still in this case and it is often followed by inhibited counter-attack (in *tetrazona*, see chapter III, section II, p. 141) or by immediate resumption of courting or competition with other males (in *conchonius* and *nigrofasciatus*, see chapter III, section II). Roll is a common activity of inferior males in 2 ♂♂

and 3 ♂♂ groups in small tanks (see chapter III, section IV). Such males often launch counter-attacks at the dominant from time to time. However, if such males are completely submitted by the dominant (see 2 ♂♂ groups of *nigrofasciatus, tetrazona* and *cumingi,* chapter III, section IV) both attack and roll are comparatively rare and the fishes often show plain flight only in response to the attacks of the dominant.

It is much more easy to associate the occurrence of roll with the presence of fleeing tendency than with aggressive tendency. Both fleeing and roll are typical of inferior fishes in small groups; they often occur together or in close temporal association; as long as the inferior fishes do not hide, the proportion of time spent fleeing and the proportion of time spent rolling are often positively correlated (see chapter III, section IV, fig. 57, 65, 73).

It may be tentatively concluded from the evidence presented so far that roll indeed is an ambivalent posture, motivated by aggression and flight with a strong emphasis on flight. (For a quantitative elaboration see KORTMULDER, in prep.). The aggressive factor in roll has to be more accurately specified as the tendency to fixate the opponent binocularly.

Though the motivations of dorsal and ventral roll probably are very much similar to one another, there seems to be a slight difference in accent. Both types of roll often accompany fleeing but dorsal roll is relatively more often performed while standing still. Dorsal roll sometimes accompanies inhibited attacks. This has never been observed for ventral roll in any of the five species discussed here. In *tetrazona* dorsal roll may alternate with frontal head-down threat through intermediate forms such as described on p. 63 These observations suggest that in dorsal roll the accent is slightly more on the aggressive tendency than in ventral roll.

(e) the structure of agonistic behaviour.

It would follow from the fore-going sub-sections that *Barbus* agonistic behaviour results from three main tendencies and their interactions: attack, flight and threat. The occurrence of threat is facilitated by a conflict between the attack and flight tendencies which leads to lateral orientation. Some kind of sensitizing of spinal reflexes seems to be necessary for threat to be elicited and to express itself as a separate tendency. A further complication is provided by swimbladder distension resulting from autonomic activity.

Though there are arguments to consider the aggressive tendency as a unitary whole, several of the components of aggressive behaviour may interact with the fleeing tendency in their own specific ways: butting interacts with flight to produce intention butting; attack may interact with flight to produce lateral orientation; binocular fixation may conflict with flight and produce roll.

The various ambivalent behaviour elements discussed in this section apparently reflect rather different balances between the aggressive and fleeing tendencies. These differences are illustrated in a simple manner in fig. 41 in which a number of ambivalent elements are ranged between the extremes of pure aggressive behaviour and pure flight. For the sake of simplicity the various forms of butting movements are lumped. It is indicated where lateral threat is most likely to occur. Frontal and head-down threat are added to the aggressive side and to the flight side of lateral threat respectively.

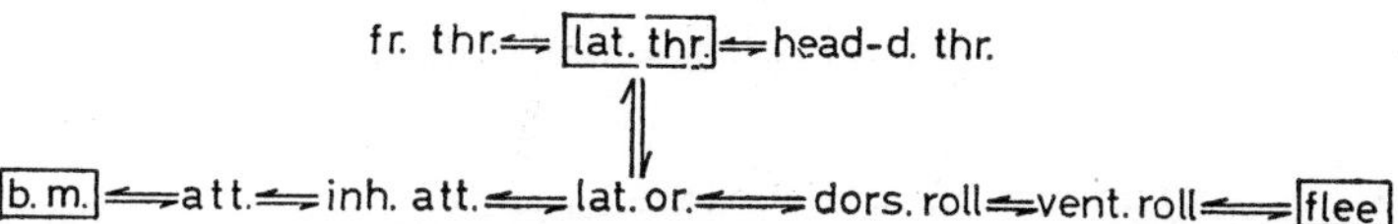

Fig. 41. Simple diagram of structure of *Barbus* agonistic behaviour. fr. thr. = frontal threat; lat. thr. = lateral threat; head-d. thr. = head-down threat; b. m. = butting movements; att. = attack; inh. att. = inhibited attack; lat. or. = lateral orientation.

Though roughly illustrative of the differences in balance expressed by the various elements, the diagram obviously is too simple. Reading from left to right binocular fixation of the opponent is found to be present from butting movements through inhibited attack; it is absent in lateral orientation, re-appears in dorsal and ventral roll and disappears again in pure flight. Sequences of attacking, lateral orientation and fleeing are typically performed with normal or somewhat above normal general muscle tone of the rump musculature [1]). General muscle tone is below normal during dorsal and ventral roll. In spite of a certain tenseness, low general muscle tone seems to be present also during inhibited attacks. The same combination of tenseness and low general muscle tone is typical for the frontal posture of *stoliczkanus* and *tetrazona* inferiors hanging in a corner of the tank. This posture may be maintained for a while on approach by the dominant male: the posture may then be followed either by inhibited attack or bite (mouth-fight) or by dorsal and ventral roll (see p. 63, 94; and chapter III, section IV, p. 253, 261). In the sequence: inhibited attack — frontal posture — dorsal and ventral roll, there is an increasing tendency to actively fold the fins, even though locomotion is often very slow or absent. On the other hand, in the sequence: attack — lateral orientation — fleeing, the fins are strongly adducted only during fast locomotion; for the rest they are kept in neutral

1) 'Normal' refers to normal locomotion outside reproductive periods.

position or loosely spread (most strongly so during lateral orientation, see sub (b); cf. also WIEPKEMA, 1961, p. 144). Rolling seems to be temporally associated with inhibited attacks rather than with (full) attacks and distance attacks; it seems to be the other way round for lateral threat (see also KORTMULDER, in prep.).

These considerations suggest the diagram of fig. 42 in which the row of ambivalent elements of fig. 41 has been split in two; frontal posture has been added as a separate element. The upper row is characterized by binocular fixation being sacrificed in favour of lateral orientation; it is the other way round in the lower row. Threat remains most strongly associated with lateral orientation and therefore with the upper row.

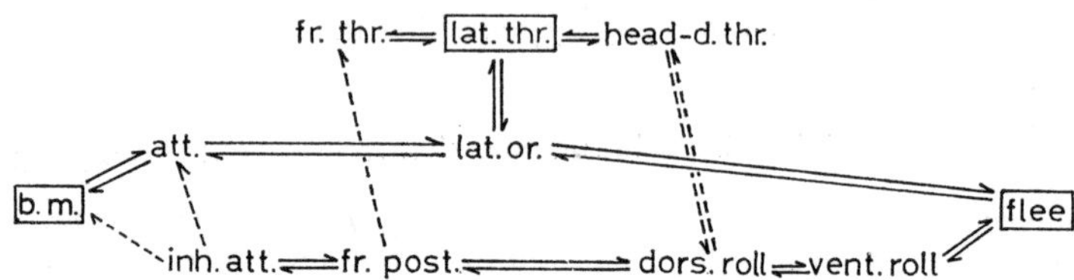

Fig. 42. Elaborated diagram of structure of *Barbus* agonistic behaviour. fr. post = frontal posture; see also fig. 41.

The two pathways between pure aggression and pure flight may be considered as the logical extensions of the inhibition types (b) and (c) discussed on p. 74. The two pathways are not absolutely isolated from one another. Transitions may occur such as indicated by the broken lines in the diagram.

A final relevant question may be what decides between the two pathways. In trying to answer this question it may be important to note that the lower pathway is typical in the first place of inferior males in small groups. The other pathway, including threat, is typical of dominant males of small groups and of males of roughly equal status (referred to together as 'self-actualizing' males hereafter) 1).

More or less continuous inferiority may obtain as a natural role in *tetrazona* (see chapter III, section II, p. 140). In most cases, however, self-actualizing males from time to time just 'dip' into the inferior role for a few seconds as long as they are attacked by a stronger male. The state of inferiority may last somewhat longer (a few minutes) in a male which has just lost a prolonged duel (see also p. 93).

In males which have been inferior in a small group for several days the state of inferiority tends to be prolonged even if the dominant fish is removed

1) The term has been borrowed from A. H. MASLOW's 'Towards a psychology of being' for no other reason than that intuitively it seems an appropriate term.

from the tank. It may last several days before the ex-inferior can again fight a freshly introduced male. A similar phenomenon was found in one of the *nigrofasciatus* males which was trained to fight models. After it had been beaten by the model (moved by hand) in a duel lasting some 10 minutes, it took some 3 days before the fish could again be induced to threaten and to attack the model. During that period it only responded with fleeing, rolling and inhibited attacks from a distance.

In the course of prolonged inferiority also other than reproductive behaviour elements seem to be affected by the general loss of tone in the rump musculature. Whereas self-actualizing fishes typically swim around actively also outside reproductive periods, or feed, inferior fishes often stand still for long periods of time, even in the absence of the dominant male. If they move, locomotion is typically wavering and slow. They often do not feed at all. The typical comfort movements of self-actualizing fishes are yawning-stretching, stretching, finflickering and chafing. In inferior fishes all comfort movements involving rump musculature often are absent. They typically show headshaking and mouthflickering. When performing yawn-stretch, only the mouth is opened and closed; the concurrent contraction of body musculature and the raising of the fins are often very weak or absent. The general behaviour of such inferior fishes is somewhat similar to the behaviour of a sick fish. This is so even in the absence of any visible trace of bodily damage.

The above observations suggest that the switch from self-actualizing to inferior behaviour marks the onset of a profound physiological change which is easily reversible when short-lasting but which tends towards irreversibility when submission is prolonged. The main feature of inferior behaviour is the loss of general muscle tone in the rump musculature. It may be hypothesized that this also makes the fish avoid normal turning movements and therefore lateral orientation. This would give another aggressive component, binocular fixation, the chance to express itself. The inferior state is apparently not favourable for the occurrence of threat, probably both because of the avoidance of lateral orientation (and therefore of the relevant stimulus for threat) and because of the low muscle tone which does not allow strong muscle contraction. Besides low muscle tone, active adduction of the fins is typical for inferior behaviour.

B. Male courtship.

Animal male courtship is often considered as a product of three main tendencies: one to attack, one to flee and one to mate (for a review see Hinde 1966, chapter 16). The three tendencies are assumed to be irreducible to each other. The strengths of the motivations under-lying the tendencies are thought to be mutually more or less independent (see *e.g.* WIEPKEMA 1961). Simultaneous presence of the three tendencies (or of any combination of two of them) may give rise to the various kinds of conflict behaviour discussed on p. 71.

The relative contributions of the three tendencies may differ in different species (MORRIS, 1957b). In some cases it may be difficult to find any unmistakably sexual components in any of the courtship elements up to ejaculation. In such cases the main role of the sexual motivation in courtship may be to influence the ways in which the other two tendencies are expressed. Such influences may explain why courtship behaviour, even if largely built out of components of aggressive and fleeing behaviour, differs from agonistic behaviour in form (see *e.g.* KRUIJT, 1964). As an extreme case it may be that the sexual motivation largely dictates also the sequence of courtship elements (*cf.* BAERENDS *et al.*, 1955). It is likely that such a directive influence of sexual motivation is to some degree present in all types of courtship because courtship of any species tends to lead to fertilization sooner or later and no doubt has been evolved with the function to do so.

Form analysis suggests that *Barbus* male courtship is almost entirely built out of agonistic components. This statement may be exemplified by the discussion of three components of *Barbus* courtship — leading, tapping movements and leaning — in the following sub-sections. No data are available to assess in how far sexual motivation determines the sequence of courtship elements (see however the influences of depth and vegetation upon the form of courting behaviour, chapter III, section II).

(f) leading (*s.l.*).

Leading *s.s.* is similar to fleeing in that the performer swims away from the other fish, in that it is often directed obliquely upwards and in that it is often directed towards vegetation. Leading differs from fleeing mainly in the stronger adduction of the fins and in the type of locomotion, *i.e.* in such characters as are typical of all courtship elements. These facts suggest that leading is caused by the fleeing motivation mixed with other factors causing courtship behaviour (which factors are not necessarily identical with the sexual tendency, see sub (i)).

All other forms of leading *s.l.* begin with an initial dash away from the female like in leading. Subsequently, however, the male turns back to the female. This indicates the presence of a balance between tendencies to move away and to approach the female. Apparently, the approach tendency is relatively weak in circle-leading, intermediate in circling and relatively strong in circle front and half eight. The fact that the latter two movements may end in frontal orientation by the male and in low intensity butting movements may indicate that it is the aggressive tendency which counter-acts the fleeing tendency in all forms of leading *s.l.*

Leading *s.l.* also differs from fleeing in that it typically occurs *not* in response to attack by the other fish. This may indicate that arousal of the sexual motivation causes a shift in balance between the aggressive and fleeing tendencies in favour of flight.

(g) tapping movements.

Tapping movements occurring during mating attempts are similar in form to the tapping movements occurring in agonistic behaviour. This may indicate the presence of aggressive motivation during mating attempt. In agonistic tapping the more or less lateral orientation seems to result from a conflict between the tendency to butt on the one hand and a tendency to turn away from the other fish or to maintain lateral orientation (as in the case of threat) on the other. The same two possibilities are present in the case of courtship tapping. The data of table 2 (p. 68) together with the observation that a male may switch to frontal orientation at low intensities of courting (see p. 68) suggest a conflict between straightforward butting and parallel orientation. The tendency to maintain lateral orientation may be sexually motivated because it is the orientation in which mating takes place. However, it is equally possible that the lateral orientation obtaining in mating attempts is itself the product of a conflict between the tendencies to turn towards and to turn away from the female. The point may be an academic one. Also in the latter case there is little doubt that sexual motivation plays an important part in adjusting the balance between the two tendencies.

C-curving during circling movements apparently serve the function of turning back towards the female. C-curving during mating attempt may be interpreted as an intention movement of tapping. In both cases it is likely that C-curving is an expression of the aggressive tendency. However, as in the case of parallel orientation an alternative or complementary explanation is possible. C-curving during mating attempt (and perhaps during other courtship elements as well) might be derived from tail bending during mating and therefore sexually motivated.

(h) leaning.

As was mentioned on p. 87 *conchonius* males often show a tendency to drift upwards as soon as they start courting. The same tendency is clearly present in *nigrofasciatus*. Like in threat in *conchonius*, the drifting tendency during courting increases with depth and disappears as soon as the fish stops courting or goes to higher water levels. As in threat the most likely source of the drifting tendency is to be found in a distension of the swimbladder.

Also during courtship the males apparently often manage to counter-act the lift or to regain lower water levels. Because of the presence of C-curve they cannot do so by pitching downward combined with forward movement — as is done in head-down threat [1]). They may, however, use the C-curve to exert a downward force by rolling towards the concave side. This technique is essentially analogous to the one used in threat while lying on one side (see p. 87) and in turning in an inclined or vertical plane. Rolling towards the concave side is exactly what the males do in courtship leaning. Therefore I hypothesize that courtship leaning is a device to counter-act the lift caused by a distension of the swimbladder. This interpretation is in accordance with the rule (see p. 67) that leaning is always strong when the male courts above the female and starts to return to lower levels. Like pitching in head-down threat, I assume leaning to be aggressively motivated.

Apart from the above rule, leaning is strongest during circling and during the circling parts of circle front and half eight. Leaning is abolished (a) in straight leading, especially if the course is obliquely upwards (b) during frontal orientation following circle front and half eight and during head-up courting posture perpendicularly below the female (see p. 66). These facts indicate that leaning and apparently the lifting force are strong during conflicts between the tendencies to swim away and to approach. Both stop as the conflict is resolved in one or the other direction. This accords rather well with what was found on the occurrence of swimbladder distension during agonistic behaviour.

Finally, one alternative interpretation of leaning should be briefly considered. Male courtship leaning is qualitatively similar to rolling during mating. This might suggest sexual motivation for courtship leaning. However, as was discussed on p. 69, rolling during mating is probably initiated by the female, the male just trying to keep parallel to the female. If the female stays in normal position, the male abolishes leaning during mating attempt. Whereas female leaning and rolling during mating are most strong when the fishes are near to the water surface, male courtship leaning is strongest when courting takes place near to the bottom (see also chapter III, section II,

1) Head-down threat with C-curving may occur during close carousselling. In this case, however, circling is always very slow and intermittent; the fishes seem to have difficulty in maintaining their balance.

p. 111). Therefore I reject the hypothesis that male courtship leaning is sexually motivated.

(i) muscle tone and fin folding.

The relative laxness of the body musculature and the tendency to actively fold the fins throughout courtship make male courting behaviour somewhat similar to inferior behaviour (see p. 97). The similarity in form may suggest a similarity in motivational state. This interpretation would be in accordance with the shift towards fleeing (leading) behaviour which marks the onset of courting behaviour (see p. 99).

In contrast to the motivational state of inferiority, the motivational state of courtship is easily reversible even after long periods (minutes) of courting. Most males can immediately switch from courting behaviour to threat and/or attack when they are approached by a rival. (For differences between individual males see chapter III, section II, p. 124 ff.).

CHAPTER III. BEHAVIOUR II

'Pictures of an exhibition'

INTRODUCTION

In this chapter a comparison is made of the reproductive behaviours of the five species: *B. conchonius, B. nigrofasciatus, B. tetrazona, B. stoliczkanus* and *B. cumingi*. Some information is added on the behaviour of *B. phutunio* and *B. gelius*.

The first section of this chapter contains descriptions of behaviour elements of the first five species insofar as these differ from the general description given in chapter II. In the second section a semi-quantitative description is given, for each of the five species, of the behaviour of a group of 12 individuals (6♂♂ 6♀♀) in a large tank (130 × 65 × 60, 300 × 60 × 50, or 400 × 100 × 60 cm). The third, and fourth sections deal with the behaviour of the five species in smaller groups in small tanks (60 × 35 × 35 cm). In these sections special aspects of the behaviour as observed in the large groups are studied in more detail. The third section deals with behaviour in groups of 2♂♂ 2♀♀. In the fourth section interactions between 2 or 3♂♂ are analyzed.

The observations of the third and fourth sections involve quantitative recordings of the behaviour of all individuals. Since it proved impossible to make complete records of more than one fish at a time, the individuals in a group were recorded one after another.

Though the natural behaviour of none of the species is known from direct observation, and though their natural surroundings are known only superficially (see chapter I, section II), the situation with 12 fishes in a large tank is assumed to resemble most closely the natural situation. Barbs are obviously schooling fish and, at least for most of the species described here, individual territories play a relatively less important role in their lives than, for instance, in sticklebacks or substrate-breeding cichlids. Therefore, much more may be learned of their natural behaviour by observing groups rather than single individuals or pairs. Even so, 12 individuals in a large tank probably represent only a fraction of the numbers of fishes and of the space available in nature, but it is about the best that can be reached under laboratory conditions.

The conditions of the experiments were the same for all species, not because the natural surroundings of the different species are supposed to be the same (they apparently are not) but in order to find such differences between the species as depend on properties of the species themselves rather than those which are merely a direct reflection of differences in natural environment.

Unless otherwise stated, all experiments were done under the same artificial light regime (see p. XI). It may be reminded here that none of the five species shows any obvious seasonal variations in behaviour under the laboratory conditions. Therefore, experiments made at different times of the year are directly compared.

SECTION I. THE BEHAVIOUR ELEMENTS OF THE 5 SPECIES

On the behaviour element level all species are very much alike. It is in the absolute and relative frequencies of the separate elements and, most impressively, in the patterning of behaviour in time and space, that the main differences between the species lie.

Two categories of differences may exist on the behaviour element level. Some behaviour elements may be shown by some species and be absent in others. Besides, homologous behaviour elements may have a slightly different form in different species. This may be partly due to more general differences between the species in locomotion. Both categories of differences are dealt with in the following paragraphs.

1. *B. conchonius.* The imperturbable.

Biting is relatively rare in this species. Attacking is mostly accompanied by relatively mild butting. Both normal and head-down threat occur very frequently. They almost imperceptibly merge into each other. All possible angles between 0° and about 60° with the horizontal seem to be equally common. More generally, *conchonius* exhibits a great plasticity in its orientation relative to the surroundings during threat. Carousselling may be performed in any plane, though horizontal carousselling is probably most common. Also during keeping-off almost any orientation relative to the surrounding may be seen, though head-up orientation seems to be avoided. Both during parallel threat and during carousselling both fighters tend to rise towards the surface. Prolonged carousselling mostly ends up with both partners touching the surface. Mouth-fighting was never observed in this species. Fin-folding during rolling is usually distinct, but not as extreme as in some other species (see below).

Also in male courtship there is a great deal of plasticity. All variations of leading and circling described in chapter II occur frequently. There is no specialization in straight leading as in *tetrazona* or in smooth circling as in *nigrofasciatus* (see below). Leaning during courtship is usually present, but not as consistently as in *nigrofasciatus*. Fin-folding during courtship is usually present as well, but not as extreme as in some of the other species.

In general, the locomotory patterns of fighting and courting are not as stereotyped as for instance in *nigrofasciatus* and *tetrazona,* but resemble more the normal locomotion of the species.

2. *B. nigrofasciatus*. The knight.

Besides butting, biting is common in this species. Head-down threat was never observed. Instead, a very slight head-up position (not more than a few degrees) is sometimes seen during threat, for instance during parallel threat. Carousselling is a very conspicuous activity in this species. It is mostly performed in a horizontal plane, often with great velocity, much faster than in any of the other species. Mouth-fighting was never observed. There is often a distinct rising tendency during parallel threat swim. Carousselling is mostly performed close to the bottom. Fin-folding during rolling is usually strong.

The most conspicuous elements in courting are all possible combinations of leading and circling. One courtship element follows another in one continuous sequence. Leaning and fin-folding are strong and continuous through most sequences of courtship.

Both fighting and courtship are characterized by a combination of elegance, strength and stability of movement. They are easily distinguished from the normal locomotion of the species, which is much more irregular.

3. *B. tetrazona tetrazona*. The vulnerable.

Fierce biting is very common in this species; butting does occur as well. Both normal and head-down threat are common. In contrast to *conchonius* horizontal threat is a more or less separate behaviour element from head-down threat. Though both may merge into each other, horizontal orientation is often deliberately maintained for some time. In head-down threat any angle up to 90° may occur, though angles between 0° and *ca.* 60° are most common. Carousselling usually is performed in the horizontal plane. It is mostly rather slow, especially when it goes with head-down threat. During head-down caroussel the fishes may get on very close quarters and sometimes they even seem to grasp each other with the bodies which are slightly curved with the concave side towards the opponent. Mouth-fighting is a very common occurrence in duels in this species. Opponents may interlock their jaws and tug at each other for considerable time. Duels between males often lead to mouth-fighting soon after they have begun. A rising tendency is often exhibited during parallel threat swim. During carousselling and especially during mouth-fighting the fighters usually sink slowly towards the bottom. Most of the duels involving mouth-fighting end up very close to or even in contact with the bottom. Rolling is typically performed with extremely folded fins.

Male courtship mainly consists of an alternation of approaching, often ending in a bite intention from below, with leading in a straight course away from the female, very often distinctly directed towards a plant. At the end of leading the male often stops just in front of the plant and stands there for a while quivering slightly, waiting for the female to follow. In this position he may sometimes gently nip at the plant, or he may occasionally chafe on it. Circling and intermediates between leading and circling are very rare. Courting posture as a sustained posture is often very brief or absent (see also p. 141). Accordingly, leaning during courtship is rarely distinct. Extreme fin-folding may occur during the approach and at the beginning of leading. During one leading movement often the ventral fins are loosely erected to some degree. This is often seen when leading follows a horizontal or more or less downward course. Concurrent with the spreading of the ventral fins the male often tips slightly backwards.

During normal locomotion *tetrazona* makes an unstable impression. Its movements are hesitant. Characteristic of all locomotion that accompanies threat is the enormous tenseness of body and fins. Also the smooth and restrained movement of leading is distinctly different from normal locomotion. During very intensive courting, when a male is not disturbed by other conspecifics, all his movements may get this same quality.

4. *B. stoliczkanus*. The wavering.

Severe biting may occur in this species, but butting is much more common. Like in *tetrazona* there is a clear distinction between horizontal and head-down threat. Also carousselling shows the same characteristics as in *tetrazona*, but mutual grasping was never observed. Mouth-fighting is not rare, but usually very brief. During parallel threat and carousselling both fighters tend to rise towards the surface and duels often end very near to the surface. Fin-folding during rolling is usually strong, but especially in ventral rolling the dorsal fin may remain more or less extended.

Courting posture, leading, circling and combinations of leading and circling are common in this species. The paths described in leading and circling are often rather irregular. Leaning is mostly distinct and finfolding is often very strong. Like in *conchonius* there is little specialization in fighting and courting locomotion compared to normal locomotion. All locomotion in this species is rather irregular and wavering.

5. *B. cumingi.* The unrestrained.

Strong butting and severe biting are common in this species. There is no clear distinction between horizontal threat and head-down threat. Especially during keeping-off threat may occur in any orientation from head-up to head-down. Compared with those of the fore-going four species, duels between males seem to be little formalized. Brief bouts of parallel threat and carousselling with considerable distance between the partners occur frequently. Also, males may exchange a few vigorous bites on each other's flanks or gill-covers. Prolonged fights are relatively rare. If they occur, they usually consist mainly of carousselling at close quarters and exchanging of bites, alternating frequently with fleeing plus head-down threat ("plough", *cf.* Dunham *et al.*, 1968) and with very brief *en passant* mouth-fighting. Carousselling in such fights may be performed with little curving of the bodies and slow turning, the opponents almost touching each other, or it may be performed with distinctly curved bodies and higher speed, though it is never as fast as in *nigrofasciatus.* During brief bouts of fighting, a strong tendency to rise is usually exhibited, both during parallel threat and carousselling. Alternatively, during carousselling both partners may spiral slowly downwards in head-down threat, apparently with much effort. During prolonged fights the partners usually rise quickly during carousselling and during exchanges of bites, and laboriously go down again during ploughing. Much time is spent near to or even touching the surface. Fin-folding during rolling is usually strong but, especially during ventral roll, the dorsal fin may remain more or less extended.

Prolonged courting posture is common in male courtship. Both straight leading (mostly upwards) and circling are common, as well as intermediates between the two. However, the paths followed are not as regular as in *nigrofasciatus,* mainly because a male, once it starts courting a female, tends to stay at a distance from the female between successive leading movements. Fig. 43 shows a typical example of the path followed by a leading male between two courting postures seen from above. Fin-folding

♀

Fig. 43. Course relative to courted ♀ followed by *cumingi* ♂ in a series of successive leading movements; top view; schematic.

is usually strong during courtship. Leaning is usually distinct during courting posture and circling.

The type of locomotion in fighting and courting is not distinctly different from normal locomotion, but both fighting and courting are characterized by extremely vigorous and unrestrained movement.

SECTION II. THE BEHAVIOUR OF THE FIVE SPECIES IN GROUPS OF 6♂♂ AND 6♀♀

(With some notes on the behaviour of *B. phutunio* and *B. gelius*)

This section contains data of roughly four categories. (1) long-term day-to-day variations in activity; (2) activity patterns within one day; (3) spatial patterning of behaviour; (4) inter-individual structure of behaviour, *i.e.* differences in behaviour displayed to different individuals. For each species the data will be presented roughly in this order.

Procedure.

All tanks used for these observations were well planted with at least a few different species of plants. Plants used were *Vallisneria, Myriophyllum, Hygrophila, Cabomba, Synnema, Ceratophyllum, Cryptocoryne, etc.* Plants were put in larger or smaller clumps with at least 10 cm between clumps. At least one area of at least 20 × 20 cm (depending on the size of the tank) was left completely bare. Part of the surface was also covered with floating plants. Water depth varied between 40 and 60 cm but was kept largely constant during an experimental series. The temperature was kept constant in each series somewhere between 20 and 25° C. The fishes were taken from smaller tanks, usually all from the same tank. As far as possible they were selected to be all of them of about the same size. All 12 individuals were introduced simultaneously into the experimental tank, mostly in the early afternoon. Each day, including the day of introduction, the fishes were fed late in the afternoon. If at the normal feeding time (between *ca* 16.30 and 17.30) intensive reproductive behaviour was going on, or when it seemed to be impending, feeding was often postponed for one or two hours. Feeding times were chosen because, at least for some of the species, they seemed to coïncide more or less with the time of maximal spontaneous feeding activity of the fishes.

Observations were started the morning after introduction. In some cases also observations were made on the introduction day. The behaviour of the fishes was recorded in a semiquantitative way, at intervals varying from a few minutes to one or two hours in such a way that a reliable picture was obtained of the changes in activity throughout the day. Observations were sustained, as much as possible, through consecutive days, often for several weeks, until an overall picture of day-to-day changes in activity patterns emerged. In *conchonius* no consistent day-to-day notes were taken. However, the day-to-day activity pattern in this species was clear enough from more casual observations. Observations on *stoliczkanus* are less complete than those on the other 4 species (see below).

1. *B. conchonius.*

(a) temporal patterning.

One group of 6♂♂ 6♀♀ of this species was kept in a 130 cm tank (water depth 40 cm) for several months. Although no consistent daily notations of the activity were made, it was clear that reproductive behaviour was shown by practically all males on practically all days, beginning with the first day after introduction. Only rarely a day was skipped, mostly without a known cause.

The daily rhythm of activity was very much the same on different days. Therefore, the activity pattern of this species may be satisfactorily described by describing an average day. The data derived from the 12-individual group are amply corroborated by many other observations under less standardized conditions, mostly in 125 cm tanks with varying numbers of individuals of both sexes.

Typically reproductive behaviour starts very early in the morning. Often the males have already taken their nuptial colourations (red bodies with black fins and completely bleached tail-spots) when the lights go on, or do so shortly afterwards. The males swim around at about 5 to 15 cm below the surface, forming a very loose school. Schooling is characterized both by staying together and by parallel orientation. Compared to non-reproductive schooling the distances between the individuals are rather large and perhaps more constant. At the same time the females move around near the bottom and often feed there. As long as the females stay there, there is extremely little aggression between the males. Attacking is practically absent and only an occasional low-intensity threat posture is seen. Within the first half hour one or more of the females start to swim upwards. As soon as they appear near the surface the males rush at them and start courting. At the same time they start to fight each other whenever two or more of them are near to a female (see, however, p. 110). The first matings occur very soon. If a female returns to lower depths it is usually followed by competing males. However, when she reaches the bottom layers, usually no more than one or two males are left with her. If one male is left he may go on courting, or he may return to the surface as soon as he is left alone with the female. Matings only occur near the surface. Gradually the males may come downwards more frequently also on their own account. Males also start to attack and threaten each other when no female is nearby, but this they only do if they meet in the bottom layers. In the surface layers the males only fight each other as long as at least one female is nearby.

For at least the first hour the females continue coming to the surface layers more frequently and stay there for longer periods. Males court more intensively and the mating frequency increases greatly. The males get more and more persistent in following the females downwards and in courting there, but matings, throughout the entire period, almost exclusively occur near the surface. The separation of males and females in two layers above each other, which was almost perfect in the beginning, becomes much more vague now and at times may even vanish completely. Mating frequency becomes very high and very often a male performs a series of matings in quick succession with the same female.

After the first hour all activities described may go on increasing or remain more or less stable at a high level, for at least another hour. The incidence of fighting between the males, especially those cases when no female is present nearby, keeps increasing all the time and they become more and more persistent in swimming around or courting near the bottom. Towards the end of the reproductive period the males seem to become relatively more interested in mutual competition for females, rather than in courting them. Often the whole group of 6 males pursues one female, trying to mate with her and to keep-off rivals.

Two or three hours after the beginning of the day, reproductive activity starts to decline rapidly. Concurrently, the males start to loose their nuptial colours. First their tail-spots get darker and gradually their red colours fade. Reproductive activity in both males and females is more and more replaced by feeding, mainly on the bottom. The observer, who was practically ignored up to this time, now begins to attract their attention. They often flutter against the front glass of the tank in his direction, hoping for food. Swimming around and feeding are their main occupations for the rest of the day. Often there is not more than one half hour between maximal reproductive behaviour and almost exclusive feeding behaviour. There is never more than one reproductive period on the same day.

The above picture is somewhat idealized. The changes of behaviour in time are not as continuous and gradual as the description might suggest. Most of the time activity oscillates back and forth every few minutes. For instance, females keep going to the bottom at intervals to feed. Even males may do so occasionally, though this is rare in a group of this size. Males alternate periods of intensive courting near the surface with periods of increased activity near the bottom. There seems to be a considerable amount of sympathetic induction in these oscillations. Both males and females obviously induce each other to display congruent behaviour. This seems to be true for both feeding and reproductive behaviour. This is particularly conspicuous in comparable groups in smaller (125 cm) tanks towards the end of the reproductive periods. In this situation often a few very deep oscillations between almost general feeding and almost general reproductive behaviour occur, before the main switch to feeding activity is made.

(b) spatial patterning.

During the reproductive periods, there is a very conspicuous spatial patterning of behaviour. Most behaviour elements may occur in principle at any place in the tank, but many elements are much more frequent on some places than on others. At least two environmental dimensions have a profound influence on the composition of behaviour: (1) the depth at which the animal swims and (2) whether the animal is between or at least in contact with plants, or in the bare spaces between the clumps of plants. Since no sufficient quantitative data were collected, only the most obvious effects will be discussed here.

As was already mentioned above, fighting between males without females nearby, occurs almost exclusively at greater depths. As was mentioned in chapter II, fleeing by males is often directed upwards.

If two or more males court one female near the surface, there is hardly any keeping-off or threat between the males. At the most their fins are slightly more raised than during solitary courtship. Each male tries as hard as he can to come alongside the female in order to mate. There is no apparent individual distance between the males. As soon as the female begins to swim downwards, the distance between the males starts to increase and they start to threaten and keep-off each other. Also the distance between the female and the nearest male increases, even more than the distances between the males. (Apparently, at least for the first male the tendency to compete with the other males increases relative to his tendency to approach the female). These effects become progressively stronger with greater depth. Concurrently, threat during keep-off is more and more replaced by head-down threat and by threat while laying on one side. The males get more and more trouble in swimming downwards against a lifting force (see chapter II, section II, sub (c)). When near the bottom the first male, in addition to keep-off and threat, may also attack the next male.

Near the surface often many males pursue one female. If the female swims downwards, some of the males may stay behind immediately; others are discouraged halfway, probably in part by the keeping-off and threat of the male(s) nearest to the female. As a result groups of males pursueing a female are on the average much smaller at greater depths, than they are near the surface. However, this cannot alone account for the differences in behaviour of the males at different depths. These differences remain the same, even if only 2 or 3 males are after a female near the surface, or if 4 or 5 males do so near the bottom.

When, as the female goes downwards, only one male is left, he often immediately approaches the female again, even if in the meantime she has reached the bottom layers, and starts courting. At the same moment his troubles with the lifting force stop, or at least become much less. He may, however, also leave the female and return to the surface layers as soon as he is left alone with the female.

If a single male courts a female near the bottom, the male almost invariably stays below the female during courting posture. Leaning is mostly strong both during courting posture and circling. Leading and circling are frequent. Leading paths are often relatively long and often directed obliquely upwards. Mating attempts are comparatively rare and mostly incomplete and/or aggressively tinged. This type of courtship is mainly performed in the open spaces. When courting near the surface the male is usually at the same height with the female or even above her during courting posture. Leaning is often absent. Even if they are still in the open space, leading and circling are comparatively rare and the male stays relatively close to the female. Mating attempts are common. Mating may occur.

In both types of courtship leading is often directed towards plants. Complete mating attempts and matings are typically performed between, or at least in contact with plants. As already mentioned above, matings occur exclusively near the surface. They are usually performed with a considerable degree of rolling over by both partners. Rolling during mating is probably a result of the female leaning away from the male (see chapter II, p. 69) and the male trying to keep parallel to her. In smaller tanks (60 × 35 × 35 cm) matings may also be performed near the bottom. The pair tends to remain more in normal position then. (See also p. 122, *B. nigrofasciatus*).

Most of the non-reproductive behaviour (swimming around, schooling and feeding) is performed preferably in the lower layers. Feeding is often from the bottom.

Independent from depth, the structure of behaviour is influenced by the vicinity of plants. These influences are of the same kind as those of depth.

Both non-reproductive behaviour and fighting between males mostly take place in the open spaces. Fleeing is often directed towards plants, though in this situation a fleeing animal almost never really enters a clump of plants. (However, in smaller groups in smaller tanks inferior males may do so). In courtship leading and circling occur most frequently in open spaces. Leading is often directed towards plants. (Leading apparently serves to induce the female to enter plant clumps). Leading *s.l.* is largely replaced by mating attempts and mating as soon as female and male enter a plant clump.

At no stage of reproductive behaviour were males seen to exhibit any territorial behaviour. In small (60 cm) tanks males may occasionally defend territories, especially if only males are present in the tank (see also section III, p. 185).

(c) inter-individual patterning of behaviour.

In the 6♂♂ 6♀♀ situation males were never seen to court males. Males may attack females, but, typically, as soon as a male spends any considerable part of his time in courting, he no longer attacks the female he courts. Also other females are attacked very rarely; they may be kept-off when coming too near to the courting male.

At least during the reproductive period, females are never seen to behave aggressively towards other females or towards males. During the reproductive period females, when very receptive, may show *vacuum* leaning and quivering. It is displayed exclusively near the surface. This is so even in small tanks. Females may get slightly bleached tail-spots when they are willing, especially during a series of matings and during *vacuum* leaning.

Discussion.

The spatial structure of behaviour may be interpreted in terms of a few simple assumptions. These are listed below. The phenomena to be interpreted by each of the assumptions are added in parentheses.

For these interpretations, the hypotheses of agonistic behaviour as resulting from the interactions of an attack and a flight tendency, and of courtship as resulting from attack, flight and sexual tendencies, as given in chapter II, have to be accepted. On the other hand, the analogy between the effects that both depth and plant density have on agonistic behaviour and on courtship, may be considered as an additional argument for the ambivalence hypothesis of courtship.

1. In the males, the agonistic system is progressively more activated at greater depths. (Fighting between males mainly at lower levels). The

sexual system, on the other hand seems to be more strongly activated at higher levels; at least there is a relative shift from sexual to agonistic tendencies with increasing depth. (Competition between males progressively stronger relative to sexual tendencies at greater depths: more keeping-off and threat at lower levels; increase of the distance between female and first male relative to the distances between the males; first male may attack second male from keeping-off position at lower levels. Accent of solitary courtship on leading *s.l.* and on aggressively tinged incomplete mating attempts; accent on more complete mating attempts and mating at higher levels).

2. The same assumptions may be made with respect to the plants-bare space dimension: agonistic elements and components are more common in the open spaces, sexual components are more common between plants. (Fighting between males and competition mainly in the open spaces; leading phase of courtship also mainly in the open spaces; mating attempts and mating mainly between plants).

3. Within the agonistic system, both attack and flight systems are progressively more activated at lower levels. (Attacking mainly at lower levels; attacking from keeping-off position at lower levels only; aggressively tinged incomplete mating attempts near the bottom. More head-down threat and threat while lying on one side relative to horizontal threat at greater depths; increasing difficulty with lifting force at greater depth; more leading in solitary courtship near the bottom. The increasing absolute distance between competing males at lower levels may be a result of a relative increase in flight tendency, but may also in part be a reaction to increasing keeping-off and threat behaviour of the opponents).

4. The same assumptions may again be made for the plants-bare space dimension. (Attacking between males mainly in the open spaces; both leading *s.l.* and aggressively tinged mating attempts typical for courtship in bare spaces).

Not in all situations (fighting males, competing males; solitary courtship) is it equally clear that both aggression and flight are activated at greater depths and in bare spaces. For instance, in solitary courtship at lower levels and in bare spaces, the high frequency of leading *s.l.* is much more obvious than the rather subtle aggressive tendencies in mating attempts. Which of the two systems is more strongly activated may be determined by the momentary motivational state of the animal.

The effects discussed refer to a situation in which the bare spaces between plant clumps are relatively small. Other observations suggest that very large open spaces elicit flight only. However, even here it is possible that the aggressive system is in principle activated, but that its out-put is almost completely inhibited by the much more strongly activated flight system.

One observation would seem to contradict the assumption that the fleeing tendency increases at lower levels. When the fishes are frightened by a disturbance outside the tank (not necessarily from above) they immediately swim towards the bottom. When willing females first start to swim upwards they are often very shy as long as they are in the surface layers. They swim around making frantic movements and soon return to the bottom layers. The same phenomenon may be seen in both males and females in some of the other species (see *e.g.* p. 122). This behaviour strongly suggests that the fishes are more afraid at high than at lower levels. However, it might be that this is primarily a reaction to predators. The fact that the reaction is stronger at higher levels would suggest that in these fishes flight for predators is based on a different mechanism than intra-specific flight.

Since females show practically no agonistic behaviour in this situation, much less can be said on the spatial structure of female behaviour. Like the males, the females swim around and feed preferably near the bottom and in the open spaces. They also display most of their sexual behaviour near the surface, and in contact with plants, *vacuum* mating movements included. As agonistic behaviour is absent, it would seem probable that in the females the sexual system is more strongly activated at higher water levels, in an absolute sense. However, like in the males, it is not certain whether this is not a question of a relative shift between agonistic and sexual activation only. Agonistic activity in females in this situation seems to be suppressed rather than not activated at all. In the absence of active males, females may display quite intensive agonistic (and male courtship) behaviour. Therefore, it still seems possible that interactions between agonistic and sexual systems do occur in females as well.

It is also possible to understand the movements of the fishes through the tank, relative to depth and to the vicinity of plants, by making the following additional assumptions.

5. Both aggression and sex systems have a positive feed-back in that an animal in which either aggression or sex is activated will go to a place where aggression or sex respectively are even more stimulated. This may be an opponent or a mate, but it may as well be greater depth or open spaces for aggression and surface layers or plant clumps for sex. The flight system, on the other hand, has a negative feed-back. This means that an animal in which flight is activated will strive for a place where flight is least activated. This may lead to swimming away from an opponent, but also to swimming upwards and/or heading for plants. These mechanisms would

explain why, whereas it is aggression and flight which are influenced in the same way by the location, it is flight and sex which bring the animal to the same places. They, of course, also make clear why leading is often directed upwards, and towards plants.

As will be shown in the next sub-section, the assumptions can also provide an explanation of some differences in the spatial structure of behaviour between *B. conchonius* and *B. nigrofasciatus.*

As relative depth and the vicinity of plants have very much parallel effects on the three systems, the question arises what makes the animal decide between going upwards or towards plants, and between going downwards and out into the open spaces. This question cannot be satisfactorily answered at this moment. It may be that, for instance, the absolute levels of activation of any of the three systems are responsible.

The temporal structure of behaviour in the males suggests one more pair of assumptions.

6. The degrees of activation of both the attacking and the sex system rise gradually throughout the reproductive period, probably independently from each other. (Increasing tendency of the males to seek lower levels, together with more fighting between males without females in the direct neighbourhood; males becoming progressively more interested in competition for females, relative to the tendency to court them. In spite of the foregoing increasing frequency of matings, becoming more or less stable on a high level).

It is not quite clear what causes the decline of reproductive behaviour at the end of the period. It seems likely that at least one important factor is the inhibition of reproductive activity by the increasing feeding activity. The following facts plead for this interpretation. (1) The deep oscillations between intensive feeding and intensive reproductive behaviour that may occur towards the end of the reproductive period suggest that a conflict is going on between the two systems. (2) Even some hours after the reproductive period is over, very brief bouts of rather intensive fighting or courting behaviour may suddenly occur. For instance, a male may suddenly rush at a female which is feeding with the snout just touching a plant (a posture somewhat resembling that of a willing female) and try to mate. These attempts are usually rather wild and not well oriented and they were never seen to lead to actual mating, but their occurrence shows that sexual motivation is probably not absent, even during the afternoon. (3) That fighting too may be elicited later in the afternoon, when 'spontaneous' fighting is low, is demonstrated by the following experiment. Two tanks (60 × 35 × 35 cm), both planted, contained 3♂♂ each; in each tank one of the males was dominant. On the third day after introduction, between 14.30 and 15.00 an encounter was arranged between the two dominant fishes. At that time there was little reproductive activity in both tanks. The colours of the dominant male in tank A were: red 3/4; tail-spot 0-1/4. Those of the other dominant: red 0-1/4; tail-spot 0-1/4 [1]). Both dominant males showed some feeding activity. After both inferiors had been removed from tank A, the dominant of tank B was introduced into tank A. Within a few seconds (necessary for the new-comer to sink to the bottom and to swim slowly towards the resident) the resident started to threaten and attack the intruder. In about a minute a complete duel was developing. This duel lasted as long as 13 minutes. It was won by the intruder (who was slightly bigger). Immediately after the fight had ended, both males swam to the bottom and began to feed intensively. The resident, immediately as he stopped his last threat in that duel, pecked at the bottom 7 times in quick succession. This contrasts with his feeding pattern prior to the encounter, in which he pecked only 7 times during 5 minutes, most of them once at a time. Soon after this brief burst of feeding agonistic behaviour was resumed (now consisting mainly of the winner chasing the other male) and feeding returned to about its original level. Immediately after the duel the intensity of the red body colouration of both males was 3/4 and their tail-spots were the same as before. Besides demonstrating that intensive fighting may be elicited in the afternoon, the experiment also suggests that feeding was suppressed during the fight and in some way "dammed up". This would mean that the inhibition between feeding and reproduction is mutual. (4) If no food is supplied in the afternoon, often the next morning there is more than normal feeding activity and the reproductive period for that day may be skipped. (5) If food is supplied on a normal day in the afternoon (or on a day on which there is no reproductive period, for instance as a result of non-feeding on the foregoing day) an initial bout of very intensive feeding is often followed by a short bout of fighting and courting, lasting a few minutes. The last two points suggest that hunger or intensive feeding activity may suppress reproduction, which may in some way become "dammed up". Point (5) shows more or less the reverse of what happened in the duel experiment.

No further details of the mechanism of inhibition between feeding and reproduction are known. It is possible that the inhibition is mainly between the

1) The system of evaluating colour intensities is given on p. XI.

feeding system and one of the reproductive systems only (for instance the attack system); or it may be between the feeding system on the one hand and the attack system and sexual system separately on the other, or between feeding and a reproductive system as a whole. More experiments are needed to decide these points.

Again, since the females show practically no agonistic behaviour, and since they seem to play a relatively passive role in courtship, little can be said about the temporal pattern of female behaviour. Like in the males, sexual activity in the females increases in the course of the reproductive period, and is replaced by feeding afterwards. No *vacuum* sexual behaviour is seen outside the reproductive period, nor has successful mating been observed outside the reproductive period or after feeding.

It must be added that in the fore-going discussions it is taken for granted that it is the males who take most initiative. This, of course, it not quite certain. The possibility should be considered that it is the females who, possibly by chemical cues, initiate much of the behaviour of the males. So far no experiments have been done to assess this point.

Whatever its mechanism, the mutual inhibition of feeding and reproduction is no doubt functional. Inhibition of feeding prevents the animals from eating their own eggs, which they readily do outside the reproductive periods. On the other hand, inhibition of reproductive behaviour during feeding hours prevents a waste of reproductive material and of energy.

In spite of the temporal separation of reproduction and feeding, the fishes often consume their own eggs in the tanks used in this study. The eggs do not adhere to the plants in which spawning takes place and sink to the bottom. There they are devoured during the afternoon or by individuals which don't take part in reproductive activity. This suggests that under natural circumstances the spawning and feeding grounds are separated also horizontally. The spatial structure of courtship in which males and females go upwards and towards plants to mate, suggests that in nature spawning takes place in shallow waters near banks with relatively dense vegetation. Feeding will probably take place in deeper waters on the bottom. Such an arrangement would protect the eggs from being eaten and also provide them with sufficient oxygen for development. Under natural circumstances the groups assembling on the spawning grounds are probably much larger than 12 individuals — the maximum group size in the experiments. Such a pattern would be not unlike that of a number of Cyprinids of temperate zones (for instance, carps, breams and roaches).

It must be emphasized, however, that the above picture of behaviour under natural circumstances is speculative. It must be considered as a hypothesis which should be tested in nature. Only then questions as to the adaptiveness of the specific behavioural structure of the species can be satisfactorily answered.

2. *B. nigrofasciatus.*

Three groups of 6♂♂ 6♀♀ of this species were observed in large tanks. One group was kept in the 400 × 100 tank (waterdepth 55 cm) for a week. This series was run under normal daylight conditions in July. Two other groups were kept in 130 × 65 cm tanks (water depth *ca* 40 cm) for at least 2 months each. Both series were run under the usual artificial 14 hours day-light regime.

The first group was introduced probably between 14.00 and 15.00. The introduction time of the second group is not known; that of the third group was 14.35. The animals for the first group were collected from three different 125 cm tanks. Each of the other groups was composed from animals from one 125 cm tank.

(a) temporal patterning.

In both the first and the third series complete reproductive behaviour was first seen on the first day after introduction. In the second series no observation was made on the first day. Many unstandardized experiments have shown that it is typical for at least ambisexual groups, in both *B. conchonius* and *B. nigrofasciatus,* that complete reproductive behaviour begins on the very first day after introduction into a new tank, even if the introduction is late in the afternoon.

Like in *conchonius,* reproductive behaviour is shown on practically all days by practically all males. This may go on for weeks or even months. Each day there is one clear-cut reproductive period. Only now and then a day is skipped. For instance, in the third series complete reproductive behaviour was shown on days 1, 2, 3, 5, 6, 7, 8, 10, 11, 12 (probably) and 13 after introduction. There was no observation on day 9. Reproductive behaviour was skipped on day 4, probably because the fishes were strongly disturbed in the early morning by noisy reparation of central heating pipes. This left them very shy for the rest of the day. More casual observations indicated that the regular day-to-day pattern was maintained for at least 3 months. These observations are amply corroborated by unstandardized observations of ambisexual groups of varying size, mostly in 125 cm or 60 cm tanks. Only in small groups (for instance 2♂♂ 2♀♀) reproductive activity may decline after some weeks.

Like in *conchonius* the activity rhythms on different days are very similar. Therefore also for this species a satisfactory picture of the species' reproductive patterns can be obtained by describing an average day. The general temporal and spatial structure of behaviour in *nigrofasciatus* is very much like that of *conchonius,* but there are some distinct differences in accent.

Unlike in *conchonius* the males have not yet taken on their nuptial colourations when the lights go on. Complete reproductive behaviour usually does not develop before one to three hours after the beginning of the day. Occasionally full courtship begins even later: between 6 and 8 hours after dawn, but these are exceptions.

At the beginning the males have little black and red colour on their bodies. The black bars may be darker. The fins may be relatively dark. Usually within the first quarter of an hour at least some males start to blacken all over the bodies. The intensity of this pattern may increase gradually during the first few hours. Maximum intensity is usually reached well before intensive courtship occurs, and even before fighting between the males is at its maximum. The red colouration develops much more slowly. When high intensities of red are reached, the beginning of intensive courtship is usually not far off.

Usually the first attacks and threats are seen within the first quarter of an hour, but the intensity, duration and frequency remain very low at first. All three parameters increase gradually during the first few hours and stay high or even increase further during and after the courtship period. Fighting always declines after some time, usually after courtship has already come to an end. Full courtship doesn't usually develop before the males have been fighting intensively among each other for at least half an hour or an hour. Fighting between males before the courtship period is quite independent from the proximity of females. Often, before actual courtship begins, males begin to compete for females and they continue to do so throughout the courtship period. However, also fighting between males without females being nearby continues to occur frequently throughout the courtship period. One frequently sees 3 or more males threatening and attacking each other, often for minutes on end. In the fighting period preceding courtship some males may establish and defend territories, but this tendency is not very clear in most males (see also homosexual males, p. 125). Centres of territories (defined as the place to which the male mostly returns after having attacked an intruder) are always near the bottom. Occasionally also a female may defend a small territory near to the bottom and display aggressive behaviour towards intruders during this period. Like in the males, the frequency and intensity of the attacks of such females increase in the course of the fighting period.

For the first hour or longer both males and females spend most of their time near the bottom. Females often feed from the bottom, though not very intensively. Males may also occasionally do so. When the courtship period is impending, in both males and females a tendency to seek higher levels becomes apparent. It seems that females mostly take the initiative in this. On their way upwards females, and occasionally males, may nip at plants. When courtship first begins to occur, it is performed mostly near the bottom, where males as well as females then spend most of their time. Most matings are also performed near the bottom during this period. As courting behaviour grows more intensive, both males and females more and more seek higher levels, courting and mating there, and matings become more frequent. The highest intensity of courting (and the highest mating frequency) are reached about an hour after the beginning of courtship. From the beginning of courtship there is very intense competition between the males. They keep-off and threaten each other and they also often attack rivals from keep-off position. They mostly keep doing this also when courtship becomes more intensive and is performed at higher levels. As soon as males start courting intensively, they give up their territories if they had one before, and move freely around the whole tank (see also homosexual males, p. 125). The severe competition between the males mostly does not prevent them from courting persistently and mating frequently. Like in *conchonius,* males, in this situation, are often seen to perform several matings with the same female in quick succession.

On some occasions an extremely intensive form of courtship was seen, in which all individuals (or almost all) courted and mated between the floating plants at the surface. In this type of courtship males no longer keep-off and threaten one another, but they just try to push aside each other in order to mate with a female. This type of courting is very similar to normal intensive courting in *conchonius.* A further interesting feature is that the black body colours of the males fade, the vertical bars included, leaving the bodies red and the fins black. Thus they look more like *conchonius,* also in colour pattern.

Usually after a few hours, courting activity begins to decline, often primarily because the females stop to be willing and start to feed from the bottom. As long as one female is willing, the males continue to court her. As already mentioned above, the males often go on fighting among each other for some time after courtship has come to an end. Males may now again stake out territories near the bottom. Later also fighting declines gradually and is replaced by feeding. Concurrently the males loose their nuptial colourations.

Like in *conchonius,* the above picture is somewhat idealized. Typically the animals switch forth and back between successive stages every few minutes. For instance, going upwards towards the end of the aggressive period before courtship and during the first stages of the courtship period, typically occurs in bouts. There is apparently some mutual stimulation in this, but not as much as in *conchonius*. Deep oscillations between general feeding and general reproductive behaviour, such as may occur in *conchonius* towards the end of the reproductive period, were never observed in *nigrofasciatus*. Rather both females and males switch from reproductive behaviour to feeding one by one (see also individual differences, p. 123).

(b) spatial patterning.

The differentiation of behaviour according to depth is not as dramatic as in *conchonius*, where some behaviour patterns occurred almost exclusively at certain levels only, at least in the 6♂♂ 6♀♀ situation. In *nigrofasciatus* all behaviour elements may occur at any level. However, the same trends that were observed in *conchonius* seem to be present in *nigrofasciatus* too. Most difficult to assess were differences in fighting behaviour and competition between males at different levels. There are several reasons for this. First, males already spend most of the time in the lower layers, where — according to the rules found in *conchonius* — maximum fighting activity might be expected. This, in combination with the fact that they do fight and compete at higher levels as well, makes it difficult to say with certainty whether fighting at lower levels is, relative to the time spent there, more common and/or more severe than at higher levels. Finally, the considerable individual differences in behaviour between the males, see p. 123 ff.) make it difficult to find comparable situations at different levels, involving the same number of males of the same type, with the same distances between them. Nevertheless I certainly got the impression that males are considerably less querulous when at higher levels than near the bottom, both in intermale fighting and in competing for females. Further, territories are always near the bottom (see also homosexual males, p. 132). Also in the extreme type of courtship, where courting and mating occur mainly in the plants at the surface, competition has become very much reduced relative to courting activity. It should however be noted that this type of behaviour is displayed mainly between plants which makes it difficult to abstract depth effects from the effects of plant vicinity.

Certainly the increasing tendency to seek higher levels just before and during the courtship period correlates with increasing courting activity and mating frequency. It also seems to be much more easy for a male to perform a series of matings with the same female in quick succession at higher levels, even if he is seriously hindered by other males.

Further, males display much more frequent and intensive leading and circling when courting near the bottom and more mating attempts and mating at higher levels. Like in *conchonius* a male's tendency to stay below the female during courtship is much stronger at lower than at higher levels, and so is his leaning. Near the surface the male is often at about the same height with the female and leaning is almost absent. Matings occurring near the surface are often characterized by strong rolling over of the partners, whereas rolling over is practically absent in matings near the bottom. Like in *conchonius* rolling over during mating seems to be a result of the female leaning away from the male and the male trying to keep parallel to her (see female sexual behaviour, chapter II, p. 69). Matings acts, which are always very brief in this species, are even quicker when performed at the surface than at lower depth. This also might be an expression of stronger sexual activation relative to agonistic activation. Since head-down threat doesn't occur in this species, no differences in the occurrence of head-down threat relative to horizontal threat can be found.

Differences according to the vicinity of plants are much the same as those found in *conchonius*. Centres of territories are mostly in the open spaces, though not very far from plant bushes. Fighting, competing and the leading phase of courtship typically occur in the open spaces too. Both fleeing and leading are often directed towards plants (and upwards). Inferior males in small tanks often hide between plants. Mating attempts and mating typically occur between or at least in contact with plants. Females may exhibit *vacuum* sexual behaviour of the same kind as described for *conchonius* (leaning over, quivering and fading of the black bars). Like in *conchonius* this is always performed near the surface and in contact with plants. In addition females may make characteristic wriggling movements, for instance between *Myriophyllum* leaves. This also is an expression of being sexually activated. This behaviour was never seen in *conchonius* females, possibly because of their greater size relative to the available plant leaves.

Also normal swimming around and feeding take place mainly in the open spaces and near the bottom. When a group of *nigrofasciatus* is introduced into a large tank, they often swim around in a school, mainly through the open paths between plant clumps. Now and then, however, they enter a clump of plants and stay there for a while, nipping at the plants. This kind of behaviour was not seen again after the animals had settled in the new surroundings. It is not known whether the same kind of behaviour occurs in *conchonius*.

Especially just before and in the beginning of the courtship period, when they start to seek higher levels, both males and females display the peculiar shyness which was described for *conchonius* females (p. 114). This behaviour is even stronger in this species than in *conchonius,* possibly because *nigrofasciatus,* also when not reproductively active, are even more reluctant to come to higher levels than *conchonius.*

(c) inter-individual patterning of behaviour.

Like in *conchonius,* in this situation (6♂♂ 6♀♀) males do not court other males (except for homosexual males, see p. 124). If two males are competing for a female they may, in their mutual attempts to keep the other off, become tightly pressed together at some distance behind the female. In this situation one of the males may clasp the other and perform a complete mating movement. This is done by 'normal' males and probably has nothing to do with homosexual preference (see below). Also like in *conchonius,* males may attack females, but only if they are not courting intensively. As soon as a male spends considerable time in courting, he hardly attacks the courted female any more. He may switch to aggressive behaviour towards the courted female, if she behaves aggressively towards him (which most females don't). Other females, when coming too near to a courting male may be kept-off and sometimes be attacked.

In contrast to *conchonius,* females *may* exhibit considerable aggressive behaviour both towards other females and towards males, but this mostly stops when intensive courtship has developed in the group. Most females, however, do not display aggressive behaviour, even outside the courtship period. In small ambisexual groups (2♂♂ 2♀♀) some individual females may display considerable aggressiveness, but they never become dominant (see below, p. 186). Like in *conchonius,* females may, in the absence of active males, display male courtship behaviour towards other females, but this was never seen in the 6♂♂ 6♀♀ situation.

(d) individual variability.

An interesting aspect of *nigrofasciatus* reproductive behaviour is the presence of considerable individual differences in behaviour. These differences are perhaps not sufficiently constant and discrete to speak of behavioural polymorphism. For instance it is not known whether different types of behaviour are properties of certain individuals for life, or whether they are to any degree genetically determined. But the types described below are at least consistent for periods of several weeks or more, and under different circumstances.

Individual differences to the degree as found in *nigrofasciatus* were not found in *conchonius.* This does not necessarily imply that the species differ on this point also in natural populations. There is a variety of possible reasons why the greater variability in *nigrofasciatus* and/or the uniformity in *conchonius* might be an artifact. For instance, it is possible that our *conchonius* came from more closely inbred captive stocks and/or from a more restricted area of their natural range; or *nigrofasciatus* might be more sensitive to abnormal laboratory conditions during ontogeny. However this may be, the types found in this study are interesting in themselves and may shed some more light on the mechanisms of *nigrofasciatus* behaviour.

Males differ in the intensity of the black pattern all over the body. Some males in a group remain consistently less black than others, throughout the reproductive periods. The same individual males show the same pattern on consecutive days. In such lighter males the red body colour is usually well developed during reproduction 1). These males show little inclination to compete with other males. If they do, they are mostly not successful. However, if they are not hindered by other males, they court very successfully. Their movements during courtship are very smooth and they mate frequently. Darker males, on the other hand, are much more successful in competing for females with other males. However, if they are left alone with a female their courting movements are much more brusque than those of the lighter males and their mating frequencies are much lower. Light males may sometimes get somewhat darker late in the reproductive periods. Concurrent with this they become somewhat more persistent in competing with other males.

The above typology refers to males which are of about equal size. Size is another factor which correlates with a male's role in the group. Larger males, even if the difference is small, are more successful in competing for females. (This in contrast to *conchonius,* where a male may easily drive away a much larger rival when courting a female). If one or two males in a group are slightly larger than the others, they may get rid of rivals during courtship without much keeping-off and threat. Such males are often intensively red with little or no black on the body. Their courting movements are smooth and they mate frequently — like the lighter males described above.

Yet another type of male is what I shall call the homosexual male. Whereas "normal" males practically exclusively court females, homosexual males are often seen to court other males, even if these are coloured intensively red and/or black 2). At least some of them may also court females

1) This sets them apart from males which just do not partake in reproductive behaviour. These males show neither red nor black colour.

2) "Normal" males may court sub-adult males, which have not yet got their nuptial markings, or adult males in which the nuptial colours are very weak. This is not

and even mate with them. One homosexual male was present in series II and two in series III. The male in series II was on several occasions seen to court a female and to mate with her, though this was comparatively rare. Most of his courting was directed towards males and he often ignored females. Three complete and one incomplete matings with a female were observed. All of these were performed within the male's territory (see below) but the number of matings is too low to know whether this is typical. The two homosexual males in series III courted females only occasionally and were never seen to mate. They often ignored or even attacked females, even if these came into their territories (see below). However, one of them was observed to court females intensively and to mate frequently with them when placed in a 2♂♂ 2♀♀ group in a 60 cm tank (see below). In the 6♂♂ 6♀♀ situation, all three homosexual males were characterized by very intensive black colour all over the body. They were not less red than other males during the reproductive periods. All three were more persistent than other males in establishing and defending territories. In contrast to the other males they kept territories throughout the reproductive periods. When courting a male, they may move through the whole tank, but they always stay near the bottom. They may also court females outside their territories, especially if this is accompanied by competition with other males. Also when courting females they stay near to the bottom. Courtship towards males is practically restricted to the leading and circling phase and is characterized by very wild movements and comparatively vigorous butting during courting posture. Courtship towards females may be of this type too. When courting males, courting posture is often interspersed with attacks.

The behaviour of one of the homosexual males of series III was studied in more detail in a 2♂♂ 2♀♀ group in a 60 cm tank. Figure 44 represents a simplified version of one half hour's protocol of this male's behaviour. The homosexual male here was dominant over the other (normal) male. The figure shows that a considerable amount of time was spent in courting the other male. About the same amount of time was spent in courting a female (mainly with one female, with which the male also mated frequently; no matings were performed with either the other female or the other male). The figure also shows that both courting the other male and courting the female occur in bouts which last one or more minutes each. Further there are periods in which the male attacks much, mostly the other male but also both females. The three types of activity seem to

considered to be homosexual behaviour, because the courted animal looks very much like a female.

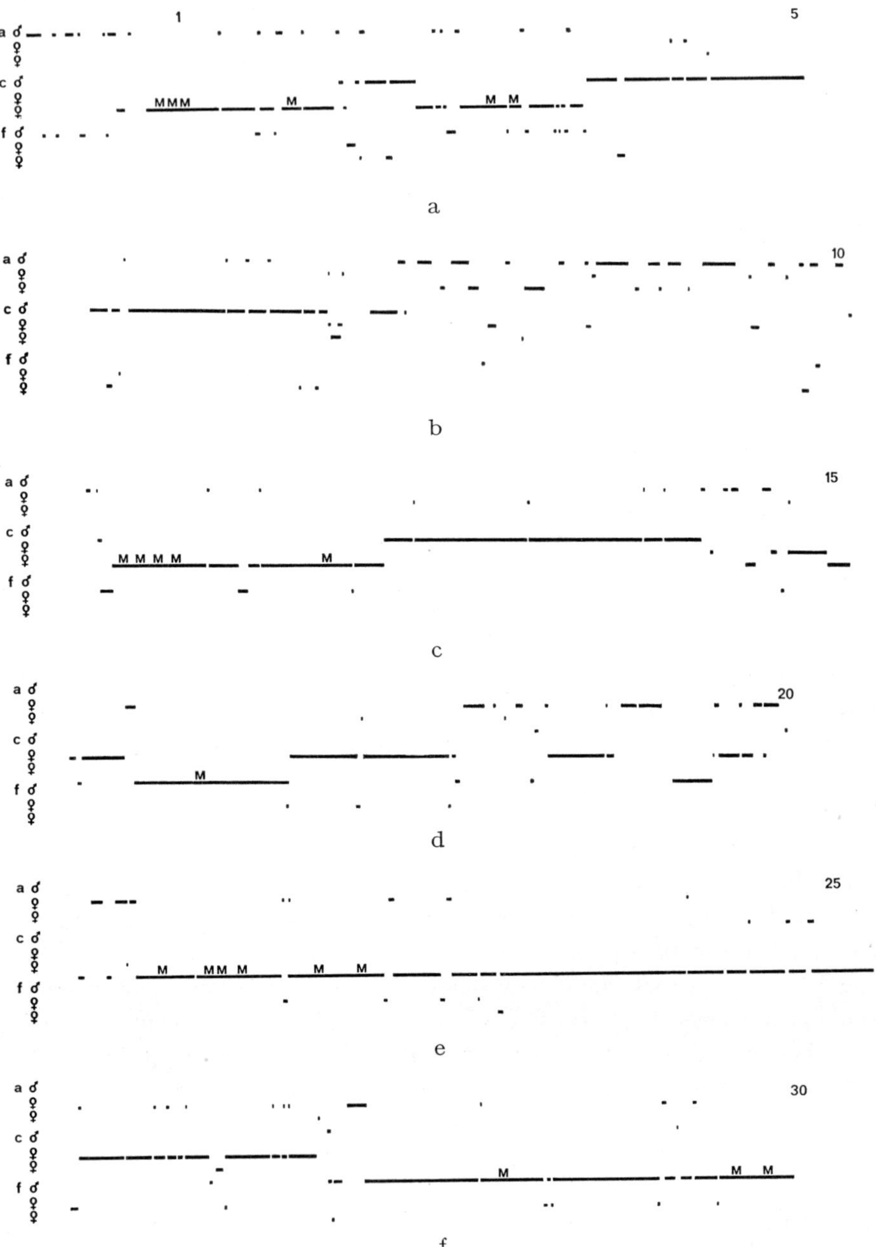

Fig. 44. Simplified behaviour record of 'homosexual' *nigrofasciatus* male (dominant of 2♂♂ 2♀♀ group); 30 minutes; a = attack; c = court; f = 'fight'; M = mate; further explanation see text.

follow each other more or less in a fixed order: 1. courting the female, 2. courting the male, 3. attacking. This cycle is run through four times in the half hour recorded. (From the end of the first minute to the beginning of the eleventh minute; from there to the end of the sixteenth minute; from there to the second half of the twenty-first minute; and from there to the beginning of the twenty-eighth minute. The protocol begins with what is probably the end of the aggressive period of the fore-going cycle, and ends with a period of courting the female again, probably the beginning of the next cycle). In addition to this overall picture there are a few more short-term fluctuations. In the first half of the third minute the male switches from courting the female to courting the male, only to return to courting the female after about half a minute. Towards the end of the first half of the nineteenth minute the male switches back from attacking to courting the male for about 25 seconds and then starts attacking again. These are the only two obvious aberrations of the general cycle. On the other hand, the second half of the second aggressive period (around the fifteenth minute) may be regarded as a short-term version of the general cycle: courting the female; courting the male; attaking. The same can be said of the second half of the third aggressive period (nineteenth and twentieth minute).

Another one quarter of an hour's record of the same male shows (fig. 45) that he may also alternate between periods of courting the male and attacking periods (again mainly towards the male, but now and then also towards the females). Courting the female is practically absent in this record.

A second homosexual male (not one of those present in the 6♂♂ 6♀♀ situations) was observed in a 2♂♂ 2♀♀ situation. Though no quantitative records of his behaviour were made, it was clear that it ran through the same cycle as the first male.

Females also show considerable individual differences in behaviour. It was already mentioned above that some females may, in the 6♂♂ 6♀♀ situation, exhibit considerable aggressiveness, whereas most others do not. Some females are much more often sexually willing than others. For several consecutive days the same few females may take part in courting and mating behaviour, while others remain reproductively inactive.

Discussion.

It seems that the same assumptions as were made for *conchonius,* can account for the temporal and spatial structure of behaviour in *nigrofasciatus,* both for males and for females. Since the time of onset of sexual activity

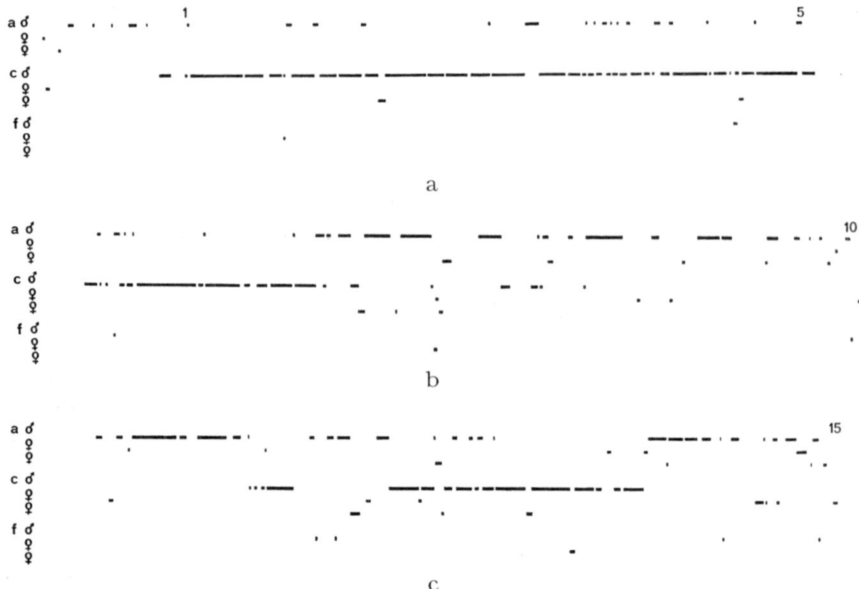

Fig. 45. Simplified behaviour record of 'homosexual' *nigrofasciatus* male (same as fig. 44) in 2 ♂♂ 2 ♀♀ group; 15 minutes; legend see fig. 44; further explanation in text.

is much more variable here than in *conchonius,* it is more obvious in this species that the daily rises in aggressive and sexual activities are mutually independent. When courting starts early, aggressive and sexual behaviour develop about synchronously (which is mostly the case in *conchonius*). When courting begins late, aggressive activity develops on its own, to be followed later by sexual activity. However, courting never develops without the simultaneous presence of aggressive activity.

Unlike in *conchonius,* some females may display aggressive behaviour in the 6♂♂ 6♀♀ situation. Its pattern closely follows that of the males: aggressive behaviour is mainly displayed at low levels and its intensity grows with time.

It seems that the temporal and spatial structures of behaviour of *conchonius* on the one hand and *nigrofasciatus* on the other differ in degree rather than qualitatively. Most of the differences may be summated by the statement that *nigrofasciatus* is in all contexts more aggressive than *conchonius.* There is more inter-male fighting and it is frequent and intensive at all levels, throughout most of the reproductive periods. Competition between males is more severe at all levels. During keeping-off the males are often very close together (so close that it may sometimes lead to male-male mating behind the female) and they are often so absorbed in the competition that they loose the female and go on fighting (even though the distance between the competing males and the female is also typically small). Often, especially in the first hour of the courting period but sometimes for the whole period, mating frequencies are kept low as a result of the intense competition [1]). At all levels, and throughout the courting period it is very common for a male to attack its nearest rival from keep-off position. In addition *nigrofasciatus* prefers much lower levels than *conchonius,* especially in reproductive behaviour. The intensity of agonistic activity and the preference for lower levels in *nigrofasciatus* seem to be associated also with the greater intensity of black colour patterns on the body in this species. The intensity of the black pattern all over the body in *nigrofasciatus* seems to be strongest during predominantly agonistic activity, especially when this occurs near the bottom (see further below: individual differences) [2]).

An interesting aspect of the individual differences in behaviour and colour patterns is that they appear as extensions of changes which may take place within one individual in time. It was already mentioned above that lighter males may become slightly darker and simultaneously start to compete more persistently. In general it seems that for each individual male there is a relation between the intensity of the black colour and the occurrence of threat. From close observation it is obvious that a male which starts threatening practically immediately grows more black. The black fades again when threatening is stopped. This is so both in theatening between males with no females nearby and in competition for females. It was also mentioned above that males which are slightly bigger than the others do not have to keep-off and threaten much to get rid of their rivals and that such males do not get very black. It is not known whether darker males prefer lower levels than lighter males, but at least it is clear that the homosexual males, which are always very black, prefer bottom layers. It is also clear that males which have just performed a series of matings in quick succession (which they usually do in the higher layers) are considerably lighter than for instance during competition with other males. When intensive courting

1) Anticipating on the comparison with the following species it must be emphasized that high mating frequences are often reached, in spite of the severity of the competition.

2) A similar bringing together of a variety of behavioural traits under the common denominator of a higher or lower species-specific threshold for aggressive or flight behaviour has been practiced by BAERENDS & BLOKZIJL (1963) in a comparative study of two *Tilapia* species. (They did not, however, take into account the differences in colour patterns between the two species, see p. 295 and appendix 5 of this study).

takes place in a 2♂♂ 2♀♀ situation in a 60 cm tank, the dominant male often has to keep-off the other male and is always more or less black. If the inferior male is removed from the tank, the dominant male's courtship becomes very intensive and his mating frequency increases sharply. At the same time his black colour all over the body fades completely. Finally, males partaking in the very intensive courtship between the floating plants described on p. 120 stop competing by keeping-off and threat and completely loose their black colours. It is possible that a male gets even more black during periods in which it both fights and courts than in periods in which it only fights, but on the other hand a male which courts in a tank with only females is not black at all. Even his bars are light and the red colour is almost absent. Such a male doesn't compete at all except for a very occasional keep-off of another female, usually without threat. He courts practically continuously and mates frequently. These observations indicate that at least the black pattern all over the body is mainly related to the agonistic systems, especially to threat.

There are some interesting points of resemblance between the structures of reproductive behaviour in "normal" and homosexual males. Figure 46 presents a simplified version of a thirty minutes protocol of normal male's behaviour. The situation during this protocol was comparable to that of figures 44 and 45: 2♂♂ and 2♀♀ in a 60 cm tank; the actor is the dominant male; the other male is also a normal male. The actor apparently alternates periods of intensive courtship (mainly on female C) with periods of increased aggressive behaviour (not only towards the other male but also towards both females; typically the other male is attacked more than the females but this is not clear in this case). The alternation is most clear in the last 20 minutes. In this period four complete cycles are shown. The lengths of the periods are of about the same order as those in figure 44. Another interesting phenomenon in figure 46 is that during the courting periods courting is practically restricted to one of the females, whereas during the aggressive periods there is also a considerable amount of courting towards the other female. This shift in preference is present in all four aggressive periods but it is particularly clear in the third period. The same phenomenon is present in figure 44 (homosexual male) in which the second female is the preferred one and courtship towards the other female is practically restricted to the first and second aggressive periods. It appears that the main difference between the homosexual and the normal male is that in the homosexual male a period of courting the male is inserted between each period of courting and the next aggressive period.

Only further experiments may decide what mechanisms are responsible

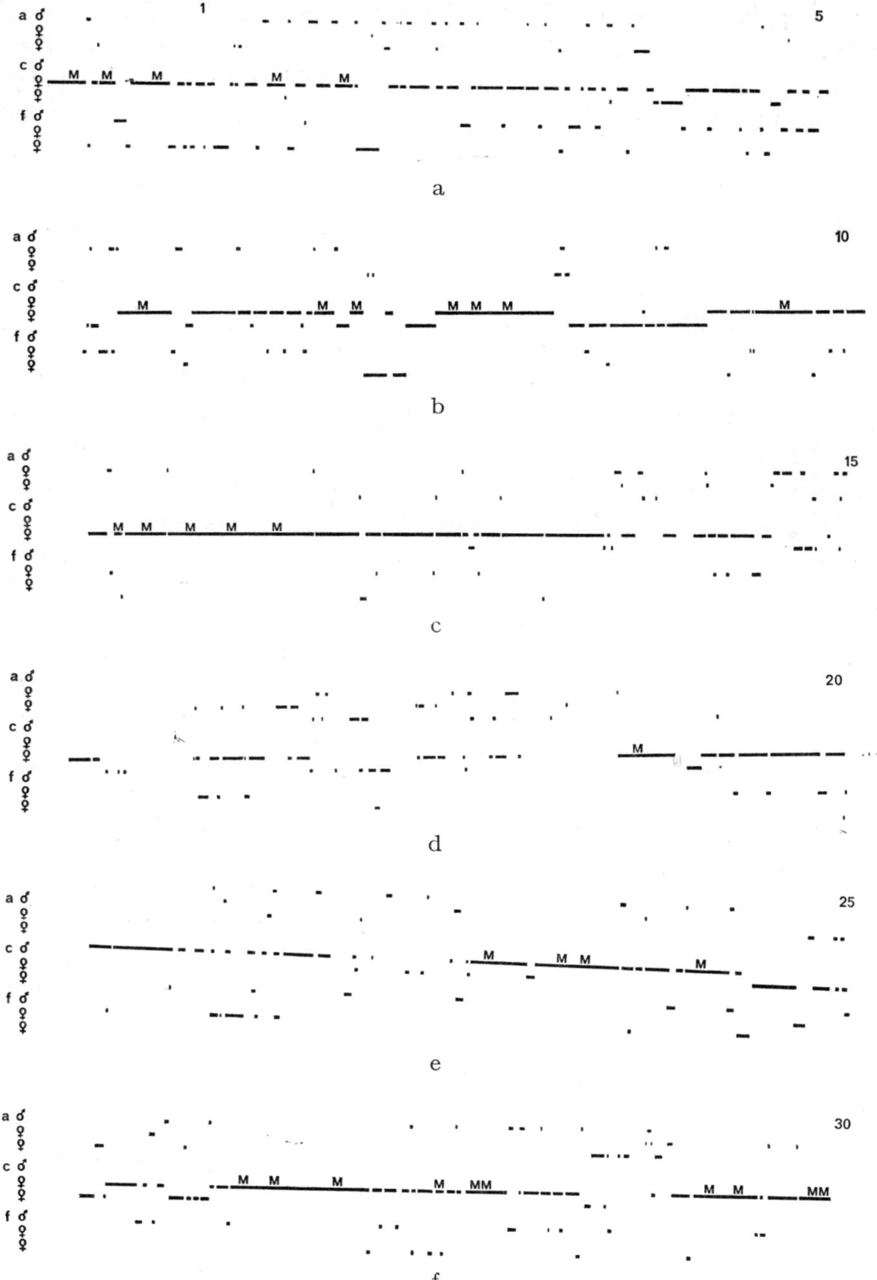

Fig. 46. Simplified behaviour record of 'normal' *nigrofasciatus* male (dominant of 2 ♂♂ 2 ♀♀ group); 30 minutes; legend see fig. 44; further explanation in text.

for the cyclic changes in the behaviour of both normal and homosexual males. However, the following speculations may be relevant. It follows from the aggression-fear-sex hypothesis of courtship (see chapter II, p. 97) that during the leading phase of courtship (much leading; mating attempts aggressively tinged) the agonistic systems are more strongly activated relative to the sexual system than during the mating phase (little leading; many more complete mating attempts and matings). During the aggressive periods in figures 44 and 46 all courtship is restricted to the leading phase. Leading is frequent, mating attempts and often courting posture are aggressively tinged and matings are absent. This is true for the courtship towards both females as well as for the homosexual male's courtship towards the other male. The restriction of courtship to the leading phase is in accordance with the higher agonistic activity during the aggressive periods. The clear-cut courting and mating preference for one female during the courting periods might suggest that different females induce different relative amounts of agonistic and sexual activation. This would mean that the not-preferred female induces relatively more agonistic activity, even though this is often not expressed in overt attacking. If this is true it is not surprising that, when the male's motivation shifts from a more sexual one to a relatively more agonistic one, a shift in preference in favour of the formerly not-preferred female occurs simultaneously. This would be in accordance with the assumptions nr 5 made for *conchonius* (p. 114) that both aggression and sex have positive feedbacks.

An important aspect of homosexual behaviour seems to be that a homosexual male just goes one step further than the normal male and periodically courts an individual which induces even more agonistic activation, *i.e.* a male. That agonistic motivation is often relatively strong in homosexual males is amply shown in the 6♂♂ 6♀♀ situation in which the homosexual males are typically very dark, even soon after dawn, always stay near the bottom and keep their territories throughout the reproductive periods. However, the question remains why a homosexual male, when agonistic motivation becomes stronger, courts males instead of attacking them (or, to put it the other way round, why normal males never court other males and only attack them). Further, the periods in which the homosexual males court another male differ from the aggressive periods of both the homosexual and the normal males in that in the former there is almost continuous courtship and a real preference for the male (about as strong as there is for the preferred female in the courting female periods) whereas in the latter the not-preferred female is only courted now and then and not necessarily more than the preferred female. Thus the period of courting the male has properties of both the courting female period and the aggressive period. It presents itself as an extension of the courting female period, while the orientation is already an agonistic one. It seems, therefore, that in the homosexual males something is wrong with the mutual exclusiveness between fighting and courting behaviour, which is so perfect in normal males.

This is about as far as the analysis can be pushed at the moment. Further questions will have to be left to future research.

The functions of the temporal and spatial structures of behaviour in this species and the mutual exclusiveness between reproductive behaviour and feeding are probably much the same as those discussed for *conchonius*. What the adaptive significance of the differences between the two species is, can only be assessed by studying natural populations.

Again, it is not known whether the various behavioural types in *nigrofasciatus* also occur in nature. It is evident that both the dark and the light type of male may be very successful, dependent on the circumstances. If courting groups are dense and if competition is heavy, the dark type will be more successful. When groups are less dense and competition is less important, the light type will have the full advantage of its better courting technique. However, it is difficult to see what the selective advantages might be for the homosexual males and for those females which behave aggressively and/or are only rarely sexually motivated, unless these characters are in some way linked with other more advantageous properties.

Finally, it must be emphasized that the laboratory conditions, even those in the large tanks, are only a very impoverished version of the natural circumstances. For instance, the distance between the bottom and the water surface was the same all over the tanks. It is not known what influence different water depths have on the behaviour. That this influence may be profound is suggested by the experiments with the homosexual male discussed above. This male hardly ever courted a female in the large tank, but at least periodically courted a female very intensively and mated with her in the more shallow 60 cm tank. Therefore the possibility must be left open that the homosexual males would display even more normal behaviour and that females would be more receptive if they could choose their own whereabouts.

3. *B. tetrazona tetrazona.*

Both the temporal and the spatial patterns of behaviour in *B. tetrazona* differ very much from those found in *B. conchonius* and *B. nigrofasciatus.*

For *tetrazona* four series of observations were made on 6♂♂ 6♀♀ groups in large tanks. One group was observed in a 400 × 100 tank (water depth 55 cm) for eight days. This group first consisted of 6♂♂ and 3♀♀ only. On the fourth day after introduction 3 more ♀♀ (from a different stock tank) were added. A second series (6♂♂ 6♀♀) was run in a 130 × 65 tank (water depth 40 cm) for four days only. 5♂♂ and 6♀♀ (from the same larger stock as the former series) were kept in a 130 × 65 tank (depth 40 cm) for 25 days in succession. This same group was then transferred to a 300 × 60 × 50 tank (water depth 45 cm) and observed for another 20 days. One male of this group died and was replaced by another male (see table 4, 28/7). One female got ill and was removed (1/8). The first series was run under normal day-light conditions in July; for the other three the usual artificial light regime was maintained.

(a) temporal patterning.

In the first and third series complete reproductive behaviour was first seen on the first day after introduction (introduction times 09.25 and 12.00 respectively). In the second series no complete reproductive behaviour was seen during the first four days (introduction time ± 13.00) but the activity developed gradually in such a way that reproductive activity could no doubt have been expected on the fifth or sixth day. However, the experiment had to be broken off due to serious disturbances in the room from the fifth day onwards. For the fourth series the fishes were transferred at 10.00. Complete reproductive behaviour was seen from 16.00 on the same day.

All series began with intensive exploratory behaviour, to an extent as never seen in the former two species. The fishes moved around through all parts of the tank, mostly in one group, but occasionally in smaller groups or alone. This exploration was the main type of behaviour seen in the second series. Many experiments with smaller groups in smaller tanks (see for instance the 2♂♂ 2♀♀, 2♂♂ and 3♀♀ expp. sections III, IV) confirm that *tetrazona* often spend much time, often as much as the first one or two days, in exploring the new surroundings. In this they differ markedly from *conchonius* and *nigrofasciatus* which — at least in ambisexual groups — practically always start fighting and courting on the very first day after introduction, even if they are introduced late in the afternoon.

Also the day to day pattern of activity is not by far as regular as in the former two species. The following accounts give an impression of the daily activities in the third and fourth series.

TABLE 4

Abstract of daily activities of tetrazona 6♂♂ 6♀♀ group

Date	Activities	Food
29/6	Introduction *ca.* 12.00, much schooling and exploring; some fighting in afternoon, courting intensions of one ♂, no territories; *ca.* no activity in evening.	D
30/6	Some schooling first; soon intensive fighting and courting with matings by *ca.* all ♂♂, first mating 09.34, incipient territorial behaviour by 2♂♂; *ca.* 11.45 much less fighting, *ca.* no courting, one ♂ territorial, territorial tendencies of at least 2 more ♂♂; from 13.45 until at least 15.12 again intensive fighting and courting with matings of *ca.* all ♂♂, one ♂ territorial.	D
1/7	Much schooling in morning; some fighting later and some courting in afternoon, almost mating 15.25, 2 territorial ♂♂; no evening observation. Following days: 2 territorial ♂♂ whenever activity is considerable.	D
2/7	Much schooling in morning; some fighting and courting later, no matings.	D
3/7	Much schooling in morning; some fighting and courting; *ca.* no activity from 16.00; probably *ca.* no activity in evening.	T
4/7	Some schooling first; fighting in morning; courting later, first mating 13.51; intensive fighting and courting until 16.00; slowly decreasing activity.	D
5/7	Much schooling first; some fighting and courting later, no matings; little activity at least until 15.00; no evening observation	T
6/7	Much schooling first; fighting and some courting later; intensive fighting and courting with matings in afternoon, first mating 13.46.	D
7/7	Much schooling, little activity at least until 14.47; No later observations.	T
8/7	No observations	D
9/7	Observation 16.50 only, moderate fighting, *ca.* no courting.	T
10/7	Schooling first; intensive fighting and courting with matings in afternoon from before 14.30 until after 16.00.	D
11/7	Some schooling first; fighting later; intensive fighting and courting with matings from at least 14.00	DT
12/7	Schooling first; much fighting and some courting later and throughout afternoon, few matings; no evening observation.	Dph
13/7	Some schooling first; some fighting from 13.45; moderate fighting at 16.00; little courting throughout the day; little activity in evening	DT
14/7	Some schooling first; more fighting later; intensive fighting and courting with matings from 13.45; gradually less courting from 15.00, intensive fighting going on.	DT
15/7	Schooling for most of the day; intensive fighting and courting with matings at 19.00.	DT
16/7	Some schooling, little activity at least until 19.15.	DT
17/7	Schooling in morning and early afternoon; moderate fighting and courting possibly with some matings (none observed) from 16.45 until at least 20.55.	DT
18/7	Some schooling for most of the day except some fighting of territorial ♂♂ (2); fighting and courting with matings at 21.00	DT
19/7	Much schooling, little activity throughout the day (evening included).	DT
20/7	Schooling first; more fighting later; intensive fighting and courting with matings from 20.30.	DT
21/7	Much schooling first; intensive fighting and courting with matings at least from 14.00; intensive fighting, no courting from 16.30; intensive courting with matings again at 20.15.	DT

Date	Activities	Food
22/7	No observations in morning and afternoon; intensive fighting and courting with matings at 20.00.	DT
23/7	No observations.	DT
24/7	*Ca.* 09.45 transferred to 3m. tank; much schooling and exploring first; increasing fighting and courting with matings in afternoon, first mating 16.09; fighting and courting going on until at least 19.45; incipient male territories in evening.	DT
25/7	Much schooling and exploring alternated with some fighting and courting with matings, first mating 09.52, incipient territoriality of one ♂; gradually less fighting and courting in afternoon and evening.	T *)
26/7	Much schooling and exploring in morning; increasing fighting and courting with matings in afternoon, first mating 13.58; territorial behaviour of 2 ♂♂, incipient territoriality of others; less activity in evening.	DT
27/7	Much schooling and exploring first; increasing fighting and courting with matings later, first mating 10.54; little activity from at least 13.45; 14.45 again intensive fighting and courting with matings of 2 ♂♂, other ♂♂ also from *ca.* 16.00; territories settled.	
28/7	Some schooling first; more fighting and courting with matings later, first mating 12.25; little activity for rest of day and evening; introduction new ♂ in afternoon (see text).	DT
29/7	Much schooling for whole day; no evening observations.	DT
30/7	No observations.	DT
31/7	Schooling for whole day and evening.	DT
1/8	Schooling first; more fighting in afternoon; much fighting and courting with matings from 16.00, first mating 16.19; activity going on until at least 19.15.	DT
2/8	Schooling for whole day and evening; some fighting and courting intentions after 17.15.	DT
3/8	Schooling for whole day and evening; some fighting after 15.00.	D
4/8	Schooling for whole day and evening; some fighting after 14.30.	D
5/8	Schooling at least until 13.20; no later observations.	T
6/8	No observations.	DT
7/8	Schooling first; more fighting later; fighting and courting with matings from 16.45; intensive fighting and courting at 20.00.	DT
8/8	Schooling for whole day and evening.	DT
9/8	Schooling first; some fighting, some incipient courting in afternoon; little activity in evening.	DT
10/8	Schooling at least until 16.30; no evening observations.	DT
11/8	Schooling first; intensive fighting and courting with matings at least from 11.30; gradually less activity later in afternoon.	DT
12/8	Schooling for whole day and evening.	DT
27/8	After long period of little activity: again intensive fighting and courting with matings at 16.30.	

Notes: 'schooling' implies little reproductive activity; 'activity' = reproductive activity; D = Dry food; T = Live Tubifex; Dph = Live Daphnia; DT = Dry food + Live Tubifex.

*): Dry food still present on the bottom.

Intensive reproductive behaviour of all males is seen only on some days. On other days reproduction is much less intensive and on several days there is practically no reproductive behaviour at all. On such days much schooling behaviour is seen but, as long as the fishes remain in the same tank, the intensive exploratory behaviour of the first days is never seen again. More and less active days follow each other in a rather irregular fashion.

It also emerges from the abstracts that there is no consistent daily pattern of activity such as is found in *conchonius* and *nigrofasciatus*. Some general features are present: There usually is very little reproductive activity during the first one or two hours of the day. Reproductive behaviour occurs mainly in bouts lasting a few hours. During those bouts feeding activity is reduced, especially in the males. A period of increasing agonistic activity usually precedes the onset of courting behaviour. Even on days with little reproductive activity, there often is some agonistic behaviour and this activity increases gradually through the late morning and the early afternoon. So far there is a certain resemblance between the behaviour of *tetrazona* on the one hand and *conchonius* and/or *nigrofasciatus* on the other. However, for the rest almost anything may happen on a particular day. Agonistic behaviour may slowly increase in intensity up to a point where courting behaviour might be expected to begin and then wane again. After an initial increase of agonistic behaviour, activity may be reduced almost completely only to revive later on the same day and be followed by courting. Courtship periods may begin at almost any time of the day, from as early as one hour after dawn to one or two hours before the end of the light period. Once courtship has begun, it may be replaced by fighting again and reappear later (this may occasionally also happen in *nigrofasciatus*). There is even one day (27/7) with two completely separated reproductive periods. On this day complete reproductive behaviour had developed around 10.45; at 13.45 all reproductive activity had stopped; at 14.45 two of the males courted and mated again and so did most of the males at *ca.* 16.00. Roughly the same happened on 30/6.

Introduction into a new tank often seems to stimulate reproductive activity after the initial period of exploration. Introduction of new individuals into a settled group also often stimulates reproduction, as is shown in series I (3♀♀ on 22/7) and in series IV (1♂ on 28/7). In the first case courting was already going on when the females were introduced, but it suddenly increased greatly in intensity immediately after introduction. In the second case complete reproductive behaviour had already occurred during the early afternoon but the activity had come to an end at least an hour and a half before introduction of the new male. Introduction was followed by a period of increased agonistic and courting behaviour, though no complete reproduction ensued. Introduction of new fishes may also lead to increased schooling of the whole group. In fact, the introduction of the male in series IV had this effect for a short period before reproductive behaviour developed.

The possibility has to be considered that the greater irregularity in activity patterns in *tetrazona* is due to a greater susceptibility to changes in external factors such as water composition, temperature or food. It also seems possible that the laboratory conditions are less optimal for *tetrazona* than they are for *conchonius* and *nigrofasciatus*. In fact, we found *tetrazona* more difficult to breed in the laboratory than *conchonius* or *nigrofasciatus*, possibly because the young of the former species need comparatively soft water for development. However, though no consistent measurements were made of water hardness, PH or any kind of pollution, it doesn't seem probable that the above factors are alone responsible for the irregularities found. In fact, very intensive reproductive behaviour was displayed on a number of days. The relative inactivity on some days may have been facilitated by very slight water pollutions (as judged by the clearness and colour of the water and by the state of plants and algae), but the same apparent conditions didn't prevent them from displaying intensive reproductive behaviour on other days. Besides, on many of the inactive days there was no sign of pollution at all. Temperatures varied between the series from 21°C (series III) to 25°C (series IV) and also this didn't make any difference for the pattern. The kind of food supplied on a certain day might influence the activity on the next day (see table 4 column 3). On the first 4 days of series III dry food mixture was supplied. Reproductive activity declined gradually from 30/6 to 3/7. On the following days live *Tubifex* and dry food were alternated regularly. Some activating effect of live food may be present here. However, activity was high on 11/7 after dry food on 10/7; activity was only moderate on 12/7 after *Tubifex* and dry food on 11/7; activity was low on 13/7 after live *Daphnia* on 12/7. Therefore, though a supply of live food or a change in kind of food may have an activating effect, such effects probably don't last longer than a few days, when food is alternated frequently. On 11/7 and from 13/7 onwards both *Tubifex* and dry food were supplied each day; the same was done throughout series IV. Yet the same irregular pattern continued to occur.

The large scale irregularity is somehow in accordance with the general explosiveness of activity in this species, on a smaller time scale.

From the above considerations it would appear that the irregularity of temporal patterning of behaviour in *tetrazona* under constant conditions, is a species-specific character rather than a laboratory artifact. This does not mean that the same patterns are present under natural circumstances. In fact, it is much more likely that in nature the fishes are again and again reactivated by confrontation with new individuals. Further, in the laboratory it was found that a temporary increase in temperature, even of one or two degrees only, often causes an increased reproductive activity lasting as much as two or three days after the temperature had returned to the normal value. This may of course easily happen in nature as well. (On the other hand slight decreases in temperature may suppress all reproduction; subsequent return to normal temperature works as temperature increase!). Finally, the fact that in nature the fishes are able to select their own food leaves the possibility that they may re-activate themselves.

(b) spatial patterning.

Also the spatial patterning of behaviour in *tetrazona* is different from that of the former two species. In all three series in which complete reproductive behaviour occurred, males were distinctly territorial throughout the reproductive periods. Only the 3 and 4 meter tanks turned out to be large enough for all males to establish territories. In the 130 cm tank only two males managed to do so permanently, though the others kept trying now and then to establish territories as well. Since apparently the 130 cm tank is too small for a group of 6♂♂ 6♀♀ to deploy their natural behaviour, most of the following descriptions are taken from the first and fourth series (4 m and 3 m tank respectively).

The territoriality of the males is distinct but not very strict. On the first few days after introduction into a new tank, territories are even often absent or at least rather vague and locations shift frequently. On later days territories are mostly given up outside the reproductive periods. Feeding and strong schooling are the predominant activities then. At least some males start to visit their territories and to defend them as soon as the school begins to break up. The other males follow as fighting activity increases. Defence of territories is not too successful. Especially if several other fishes enter a male's territory simultaneously, his success in chasing them out is rather poor.

Neighbouring males often are not equally strong. Usually one of them may enter the other's territory comparatively easily and he may even temporarily defeat the owner, who then flees and/or rolls. Only because the stronger male soon returns to his own grounds, the weaker can again and again recover and re-establish himself.

The same males tend to occupy territories at the same places on consecu-

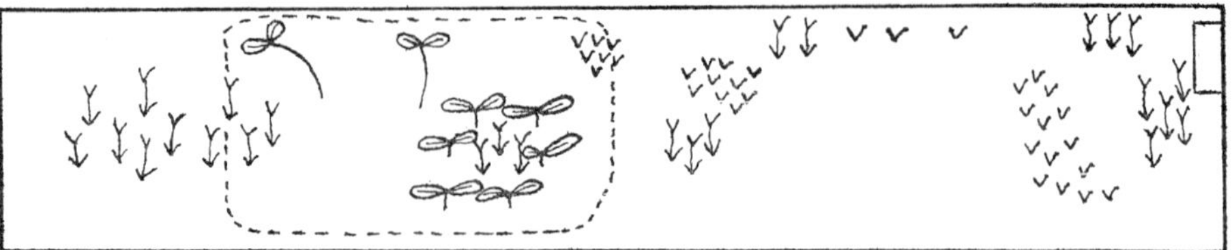

Fig. 47. *Tetrazona*. Schematic top view of 3 m tank; locations of filter (right) and rooted vegetation of various kinds; area enclosed by broken line: floating plants.

tive days. However, locations may be shifted rather often and the same location may be occupied by different males in succession within a few days. The following abbreviated accounts from the fourth series may serve as an example. The scene is the right end of the 3 m tank. Figure 47 gives a schematic top view of this area. The bare space to the left of it acts as a kind of natural barrier reducing the number of intruders. On the other hand, these qualities apparently make the area a particularly attractive one for territory-seeking males. The main actors are three males: ♂1 one of the larger males [1]), ♂3 slightly smaller than ♂1, and ♂6 about the same size as ♂1. ♂6 is the male which was introduced on 28/7. On the day of introduction (24/7) ♂1 and ♂3 are the first to break the school. ♂1 is dominant over ♂3 and establishes a territory at the right end of the tank. ♂3 tries to settle on several places in succession, but is defeated by other larger males. He ends up with a territory to the left of that of ♂1, at the other side of the bare space. He and ♂1 fight each other across the bare space. The other males display only very weak territorial tendencies. On 25/7 ♂3 is the most active of the males. He exhibits territorial tendencies, but again he cannot decide as to the location. On 26/7 he establishes a territory at the right end of the tank. He is dominant over ♂1 now. At first he is the most active of the males. After 16.50 ♂1 again claims the territory at the right end of the tank and defeats ♂3. The other males do not yet clearly choose territories. On 27/7 ♂1 is clearly in command of the territory at the right end of the tank. He defeats ♂3 again and again. ♂3 flees with strong rolling. However, he keeps trying and intruding ♂1's territory and interfering with ♂1's courtship (see p. 142). If ♂1 is absent for brief periods, ♂3 even defends the territory against intruders. On this day also the other males settle. The situation on 28/7 is essentially the same as on 27/7: ♂3 taking the territory at the right end as long as ♂1 is still reproductively inactive, ♂1 taking over later in the morning. In the afternoon of this day ♂6 is introduced. On the next observed active day

1) All ♂♂ were approximately the same size but two of them were slightly but distinctly smaller than the others.

(see table 4: 1/8) ♂6 takes the territory at the right end and defeats ♂1. ♂1 moves to other parts of the tank but ♂3 from now on plays the same role with respect to ♂6 as he did before to ♂1. Though he is again and again chased away by ♂6, he keeps intruding into his territory. (♂1 doesn't loose his interest in the area. On later days he often visits it whenever ♂6 is not (yet) reproductively active. Sometimes he meets ♂3 there and fights and defeats him).

In this situation the efficacy of rolling is particularly clear. As soon as ♂3 rolled, ♂1 or ♂6 stopped their attacks and went back to their territory centre. ♂3 then also returned at once and even now and then showed inhibited attacks over long distances at the territory-owner. Long inhibited attacks immediately following strong rolling were also seen in the more inferior of two territorial males in the 4 m tank (series I), which had territories at either side of a large bare space. These examples beautifully illustrated that rolling, just like threat, may occur at a point between flight and renewed attack (see chapter II p. 93).

Courtship and mating are mainly performed inside a male's own territory, though males now and then leave their territories to court females and to compete with other males far away from their home grounds. In the first few days after introduction, when locations of territories may shift frequently, a male may court and mate in each of his successive tentative territories. For instance, in series IV on 25/7 ♂3 mated at various places ranging from the extreme right to the extreme left of the tank. Also during the first few days and later at the beginning of reproductive periods, courting and mating may take place without any sign of territorial attachment, usually with extremely intensive competition between the males. In this type of group courtship, even when only one male is courting and therefore competition is low, leading is very rare. Courtship mainly consists of direct mating attempts alternated with courting posture. Mating frequency is very low. In the same situations, when courting activity increases, the group may be broken up and often several pairs of one male and one female move around separately and court intensively. Even in this situation, however, mating success is very poor. This type of courtship in pairs, interrupted every now and then by competition between males, also occurred in the 130 cm tank. Here it was the normal type of courtship for all non-territorial males. In the 3 m tank, as soon as territoriality has become settled, a male's territory is the only place where he may court relatively successfully, where he may for instance perform more than one mating with the same female in quick succession and where he is not continuously forced to compete with other males. Even so, however, competition remains severe and mutual penetration into each other's territories remains frequent. This, added to interference by other females, keeps mating frequencies always low compared to those found in *conchonius* and *nigrofasciatus* in comparable situations. When competitive interaction is very high, territories may become vague again and groups of 3 or 4 males and one or a few females may move across the tank, the males fighting and keeping-off each other.

In series IV the male which for the moment held the territory at the right end of the tank, invariably was the most successful in courting and mating. ♂3 on 26/7, ♂1 on 27/7 and 28/7 and ♂6 on 1/8 and consecutive days regularly managed to perform from 2 to 4 matings with the same female in quick succession. This remained rare for all other males throughout the experiment. It was rare for all males in series III (130 cm tank), though also in this tank the territorial males were relatively more successful. In the 3 m tank ♂1 and ♂6 would no doubt have been even more successful if ♂3 had not kept intruding and interfering with their courtship. ♂3 now and then managed to mate inside the territory of ♂1 or ♂6, when the resident was absent for a while or himself courting in a different part of his territory. When ♂1 or ♂6 courted in their territory, ♂3 often approached immediately and interfered with their courtship. In the 130 cm tank ♂3, who was there one of the non-territorial males, was once seen to approach suddenly one of the territorial males which was about to mate, and try to mate at the other side of the female. This mating was unsuccessful for both males. However, this kind of "sneaking" has been observed to be successful in other situations.

Summarizing the above observations we may say that the only situation in which a male may have a reasonable mating success is the territorial one. Most favourable are such locations as are "naturally defended" against intruders by being situated in a corner of the tank (or near a bank in nature) and by open spaces or dense vegetation at the open side. The fact that all males in our experiments settled in more or less permanent territories once they had become accustomed to the environment and provided that there was sufficient room for all, suggests that the territorial situation is also the natural one (see further discussion p. 148).

The influence of depth on behaviour is much less clear in *tetrazona* than in *conchonius* and *nigrofasciatus*. In general, any type of activity may be performed at any level in the tank. *Nigrofasciatus* typically explore new surroundings swimming near to the bottom, (though a certain shyness in the unfamiliar environment may play a role in this preference). *Tetrazona,* on the other hand, may do so at any level. Only if they are very shy they swim around near to the bottom. Typically, on introduction into a new tank, they swim around at about 20 cm from the bottom and now and then make excursions to higher and lower levels. Feeding may occur high up in the plants as well as from the bottom. The same indiscriminatory tendency as to depth is found in reproductive behaviour. Fighting is common at all levels and so are competition and courtship. Mating may take place close to the bottom as well as near the surface. Leading may be upwards, horizontal or downwards. At all levels it is very common for a male to attack a rival suddenly from keep-off position.

However, there are some differences in behaviour according to depth which suggest that the same mechanisms are at work as in the former two species, though the influence of territories seems to be relatively much more important in *tetrazona*. Fleeing is often upwards. Leading is horizontal or even downwards when a male is courting at relatively low levels, but if the female is more than about 15 or 20 cm from the bottom, leading is typically upwards. Not much difference in the degree of rolling-over during mating may be found between 0 and ½ or ¾ of the total water height, but rolling-over is stronger in matings near the surface (see table 5; *cf.* chapter II, table 3). Fighting between males may begin at any level, but mostly in the lower half of the tank. During prolonged fighting they tend to seek progressively lower levels. Duels including mouth-fighting always end up very close to the bottom. Territorial males, when undisturbed, swim around or stand still at various levels, but always in the lower half of the tank. Males leaving the school to establish territories typically go downwards relative to the school. In smaller groups dominant males are always nearer to the bottom than inferior ones. The territorial female in series III (see below) was always near to the bottom.

TABLE 5

Relation between degree of rolling-over during mating and depth of pair in water (proportion of total water height above bottom). B. tetrazona

height	mean angle *)	n
0–1/10	85°	10
1/6 –1/2	82°	43
3/5 –5/6	86°	3
9/10–1	76°	21

*) 90° = vertical.

Tetrazona display about the same differences in behaviour relative to plants as do *conchonius* and *nigrofasciatus*. Non-reproductive swimming around is mainly through the open paths between plant clumps. Only during the exploratory phase they (as do *nigrofasciatus*) now and then enter plant clumps in a group, often gently nipping at the plants. Feeding may be from the bottom in the open spaces, but also sometimes inside plant clumps, usually when they are in a group. They often nip at plants at higher levels. Much nipping at plants, also inside plant clumps occurs during courtship (see below p. 147). Fighting and the leading phase of courtship are mainly performed in the open spaces. Leading is almost always very clearly directed towards plant clumps (see also p. 105). Mating typically takes place between or at least in close contact with plants. Territorial males, when undisturbed, stand out in the open, but never very far from plants; the large open areas of 20 cm or more were never chosen as a site for a territory. In smaller groups dominant males typically swim around or stand still in the open, whereas inferior males tend to hide between plants or in the corners of the tank (see also sections III, IV).

(c) inter-individual patterning of behaviour.

Inter-male fighting without females being nearby is extremely frequent and intensive both during aggressive and courtship periods. It sometimes culminates in duels which may include mouth-fighting. In the 3 m tank most of the duels took place when territories started to be established and later when a male tried to enlarge his already existing territory. Duels seemed to be more frequent in the 130 cm tank than in the 3 m tank. They were also very frequent on the last day of series II (130 cm tank). The duels in this series were rather brief but included mouth-fighting. They sometimes involved 3 males at a time.

In all situations males frequently attacked females, even during periods of intensive courtship. Also the males in series IV which held the territory at the right end of the tank often attacked females which entered the territory or chased them away after one or a few matings had been performed. Competition with other males often leads to attacking and chasing away of the courted female. This often happened when ♂1 or ♂6 were hindered by ♂3 in series IV. However, also if no rivals are nearby courting is frequently alternated with attacks on the courted female. Even if a male doesn't overtly attack the female for a while, he often courts with rather wild movements. This is apparently unfavourable for successful mating. It is strongly reminiscent of the type of courtship displayed by the darker type males in *nigrofasciatus* (see p. 124). Only in one particular situation was a male seen to court a female continuously for considerable time. This was during casual observations of one male and one female in a very small tank (36 × 22 cm). This male didn't attack the female any more and mated frequently. This observation suggests that it is the recurrent fighting and competition between the males in the large tanks which keep them constantly aggressive, also towards females. In the large groups it is obvious that a female is more likely to be attacked if a male has just been fighting another male. Apparently fighting leaves a positive after-effect on aggression lasting a few minutes at least and probably much longer than that (see further discussion p. 148).

Not only do males frequently attack females, but they also occasionally court other males. This was seen several times in series III (130 cm tank) mostly by one of the territorial males. It could occur even when females were very willing during intensive courtship periods. Males courting males were not observed in either series I or series IV (4 m and 3 m tanks respectively). There is no indication that homosexual courting is a property of particular individuals, like it is in *nigrofasciatus,* nor does it seem to be a matter of a specific preference for males. The fact that it was not seen in the 3 m and 4 m tanks strongly suggests that it is an artifact of relatively crowded conditions (see further section III). Like in *nigrofasciatus* two males may suddenly mate when they are competing for the same female and trying to keep-off each other.

Males which are strongly sexually motivated are sometimes seen to creep through dense plant clumps in a peculiar fashion. They may do this when courting a female either at a distance from the female at the end of leading or close to the female together with her. They may also show this behaviour when no female is nearby. The type of movement is similar to that of the later phases of courtship. This, together with the fact that contact with plants is sought, suggests that it is a kind of *vacuum* courtship.

Females may display considerable aggressive behaviour, even more than *nigrofasciatus* females. In all series they were rather often seen to attack and threaten other females or males. In series III one female held a small territory for a number of days. She attacked intruders and sometimes had boundary fights with one of the territorial males. In such fights she was often quite successful. In series I the females which where introduced later displayed much threat, both towards males and towards other females. One of them had a complete duel with a male. This fight lasted as long as 10 minutes at least! At least one duel including mouth-fighting between a male and a female was seen on the last day of series II (130 cm tank). No territorial females were seen in series other than series III. Female aggressiveness was usually suppressed whenever males were actively courting.

In small ambisexual groups (2♂♂ 2♀♀) females often display considerable aggressiveness, though they only rarely become dominant (see below, p. 186).

Females may also play a more obviously active role in courtship than do *conchonius* or *nigrofasciatus* females. When very willing they may wriggle through plants in a characteristic manner. This is usually performed at relatively high levels in the water and they often go slowly upwards in the process. Wriggling is often almost continuously accompanied by vacuum leaning. Females often very actively follow leading males. During the last stage of courtship, if the male hesitates to mate, the female often actively sidles towards him until their sides touch and mating ensues. Females, like males, display a vivid interest in courting pairs or groups. They often swiftly approach a pair about to mate. One female was seen to try and take part in the mating of a male and another female. This mating was abortive for all three.

Even during fully developed courtship periods, when males are completely dispersed, females always have a certain tendency to cluster together.

(d) colour changes.

As was described in chapter I, male and female *tetrazona* have very much the same colour patterns when they are inactive reproductively. Only males have a couple of red streaks in the tail fin and all reds are on the average slightly more intense than in the females. However, when reproductively active and especially when courting, males get much more intensive red colours. Even the whole body may bear a faint but distinct rosy hue. The black body bars fade. The bar at the base of the tail fin often disappears completely. The pectoral bar and the first tail bar also fade partly and get a greenish colour. The dorsal and ventral ends of the tail bar may fade completely, leaving only a rounded eye-like spot in the middle. This is typically seen in territorial males. The whole body often gets a shiny silvery appearance. In the females, on the other hand, all colours become more dull. The red colours fade, also those in the fins. The body bars fade too. Often the black in the dorsal and anal fins fade as well, but the base of the dorsal fin may remain comparatively dark. The whole body often gets a slight greyish hue. As a result of these colour changes the difference between males and females becomes much greater, though it never gets by far as spectacular as in *conchonius* and *nigrofasciatus*.

(e) feeding.

Feeding activity is reduced in those individuals which are reproductively active. However, in our situations, feeding never stopped completely, especially not in the females. Even territorial males, when they are courting and competing intensively, now and then show some feeding from the bottom or from plants. As was described above (p. 105) a courting male may at the end of leading nip gently at a plant. Courting males and females often nip at plants frequently during the final stages of courtship. Females often nip frequently at the plants during wriggling. It is not certain whether during this nipping food is actually taken in. It is practically certain, however, that all eggs are devoured soon after spawning (see further discussion p. 149 ff.).

(f) individual variability.

Like in *nigrofasciatus* there are considerable individual differences in behaviour, both in males and in females. However, in *tetrazona* it is much more difficult to classify individuals in well-defined types. This is partly due to the greater plasticity of behaviour (males attacking females as well as other males; territorial as well as extra-territorial behaviour; irregular temporal patterning of behaviour). Some males are consistently quicker in leaving the school and in starting reproductive behaviour than others. Some are much quicker in establishing a territory in new surroundings than others. Individual territories differ considerably in size, though these differences may be partly determined by the spatial set-up of the tank (walls, corners and vegetation). Part of these differences is apparently independent from the relative sizes of the males (see ♂1, ♂6 and ♂3 in series IV). A very special role is played by ♂3 in series IV. In series III this male didn't manage to establish a permanent territory, but also there he was active in breaking the school and displayed considerable agonistic and courting behaviour. He was seen several times to interfere with the courtship and mating of one of the territorial males in much the same way as he did in series IV with ♂1 and ♂6. ♂3 was a bit small for his age group; his body bars had slightly irregular shapes and his red and green colours were more strongly developed than average. In the course of this study I have seen some other males of the same morphological type. No special experiments have been done with these males, but in the stock tanks their behaviour often seemed to be on the cheek side. Females differ considerably in the amount of agonistic behaviour shown. The one female which established a territory in series III also showed more than average aggressive behaviour in series IV. Even more than in the males, however, it is difficult to assess in how far behavioural differences between females are due to individual properties or to direct external conditions.

(g) schooling and mobbing behaviour.

Strong schooling is another characteristic activity of *tetrazona*. As mentioned on p. 60 all species of *Barbus* involved in this study do form schools, especially when they are not reproductively active. However, the cohesiveness in schools is much stronger in *tetrazona* than in the other species. When inactive reproductively they often stand in a group in mid-water, each individual in a slightly head-down position. When moving around they stay closely together in a group or in a file, maintaining almost perfect parallel orientation. This behaviour, in combination with their contrasting colour patterns, makes a school of *tetrazona* particularly conspicuous. The strong schooling tendency of the species, combined with extremely strong agonistic tendencies, lends a pronounced acuity to the conflict between schooling and reproduction, which is no doubt in part responsible for the typical explosiveness of *tetrazona* agonistic behaviour (see further discussion p. 149 ff.).

Lorenz (1963; p. 44-5) reports "mobbing behaviour" of *B. tetrazona partipentazona* towards large Cichlids. Our *B. tetrazona tetrazona* showed similar behaviour towards a middle-sized *Cichlasoma severum*. As soon as the cichlid was introduced in a stock tank containing *tetrazona,* the latter flocked together. When the cichlid started to swim around, they crowded behind it and now and then bit and tore at the edges of its fins. If the cichlid turned around to face them, they dispersed swiftly and reassembled behind it at the other side. The cichlid apparently "disliked" the treatment. It got dark all over and stood still with its head steeply upward or tried to hide between dense vegetation.

The same kind of behaviour was shown towards a large *Barbus schwanenfeldi,* though much less intensively and without biting. It is difficult to say whether the difference in intensity was due to the difference in appearance of the introduced fishes (*B. schwanenfeldi* is certainly not a customary predator and it doesn't look like one) or only to the fact that *schwanenfeldi,* even when standing still, causes a lot more turbulence with its caudal fin, thus discouraging the *tetrazona* to approach closely and to bite. The fact, however, that the same kind of response was shown leaves the question whether "mobbing behaviour" in *tetrazona* is essentially different from the more general response to an arbitrary new object. In general *tetrazona* do show an intense interest in any novel object. This matter will have to be investigated further, preferably using the natural predators of *tetrazona.*

No "mobbing" responses could be elicited from *B. conchonius, B. nigrofasciatus, B. stoliczkanus* or *B. cumingi.*

Discussion.

Tetrazona behaviour differs from that of *conchonius* and *nigrofasciatus* in some essential characters. This suggests that also in nature their pattern of reproduction is quite different from that of the other two species. Their extremely poor mating success in competitive situations would suggest that they do not court in dense groups in the way the other two species probably do. Their territoriality and the fact that only males courting in comparatively isolated territories may have a reasonable mating score (p. 141, 142) suggest that in nature males stake out territories well apart from each other and outside the main range of the schools of non-reproductive individuals. More or less "naturally defended" nooks near a bank, such as were described on p. 142, might be favourable for the establishment of territories.

In a suchlike situation a male could, once a willing female has entered his territory, develop intensive un-aggressive courtship and mate frequently, without being too often disturbed and made aggressive by other males or females (p. 145).

All this still leaves the problem of egg-eating. If intensive reproduction doesn't effectively suppress feeding (p. 146) the resident male and the female he admits into his territory, would eat their own eggs, even if no other males or females would be able to do so. Also if the eggs adhere to the plants they wouldn't escape consumption, because *tetrazona* apparently feed frequently from plants at all depths (p. 143). However, it is possible that feeding *is* effectively suppressed when reproduction proceeds to more continuous courtship without much competition or aggression towards the courted female. Further, the strong schooling tendency (p. 148) may play an important role in "retrieving"' the animals from their territories back to the main school, as soon as the reproductive motivation begins to decline and feeding could be resumed.

Of course, the above picture is highly speculative and should be further investigated in natural circumstances.

4. *B. stoliczkanus.*

(a) temporal patterning.

Data on the behaviour of *stoliczkanus* are much less complete than those of the other four species. One group of 6♂♂ 6♀♀ was kept in a 130 cm tank (waterdepth *ca* 50 cm) for some days. This group was composed out of individuals from a number of different tanks. On the first day after introduction very intensive fighting and courting behaviour was seen, but no extensive notations were made. On the following 3 days only some fighting and very little courtship was displayed. More complete notations of the behaviour on these days are available. Further some observations were made of a large group of males and females which were fighting and courting intensively in a 125 cm stock tank.

Though incomplete, the data available, in combination with many observations in smaller groups (see sections III, IV) give a fair picture of the species' behaviour and allow comparison with the other species.

The fact that in the 130 cm tank no complete reproductive behaviour occurred on the 2nd, 3rd and 4th day after introduction suggests a *tetrazona*-like irregular day-to-day activity pattern. In this experiment there wasn't any sign of water pollution or of any other external disturbance which could have caused the relative inactivity. Strong irregularity of the day-to-day activity pattern was also found in many standardized and casual observations of smaller groups in 60 cm tanks.

Like *tetrazona, stoliczkanus* may be found fighting and courting intensively at almost any time of the day. Males are often seen to fight intensively, without much courting in the same period. However, it is difficult to assess whether these males are not at least sexually *motivated* at the same time, because more than in any of the other 4 species it seems to be the female's behaviour which determines whether courting will take place or not. If a female doesn't react when approached by a male in courting posture, the male almost immediately stops courting and either attacks the female or swims away. Therefore it is difficult to say whether in this species clear-cut purely aggressive periods do occur.

The reproductive behaviour, both of males and of females, is typically explosive with sudden outbursts of fighting or intensive courting, alternated by periods of relative inactivity. In this aspect it also closely resembles the behaviour of *tetrazona.*

(b) spatial patterning.

The following aspects of spatial patterning of behaviour are clear from the observations. Normal swimming around is performed mainly in the lower layers, at least in the lower 15-20 cm of the tank. Feeding also is mostly from the bottom, but feeding from the plants may also occur at higher levels (see also below). Fighting between males without females being nearby is frequent; it is mainly performed at lower levels. Competition between males may occur at any level. Courting may be initiated at any level, but intensive courtship including matings is practically only seen near to the surface. Like in *conchonius* and *nigrofasciatus* the male, when courting near the surface, is typically preferably at the same level with the female or even above her during courting posture and circling, whereas he is mostly below her when courting at lower levels.

Stoliczkanus displays about the same variation in behaviour relative to vegetation as do the other species. Normal swimming around and feeding from the bottom are mainly performed in open spaces and so is fighting between the males. The leading phase of courtship is mainly performed in open spaces, whereas mating preferably takes place at least in contact with plants.

Stoliczkanus males are often strongly territorial during fighting periods, also when, during these periods, they frequently try to initiate courting. Centres of territories may be near the bottom, but also higher: between 10 and 30 cm from the bottom. Territorial males typically swim around in the open spaces. It is not clear whether territories are maintained during intensive courtship.

Females also mostly stay close to the bottom when swimming around and when feeding. When sexually willing, however, they go upwards, either slowly along plant stems, nipping gently at the plants on their way, or in sudden dashes. Willingness may also be displayed in *vacuum* leaning and quivering, typically near the surface. When going upwards slowly, females are often accompanied by the males which go upwards in the same way, also niping gently at the plants. Close clustering of several males and females may occur during this behaviour (see below p. 158). Like in *conchonius* and *nigrofasciatus,* females often show shyness when coming to higher water levels.

(c) inter-individual patterning.

Even during intensive courtship including matings, the males keep attacking the courted female frequently and intensively. Males may court other males, but the observations are too few to know how frequently this occurs.

Females often display considerable aggressiveness both among themselves and toward males. This is frequently seen throughout periods of male fighting, at least as long no intensive courtship occurs. In small groups females display considerable aggressiveness; they even often become dominant, thus precluding any intensive male courtship (see further p. 186).

(d) colour changes.

In both males and females the black body markings fade markedly during fighting and courting. When fading the tail spot gets a deep green colour. In the males the whole body gets a silvery shine and a slight yellowish hue during intensive reproductive behaviour. The red and black colours in the dorsal fins of the males may become slightly more intensive during reproductive behaviour, though the colours are usually rather brilliant even outside reproductive periods. The ventral fins may become more or less reddish. The difference in colour between males and females during reproduction about equals that found in *tetrazona*. However, whereas in *tetrazona* the sexes are much more alike outside reproductive periods, in *stoliczkanus* the males remain more distinguishable almost always, because of the persistence of the dorsal fin colours.

(c) schooling.

Schooling tendencies are strong in *stoliczkanus*. Parallel orientation is not by far as perfect as in *tetrazona,* but strong clustering of both males and females may occur during almost all phases of activity. Even during fully developed courtship including matings, periods of more complete dispersal alternate frequently with periods of close clustering of four or more individuals. This apparently leads to repeated and sharp conflicts between schooling and reproduction, reminiscent of those found in *tetrazona* at the beginning of reproductive periods.

Discussion.

The behaviour of *stoliczkanus* resembles the behaviour of *tetrazona* (and differs from that of *conchonius* and *nigrofasciatus*) in the following characters:

(1) the irregularity of the day-to-day activity pattern;
(2) the variability of daily activity patterns;
(3) the general explosiveness of reproductive behaviour;
(4) the overt conflict between schooling and reproductive behaviour, (the same antagonism of schooling and reproduction is, of course, present in *conchonius* and *nigrofasciatus,* but in these species it doesn't lead to such acute conflicts; the school is dissolved rather smoothly as reproductive behaviour starts to develop. Such schooling behaviour as persists during reproduction (*e.g.* the loose schooling of *conchonius* males when the females are still near to the bottom (p. 113) is more or less integrated in the reproductive behaviour pattern);
(5) the strong territoriality of the males during reproductive behaviour (though it is not yet clear whether *stoliczkanus* males maintain their territories during intensive courtship);
(6) the fact that the males, even when courting intensively, keep attacking the courted female.

To these characters we may add some on the behaviour element level (see section I):

(7) the clear distinction between horizontal and head-down threat;

(8) the presence of mouth-fighting in duels;
(9) a relative accent on severe biting rather than on prolonged attacking and threat.

And finally:

(10) the "emancipated" role of the females. At least outside intensive courting periods, they take actively part in fighting behaviour and may even fight duels with males; in small groups they may become dominant (though full dominance is rare in *tetrazona* females; see further p. 186).

Stoliczkanus behaviour differs from *tetrazona* behaviour in the following characters:

(1) in *stoliczkanus* fighting between males when no females are nearby occurs mainly in the lower layers of the tank, whereas in *tetrazona* it may occur at any level (though also with a preference for lower layers, see p. 143);
(2) in *stoliczkanus* intensive courting is most often performed near to the surface, whereas in *tetrazona* it may occur at any level.

In the first two characters *stoliczkanus* resembles *conchonius*.

(3) whereas in *tetrazona* the conflict between schooling and reproductive behaviour is most acute during the beginning stages of the reproductive periods, in *stoliczkanus* it seems to persist throughout courtship periods;
(4) in *tetrazona* prolonged mouth-fighting, often lasting several minutes, is a typical part of many if not most duels. In *stoliczkanus* it occurs frequently but is always very short;
(5) *tetrazona* males fighting a duel typically sink towards the bottom during carouselling and mouth-fighting. *Stoliczkanus* males typically rise towards the surface in duels, also during carouselling (see section I);
(6) in general, *stoliczkanus,* even though it may at times indulge in severe biting, is much more gentle in its aggressive behaviour and in the agonistic components in courtship.

All of these differences, except (2), point to the conclusion that *stoliczkanus* is a less aggressive "version" of *tetrazona,* more or less in the same way as *conchonius* differs from *nigrofasciatus.*

It is not quite clear what the difference in depth preference during courtship means (point 2). The same difference was found to exist between *conchonius* and *nigrofasciatus.* One might surmise that in *nigrofasciatus* and in *tetrazona* also the sexual motivation is stronger than in their less aggressive relatives, in the sense that courting may be more easily elicited also at lower depths in the former two species (*cf.* p. 112 for the dependence of sexual motivation on depth). One might speculate that in *nigrofasciatus* and *tetrazona* a stronger sexual motivation is a functional necessity to compensate for the stronger aggressive tendencies. Much more research will be necessarry to decide upon these points.

It follows from the listing of differences (p. 128 ff.) between *conchonius* and *nigrofasciatus* on the one hand and between *stoliczkanus* and *tetrazona* on the other and from the differences between the two groups listed on p. 152, that it is very difficult indeed to compare degrees of aggressiveness between the two groups. The two groups differ in the relative frequencies of the various aggressive elements and components (see also sections III, IV) as well as in the situations in which aggressive behaviour is displayed.

In view of the paucity of data on the behaviour of *stoliczkanus* in large groups it doesn't seem warranted at this moment to speculate on the way the behaviour fits into the natural circumstances of the species.

5. *B. cumingi.*

(a) temporal patterning.

One group of 6♂♂ 6♀♀ of this species was kept in a 130 cm tank (water depth *ca* 50 cm) and regularly observed for 36 days. One female was found dead on the 35th day and must have been dead for several days. The day-to-day activity pattern found in this experiment is given in table 6.

Observation times didn't extend from the morning until the evening on all days. During the first ten days no observations were made during the late afternoon and evening. From the eleventh day onwards evening observations were made as well on most of the days, especially if no intensive reproductive behaviour had occurred in the morning and afternoon. In column 6 of table 6 the time of the last observation for each day is given as well as the time of the first observation if this fell more than two hours after the beginning of the day.

Complete reproductive behaviour was seen on the first day after introduction (introduction time ± 12.00). Casual observations on small groups confirm that complete reproductive behaviour is usually shown on the very first day after introduction in a new tank, provided that the groups are ambisexual (see also p. 267). In the course of the experiment the temperature (which was normally kept constant at about 23°C) underwent a change on two occasions (table 6, column 7). A temporary rise of temperature to 25°C occurred on the 11th day (5/5). This change appeared to have a stimulative effect on reproductive activity for about three days (5/5, 6/5, 7/5 even though the temperature was normal again on 6/5. A fall in

TABLE 6

Daily activities of cumingi 6♂♂ 6♀♀ group

Introduction 24/4 *ca* 12.00; no reproductive activities on that day.

I	II	III	IV	V	VI	VII	VIII
date	fight	court	mate	1st mate	(first) last observation	temp.	nr ♂♂
25/4	+	+	+	⩽10.00(08.15)	14.00	23°	5 or 6
26/4	+	—	—		10.45		5 or 6
27/4	no observations						
28/4	(—)	—	—		13.55		—
29/4	+	+	+	09.55	14.36		⩾3
30/4	+	—	—		(12.15) 16.55		⩾3
1/5	+	+	+	⩽09.15(08.15)	13.50		2
2/5	+	+	—		15.40		2
3/5	—	—	—		(13.37) 13.37		—
4/5	no observations						
5/5	+	+	(+)?	⩽10.30(09.35)	19.00	25°	3
6/5	++	++	(+)+	⩽19.45(16.00)	19.45	23°	⩾4 ⩾4
7/5	++	++	(+)?+	<20.00(18.00)	19.56		⩾4 ⩾3
8/5	+	—	—		19.45		2
9/5	+	+	(+)?		19.28		3
10/5	+	+	+	⩽12.00(08.15)	(12.00) 12.00		3
11/5	+	+	(+)?		(12.15) 12.15		3 or 4
12/5	+	—	—		19.40		3 or 4
13/5	+	+	(+)	<20.10(16.00)	20.10		⩾4
14/5	+	++	+(+)?	<11.55(08.15)	(11.55) 20.00		3 3
15/5	+	+	+	<10.00(08.15)	20.00		2 → 4 → 6
16/5	(+)	(—)	—		22.15		—
17/5	+	+	+	⩽12.00(08.15)	(12.00) 19.40		⩾4
18/5	+	(+)	—		(11.00) 19.25		⩾4
19/5	—	—	—		16.00	21½°	—
20/5	(+)	—	—		19.12	23°	⩾4
21/5	+	+	(+)	⩽13.15(11.45)	15.45		⩾4
22/5	+	+	+	⩽10.00(09.40)	15.45		⩾4
23/5	—	—	—		19.00	24°	—
24/5	+	+	+	⩽12.00(08.15)	(12.00) 12.00		5 or 6
25/5	—	—	—		(11.50) 19.30		1
26/5	+	+	(+)?		(12.08) 16.30		3
27/5	+(+)	+(+)	+—	<08.55(08.15)	19.58		2 → ⩾4 → 3 6
28/5	+	++	—+	12.00	17.00		2 → 3
29/5	(—)	—	—		20.00		—
30/5	+(+)	+(+)	——		15.45		2 2

+ = one bout; ++ = two well separated bouts; (+) = less than maximum activity; further explanation see text.

temperature to 21½°C occurred on the 25th day (19/5). This led to an almost complete suppression of reproductive behaviour for two days (19/5, 20/5) even though the temperature was back to normal on 20/5. 19/5 was almost exclusively spent with schooling behaviour. Much schooling also occurred during most of 20/5 but in the morning of that day repeated and fierce duels were fought by two of the males, joined later by a third one. Normal reproductive behaviour was resumed on 21/5.

Leaving aside the consequences of the changes in temperature, one can see from table 6 that intensive agonistic activity (mainly between males) occurred on most of the days (28 out of 34 observation days; column 2). Courting was somewhat less common (22 days, or about 2/3 of the total number of days; column 3). Intensive courting including matings was observed on 15 days. On another 4 days (5/5, 9/5, 11/5, 26/5) no matings were observed, but probably did occur (though infrequently) judging from the intensity of courtship. Casual observations through the two weeks following 30/5 showed that during that period the same pattern persisted: fighting on almost all days and courting on roughly 2 out of every 3 days.

On the days marked + in the 4th column of table 6 high mating frequencies were seen in at least one of the males. Mating frequencies were not assessed quantitatively but on the + days they were obviously much higher than were ever found in *tetrazona* in comparable situations. The highest frequencies were of the same order as those found in *conchonius* and *nigrofasciatus*. Series of several matings of a male with the same female in quick succession were common.

Mating frequencies were low on a number of days (marked (+) in table 6). This was often due to severe competition and aggressiveness of the males. However, mating frequencies could also be high in spite of severe competition (see below, p. 158).

On some of the days all or nearly all males participated in the reproductive activity but often one or a few males were much more active than the others. The number of males taking part in the highest activity noted for a day is given in column 8 of table 6.

In general, the day-to-day activity pattern in *cumingi* seems to be much more regular than it is in *tetrazona* and *stoliczkanus*. On the other hand, it doesn't seem to be as regular as in *conchonius* and *nigrofasciatus*. At least there are less days on which both intensive fighting and courtship are shown.

Also the timing of activity on each separate day seems to be less consistent in *cumingi* than it is in *conchonius* or even *nigrofasciatus*. Column 5 of table 6 gives the time of the first mating observed on each day. Because observations were not continuous throughout the day, the first mating may in most cases (marked ⩽) have occurred earlier than the time given in the table. Therefore the time of the last preceding observation is added in parentheses (or the time of the beginning of the day if there was no preceding observation). In spite of this incertainty it is clear that there is an enormous variation in timing between days. Similar differences in timing of aggressive and courting periods were present. Further it is not uncommon that two reproductive periods occur on the same day. There are at least four days with two complete reproductive periods both with fighting and courting (not all of them including matings) separated by a period of relative inactivity (6/5, 7/5, 27/5, 30/5). In two other cases (14/5, 28/5) two courting periods are present with at least some fighting activity going on between them. The last pattern does occasionally occur in *nigrofasciatus* too, but two completely separated reproductive periods were never found in that species. (Since observations were not continuous throughout the day on all days, it is very well possible that there were even more days with double periods).

It was mentioned above that the number of active males may vary between days. It may also vary within one day. On days without courtship (or in aggressive periods preceding courtship: see below) several or all males may start to be aggressive more or less simultaneously and get more and more aggressive while remaining on a more or less equal footing. Another possibility is that two of the males become aggressive earlier and remain much more aggressive than the others throughout the fighting period. Such males are territorial (see below, p. 158). They divide the tank between them and apparently suppress most reproductive behaviour in the other males. The other males may become slightly more aggressive among each other as time proceeds and they may even show some incipient courting, but they never get on an equal footing with the first two males.

The first situation (several aggressive males) may develop into one in which several or all males court and mate successfully. Starting from the second situation (two territorial males) the same two males may start courting and mating successfully, all the while keeping reproductive activity low in the others. (This situation is reminiscent of the one typically found in *tetrazona* in the 130 cm tank (p. 139). Alternatively, other males may become progressively more active in both courting and fighting. This may lead to a situation similar to the one described above, in which all males had been equally active from the beginning. For instance on 15/5 at 10.00 two males were already fighting actively. One of them courted intensively and mated. The other often penetrated into the territory of the first and competed with him for a female. A third male was moderately aggressive too but much less so than the first two. At 10.37 four males were fighting actively. At least three of them courted intensively. Probably all four courted by 11.00. At 11.10 at least five and probably all six males fought, courted and competed intensively. This went on until at least 16.40. At 20.00 all males were exceedingly aggressive. Some courtship still occurred, but successful mating had become practically impossible because of the severe aggressiveness and competition of the males.

In spite of the great variability in timing of reproductive activities and in the number of males participating, some temporal regularities, similar to those found in the other species, are present in *cumingi* too. Reproductive behaviour occurs in rather clear-cut periods which typically last several hours. Oscillations in activity may often occur within such periods, for instance between fighting and schooling or between courting and fighting. The "wave length" of these osciliations is of the order of minutes. Further, reproductive activity is usually low for at least the first hour of the day. The intensity of agonistic behaviour typically increases more or less gradually during at least the first few hours of the day. This occurs both on days with and those without courting. The rate of increase, however, may be very different on different days. High intensity fighting may be reached at about one hour after dawn or late in the afternoon or in the evening (if it is at all reached on a certain day). Courtship doesn't usually begin before a period of intensive fighting has occurred. Once courting has started, there is no distinct consistent increase in courting intensity in the course of a courting period such as was found in *conchonius* and *nigrofasciatus*. Courting may be considerably suppressed when fighting and competition between males are severe (see also p. 156). Courting may be intensive at the beginning of a courtship period and be suppressed later by growing aggressiveness of the males, or aggressiveness and competition may be severe right from the beginning of a courting period and be relaxed later. Changes in the number of active males may also profoundly influence the courting success. Fighting between males often goes on (up to several hours) after courting has come to an end. It usually also wanes after some time. All these regularities may be considerably obscured as a result of the lack of synchronization between the males.

(b) spatial patterning.

Like *tetrazona, cumingi* males are territorial throughout the reproductive periods. Agonistic behaviour of comparatively low intensity and even some courting may be performed in the school, but as reproductive activity increases in intensity the males disperse and establish individual territories. Courting and mating are also performed mainly inside a male's own territory. However, males penetrate into each other's territories to an enormous degree. Penetration may be one-sided, one male being able to intrude into his neighbour's territory rather easily and even to defeat him there repeatedly though not decisively, or it may be more mutual. Also during courtship males very frequently and persistently come and compete with other males in their territories. So strong is this tendency that often the terrioriality of the males is easily overlooked at first sight. The very general type of movement of the fishes through the tank during courting periods superficially looks much more like that of *nigrofasciatus* than like that of *tetrazona.* This is not only due to territorial penetration but also to the type of locomotion which, though not as smooth and continuous as that of *nigrofasciatus,* is much less jerky than in *tetrazona,* and to the presence of circling beside straight leading in courtship. In addition *cumingi* males, like *nigrofasciatus* and in contrast to *tetrazona,* often manage to mate frequently even in the presence of severe competition. This is certainly not so because competition would be less fierce in *cumingi* than in *tetrazona* (it isn't) but because they are in some way more immune to competition. It was stated above that courting may be suppressed if fighting and competition between the males is strong. This is probably due to a growing aggressiveness of the males towards females (see below p. 160) and to a shift in interest for attacking males relative to courting females, rather than to the actual disturbance of competition itself.

Like in *tetrazona,* territories were not yet established on the first day after introduction. Yet, on that day there was intensive fighting and courting of all males and mating was frequent. Territories started to be established on the second day after introduction (26/4). On 29/4 all males that were reproductively active had territories. Later, if there was very intensive fighting and courting of many males, territorial attachment sometimes became less distinct again. It could not be established whether the same males held territories in the same locations on consecutive days. The observations had to be made from a considerable distance due to the shyness of the fishes. This very much hampered individual recognition.

The composition of behaviour in *cumingi* is strongly correlated with depth. Normal swimming around and feeding are mainly performed near to the bottom, at least in the lower half of the tank. Fighting between males may occur at all levels, but it is mainly performed in the lower half. Only during duels they typically rise towards the surface; they may go on fighting in contact with the surface for considerable time (see also section I). (In small groups in small tanks dominant males typically swim around in the lower half of the tank, whereas completely subordinated males stand still near to the surface). Courting, on the other hand, is almost exclusively performed in the top half of the tank. Only low intensity courting, such as may occur in the school (see p. 158) may be performed at lower levels. Typically, leading starts approximately half-way up between the bottom and the surface and goes obliquely upwards to the surface in one stroke. Mating almost exclusively takes place in the upper 10 or 15 cm of the tank. Competition is fierce at all levels where courting occurs. Territorial males, when undisturbed, typically stand or swim around about half-way up between the bottom and the surface. When a courting period is impending males may already start to visit the upper layers now and then.

Cumingi males exhibit about the same differences in behaviour in relation to the vicinity of plants as do the other species. Normal swimming around and feeding take place mainly in the open and so does fighting between males. Also duels occurring near the surface take place in the open spaces. Territorial males stand out in the open, though in the vicinity of plants. (In smaller groups in small tanks dominant males typically swim around in the open, whereas inferior males hide between plants or in the corners of the tank. During the early phases of the establishment of dominance inferior males may also stand still very near to or pressed against the bottom; see also p. 265). The leading phase of courtship is mainly performed in the open spaces, whereas mating typically takes place at least in contact with plants.

(c) inter-individual patterning of behaviour.

Inter-male attacking is very common and intensive throughout the reproductive periods. Prolonged duels between males occur occasionally. One duel, lasting at least 5 minutes almost continuously, was seen in the evening of 13/5. Four or five males were very active then, mostly fighting and competing but also courting intensively now and then as far as competition allowed. Duels in the morning of 20/5 were already mentioned above. One of these duels lasted almost continuously for at least 25 minutes. A second duel lasted 7 minutes at least, after which time the fighters were joined by a third male. The incidence of fierce duels on 20/5 suggests that in *cumingi*, like in *tetrazona* (see p. 144) duels occur especially during phases in which territories are (re-)established.

Even during periods of intensive courting and successful mating males keep attacking the courted female frequently and strongly. Even obviously willing females are often attacked and chased out of the territory. Attacks on females are hardly less fierce than those on other males. A phase of frequent and persistent chasing of females may precede impending courting periods.

Males have not been observed to court other males in this series. However, I can't be absolutely sure that it didn't occur, because observations had to be made from a considerable distance and because there is little difference in colour between especially non-reproductive males and females. In fact, males were observed to court other males now and then in another group of about 6♂♂ and 6♀♀ in a 125 cm stock tank.

Females may exhibit considerable aggressiveness. Attacks by females are rather common in the school. One female held a small territory for some days in succession She attacked both males and other females entering the territory. Her attacks, like those of other females, were on the average shorter and slightly less vigorous than those of reproductively active males. The frequency and the intensity of her attacks increased in the course of the day, but she always stopped attacking and gave up the territory at some time during the day. In the course of the several days that she held the territory she became progressively more aggressive, her attacks becoming longer and more frequent. The centre of her territory was near to the bottom and more or less between plants. In small groups in small tanks females may behave very aggressively; they rather easily become dominant, thus precluding any intensive male courtship (see further p. 186).

The female's role in courtship is a rather passive one. For mating to occur she has to follow a leading male, to enter a clump of plants and to allow the male to mate. Active mating attempts on the part of the female, such as were observed in *tetrazona* (see p. 186) were never seen in this species. Females may exhibit *vacuum* leaning and quivering. When a courting period is impending females may express incipient willingness by visiting higher layers. When doing so, and also later during courtship near to the surface, females may display the same peculiar shyness as was described for *conchonius, nigrofasciatus* and *stoliczkanus*.

(d) colour changes.

Outside reproductive periods male and female *cumingi* differ very little in colour. Males are slightly more yellowish (or reddish) on the body and the fins. Some males of the yellow form (see chapter I, section I), but not all of them not even in wild imports, have somewhat more pronouncedly yellow dorsal fins with a number of small black spots and a very thin black rim. During reproductive behaviour the black body markings of the males may fade markedly, the tail mark assuming a deep green colour. In the 6♂♂ 6♀♀ experiment complete fading of the body markings was only seen on the first day after introduction. Later the body markings always remained darker than the background, even during very intensive reproductive activity. The males' yellow or reddish colours on the body and the fins are slightly more pronounced during reproduction and the body is often slightly darker than normal.

The females hardly change colour during reproduction. Even the black body markings remain almost as dark as normal. The resulting difference in colours between males and females during reproduction is distinct but yet rather subtle. It is possible to tell the sexes apart from a few meters distance but the differences are not by far as dramatic as in *conchonius* or *nigrofasciatus* and not even as large as in *tetrazona*. Even in *stoliczkanus* the males are more distinct from the females by the colours of the dorsal fins.

During prolonged duels the bodies of the males, especially the dorsal halves, may become considerably darker than normal (up to 1/4-1/2; see p. XI). The body markings may fade partly or completely, but they are never lighter than the background.

(e) schooling.

Schooling is of the same casual type as it is in *conchonius* and *nigrofasciatus* No acute conflicts, like those obtaining in *tetrazona* and *stoliczkanus,* between schooling and reproductive behaviour have been observed in this species. Reproductive behaviour is not characterized by explosiveness, like it is in *tetrazona*. Reproductive periods develop quite gradually.

(f) individual variability.

Little information is available on individual differences in behaviour. The two males which in the 6♂♂ 6♀♀ experiment of ten dominated the scene probably were the same individuals throughout the series. However, this may be due to the fact that these males were slightly larger than some of the others. No specially homosexual males have been found in this or in any other groups of the species. Except for the one territorial female no great individual differences in aggressiveness or sexual willingness have been found in females in the 6♂♂ 6♀♀ experiment. It is possible that the distance at which the observations had to be made was partly responsible for the apparent lack of individual differentiation. Nevertheless I feel that there is less individual variation in behaviour in this species than there is in *nigrofasciatus* or *tetrazona*. One reserve must be made here: more than in the other species the selection for the experiments of distinctly male and distinctly female individuals may have caused exclusion of certain types of individuals.

Discussion.

The behaviour of *cumingi* doesn't seem to fit into either the *conchonius-nigrofasciatus-* or the *stoliczkanus-tetrazona* type. It combines some characteristics of either type and displays some characteristics of its own.

The behaviour of *cumingi* is like that of *stoliczkanus* and *tetrazona* (and unlike that of *conchonius* and *nigrofasciatus*) in the strong territoriality of the males, in the fact that the males keep attacking courted females, even during very intensive courtship, and in the relatively 'emancipated' role of the females. It shares with the behaviour of *conchonius* and *nigrofasciatus* (and differs from that of *tetrazona* and *stoliczkanus*) the absence of very strong schooling tendencies and the resulting acute conflicts between schooling and reproduction and the absence of explosiveness in reproductive behaviour, the invulnerability of courting and mating behaviour against competition, the absence of a clear distinction between horizontal and head-down threat (this refers especially to *conchonius* as head-down threat doesn't occur at all in *nigrofasciatus*), the practical absence of mouth-fighting in duels and the general type of movement during reproductive behaviour [1]).

The behaviour of *cumingi* is intermediate between the two types in the degree of regularity of the day-to-day activity rhythm and in the degree of homogeneity of daily activity patterns. It differs from both in the plasticity of threat posture orientation (head-up as well as head-down and horizontal), in the relatively low degree of formalization of duels and in the general unrestrained nature of all reproductive behaviour. Finally, it shares with both types the occurrence of reproductive behaviour in more or less clear-cut periods and the dependencies of behaviour composition on depth and plant vicinity.

It is known that *cumingi* lives in rapidly flowing waters (p. 7). It seems plausible that the vigorous and unrestrained character of the reproductive behaviour of the species is an adaptation to this special condition. Swift and straightforward movement seems to be the only way to make attacking and courting effective in a strong current. The territoriality of the males during courtship and the only partial synchronization of repro-

---

1) Some of these statements are more or less tentative in that in *stoliczkanus* the territoriality of the males during courtship and the interaction between courting and competition have to be further investigated.

ductive activity might suggest that in this species there are no cyclic aggregations on common spawning grounds like were surmised for *conchonius* and *nigrofasciatus,* but that the males are more widely spaced in more or less permanent territories either near a bank or on the lee-side of stones mid-stream. Yet, the fact that males are much less vulnerable to competition than for instance *tetrazona* males might suggest that their typical pattern is more like a 'territorial society' (BAERENDS & BAERENDS-VAN ROON, 1950) rather than an array of well-separated solitary males, such as was supposed to obtain in *tetrazona*

It seems possible that reproductive mass aggregations without territorial attachment, more or less in the way as was found on the first day of the 6♂♂ 6♀♀ experiment, may occur for instance with the onset of the rainy season, when previously dry areas become available

Again, it must be emphasized that these are only tentative speculations. They should be tested through observations in the natural habitat of the species.

BEHAVIOUR OF *B. PHUTUNIO* AND *B. GELIUS*

In the following paragraphs brief accounts are given of the reproductive behaviour of *B. phutunio* and *B. gelius.* No extensive observations, comparable to those of the foregoing five species, were made of these two. Nevertheless, the casual observations that were made provide an interesting picture of their reproductive behaviour.

6. *B. phutunio.* The swift.

Phutunio behaviour was observed in a number of different groups under a variety of conditions (different numbers of individuals in different sizes of tanks). In all cases the reproductive males were strictly territorial. Centres of territories could be near to the bottom, but also higher, up to about 25 cm from the bottom (water depth 45 cm). Males stand out in the open, but in the vicinity of a plant. Neighbouring males are often not equally strong, especially if they are of unequal size. In this case the larger of the two may intrude rather easily and repeatedly into the territory of the other and even temporarily defeat him there. However, the weaker usually recovers as the intruder returns to his own territory. In general the males are much more efficient in defending their territories than in any of the species discussed above.

Attacks of males are very intensive and unrestrained. Attacking males seem to be little impressed by threat on the part of the attacked. No formalized duels were ever observed in this species. Males may display mutual lateral threat and neighbouring territorial males may dash back and forth a few times taking turns in attacking and fleeing, or they may exchange bites at close quarters near to the boundary of the two territories, but no ritualized forms of parallel threat swim, caroussel or mouth-fight were observed.

Willing females now and then enter the territory of a male. In doing so they swim slowly with folded fins and with the head tipped slightly upwards. She is usually met immediately by a flashing attack from the male. In response she may flee if the male's attack is too strong or if she is not quite willing enough. Often, however, the male stops just short of her, turns sharply and leads very swiftly toward some plant near to the centre of his territory. The female may then follow very quickly, after which the male turns around and mates from one to three times in very quick succession. Immediately after the last mating the female is chased away out of the territory. The entire procedure doesn't take more than a few seconds. So swift are the movements that for instance it is impossible to see whether the final chasing away of the female is induced by a change in the female's behaviour or by the male getting more aggressive on his own account. The whole scene may be repeated again and again with the same or with different females. Like in the other species, the female releases a few eggs at each mating act.

Males seem to stay in their territories all day round for at least several days in succession. Not much of a daily rhythm in activity was found. Males apparently do not eat their 'own' eggs, though they may feed (mainly on floating particles) when in their territories.

In one of the observations a male had his territory around a clump of filamentous algae growing from the wall of the tank at about 10 cm from the bottom (water depth *ca* 30 cm). Though some other plants were present in the same vicinity, the male always led the female to just above the algae and all matings took place there. The eggs stuck between the threads of algae and in this way got collected in the clump. When he was not courting or actively defending his territory, the male often manoeuvred around the 'nest' and now and then touched it with the snout. In doing so the type of locomotion was often more or less hovering (see below). This type of behaviour was strongly reminiscent of the behaviour of parental fishes, for instance a stickleback, about to start fanning movements. However, no movement was seen which could with any possibility be identified as fanning. Yet, the manoeuvring probably did contribute to some extent to the ventilation of the eggs. Another male was seen to 'collect' the eggs at one place on the bare bottom and manoeuvre around this spot in much the same way as did the male described above (PUTTERS, pers. comm.).

Several attempts were made to induce males to build nests. To this purpose they were provided with more or less transparent plastic watch-glasses (ø *ca* 10 cm; depth *ca* 1 cm) which could be fixed, with the concave side up, at different depths on a thin vertical plastic pole. Further, small clumps of algae were provided scattered on the bottom of the tank. None of the attempts had any shade of success.

Further experiments with more individuals and under various conditions will be necessary to assess how typical the behaviour of the 'nest-owning' males was. However, even without this feature *phutunio,* with its strict territoriality and the apparent lack of daily rhythm in its activity, contrasts strikingly with the species described so far. It is interesting that, even without the presence of a distinct 'nest', the pattern of male courtship and the way in which the female enters the male's territory are at least superficially reminiscent of stickleback behaviour.

The entire reproductive behaviour of *phutunio* is characterized by an extreme swiftness, staightforwardness and abruptness of movement. To some extent this is also true for its normal locomotion. Sudden forward dashes alternate with brief periods of hovering. This hovering movement is another characteristic feature of *phutunio.* In the other species described above, the equilibrium during standing still is maintained by fin movements (mainly of the caudal and pectoral fins) of a comparatively low frequency (roughly 2 or 3 in a second). As a result of this, weak hopping movements are always obvious during standing still. In *phutunio* the frequency of the fin movements during standing still is much higher (and possibly involves more fins) which results in a much more stable position. This kind of hovering was also frequently displayed by the male manoeuvring around his 'nest' (see above).

The sexes in *phutunio* are very much alike each other in colour patterns. During reproduction the black body markings of the males may fade to varying extents. The whole body becomes slightly brownish (especially along the rims of the scales) and the dorsal and ventral fins become grey. The females change very little in colour during reproduction. As far as observed, willingness is only expressed by entering a male's territory in the manner described above.

The resulting differences in colour between males and females during reproduction are usually distinct but still very subtle.

7. *B. gelius*. The hovering.

Several groups of *B. gelius* were observed in the laboratory, but complete reproductive behaviour was seen in one group only. This group consisted of *ca* 20 individuals (slightly more males than females) and was kept in a 125 cm stock tank (water depth *ca* 40 cm). Reproductive behaviour of this group was observed at several occasions.

Fighting between males is frequent and intensive. Both attacks and biting are typically very fierce and unrestrained. Though lateral threat is a common occurrence during fighting, in many cases it doesn't seem to make the slightest impression upon the opponent, who just goes on attacking and biting without any sign of inhibition. No formalized form of duels was seen. Fierce fighting is most often seen near to the bottom. Males which are fighting and not courting may hold (usually small) territories close to the bottom or above a clump of filamentous algae growing on the bottom. No sign of territoriality was seen in courting males. Nor does there seem to be any tendency to form aggregations such as found in *conchonius* and *nigrofasciatus*.

Courtship consists of the usual elements; courting posture, leading (*s.l.*) and mating. Leading may be in a more or less straight line over short distances towards plants, or it may follow very erratic courses with strong vertical components, the male staying near to the female. Both types of leading are comparatively slow and are accompanied by strong fin movements which give it a hovering or vibrating appearance. Matings take place in the usual manner: in contact with plants or between filamentous algae. Courtship was performed at a somewhat higher level in the water than was fighting between males. However, it was mostly displayed at less than 25 cm from the bottom and never near to the surface. It must be emphasized that this may, at least in part, have been an effect of the distribution of vegetation in this particular situation. The vegetation was mostly relatively low (20-25 cm) and nowhere reached to the surface. There were almost no floating plants.

Competition between the males is extremely fierce. Often as many as four or five males, in a very dense cluster, pursue one female. Keeping-off is often seen when a courting male is approached by or approaches other individuals. Keeping-off has a very characteristic form in this species. If the rival stays on more or less the same place, the male, hovering in threat posture, typically sidles slowly towards the rival fish. Only after the rival has left, the male resumes courtship. If the rival moves and tries to get around the defending male, the latter displays its perfect ability to manoeuvre in any direction, to the side and vertically as well as forward and backward. This results in rather erratic paths followed by the defending male, somewhat resembling those displayed during leading. As long as the male is engaged in keeping-off, the female often stays put, hovering in one place, until the male has finished keeping-off and resumes courtship. This seems to be an expression of willingness on the part of the female. It is not seen in females which are pursued by a cluster of males. In this situation the female usually tries to escape from the mêlée. Willing females, when swimming towards plants in response to a male's courtship, often tend to seek a position under the concave underside of a small horizontal leaf. This was not seen in any of the other six species described above, but it was also observed in *B. oligolepis* and *B. pentazona.* As a result of the fierce competition and of the apparent vulnerability of courting males for this competition, very few successful matings were observed. This makes it difficult to say whether the preference of willing females mentioned above coincides with preferred mating places.

Even when courting intensively, males often attack the courted female. It also often occurred that a male which was courting a female, when approached by a second male or female, attacked the courted female 'instead of' keeping-off the intruder. In general the attacks on females are hardly less fierce than those on males. No males were seen courting other males and little aggressive behaviour was seen on the part of females. However, more observations are necessary to be sure that this is typical.

No indication was found of the existence of any daily rhythms in activity, but more observations are needed to assess this point more clearly too.

The hovering movements, which play such an important role in leading and keeping-off as well as in the behaviour of willing females, are also characteristic of the normal locomotion of the species, even more so than in *B. phutunio.* Relatively short forward dashes alternate with periods of hovering. Prolonged hovering (for several seconds at least) on one place is quite common. A hovering fish was on some occasions seen to swim suddenly backwards smoothly and quickly for a short distance.

The sexes in *gelius* are very much like each other in colour patterns. During reproduction the black body markings of the males (including the internal pigmentations, see chapter I, section VIII) often fade almost completely. In some courting males all black markings had faded except the central area of the tail bar which thus appeared as a dark spot, closely resembling the tail-spot of *conchonius, stoliczkanus* and *phutunio.* During reproduction the bodies of the males become slightly yellowish and a faint copper-coloured longitudinal stripe appears on the side of the body. This stripe is sometimes difficult to discern if the yellowish body colour is most intense. It is practically absent outside reproductive periods. The fins of the males do not obviously change in colour. The black body markings of the females may fade a little during reproduction, but much less so than in the males.

The resulting colour differences between the sexes during reproduction are usually distinct, but still even more subtle than in *phutunio.*

Discussion.

It doesn't seem warranted at this stage to speculate on the natural circumstances of either *B. phutunio* or *B. gelius.*

It was suggested in chapter I that *cumingi, phutunio* and *gelius,* in that order, represented stages of neoteny derived from a *nigrofasciatus*-like ancestors. It is difficult to assess, at this stage, whether arguments for this theory can be drawn from the behaviour of these species, mainly because the ontogeny of behaviour in *nigrofasciatus* and other species is imperfectly studied so far. More detailed studies in behaviour ontogeny in *nigrofasciatus* have recently been started in our laboratory by Mr. F. PUTTERS. From these studies it might be, tentatively, surmised that the unrestrained character of reproductive behaviour in all three species, the poor development of formalized duels in *cumingi* and the probable absence of them in *phutunio* and *gelius,* the probable absence of distinct daily activity rhythms in *phutunio* and *gelius* and the hovering type of locomotion in the same two species are related to certain stages of juvenile behaviour in *nigrofasciatus* (PUTTERS, pers. comm.). However, in order to assess these points, much more work has yet to be done both on the ontogeny of behaviour in *nigrofasciatus* (and other species) and on the adult behaviour of especially *phutunio* and *gelius.*

SECTION III. BEHAVIOUR OF THE 5 SPECIES IN 2♂♂ 2♀♀ GROUPS

Procedure.

In this section some of the semi-quantitative observations of section II are further tested with quantitative behaviour records of individual fishes in 2♂♂ 2♀♀ unispecific groups. The groups were kept in 60 × 35 × 35 cm tanks with asbestos-cement sides and bottom and a glass front. The bottom was covered with a layer of sand. Some scattered vegetation was supplied both growing from the bottom and floating at the surface. Observations were made of settled groups which had been together in the same tank from at least the day prior to observation and often longer. As a rule all 4 fishes came from the same stock tank and were selected to be of approximately the same size. Often several observation series were taken from the same group but for each species several different groups were used (*conchonius* 5, *nigrofasciatus* 6, *tetrazona* 5, *stoliczkanus* 4, *cumingi* 6).

In the groups usually one individual was clearly dominant, most often a male. (If assessing dominance was difficult the male rolling towards the other male was taken as the inferior one; rolling was never displayed by both males towards each other). In some of the species a female might be the dominant individual (see section II and below for species differences on this point). Such groups were not used for the observations. In the groups used females were usually much less active than males; often one female was clearly dominant over the other one. The more dominant female might also dominate the inferior male but usually the inferior male was dominant over both females.

Observations were spoken into a tape recorder and analyzed later. The occurrence and duration of each reproductive behaviour element was recorded. As it proved impossible with this technique to produce complete records of more than one individual at a time, individuals were observed one by one in direct or near-direct succession, each for a period of 15′. The order in which the different individuals were observed was rotated from observation to observation in a more or less random way. If too great a change occurred in the general activity pattern within the joint observation periods of the separate individuals, the entire observation series was discarded. Not in all cases were all 4 individuals recorded. As a result the numbers of points in different graphs of the same species shown in this section may differ from each other.

No standard time schedule was maintained for the observations. Rather, observation times were chosen opportunistically in order to record as many different states of activity as possible for each of the species. This selection mainly involved the collection of as many different amounts of courting activity in the male(s) as possible. Therefore the data of this section should not be taken as statistically representative of the incidence of different degrees of courting activity in the different species. As no other principles of selection of observation times were applied, all other differences between species discussed in this section are representative of species differences, with the possible exception of aggressive behaviour scores, see sub (d). A considerable part of the observations was made under natural light conditions.

Behaviour characteristics analyzed in this section involve (a) the probability of courting males attacking the courted female; (b) the competitive relations between the two males; (c) the degree of 'emancipation' of the females (see section II); (d) differences between species in the degree and kind of aggressiveness of the males. The last point can and will be more accurately assessed in the next section.

Results.

(a) courting males attacking the courted female.

Figg. 48a-e show — for each of the 5 species — the amount of time spent by the dominant males in attacking and courting the inferior male and the two females. Each point in the graphs corresponds to the total number of seconds of attacking (ordinate) and courting (abscissa) displayed towards each of the other 3 fishes in one period of 15′. So every 15′ period of the dominant male yields 3 points in a graph. Activities of the dominant male directed at the other male are represented as crosses, activities towards the females as circles. In the *nigrofasciatus* graphs the activities of a homosexual male (the same individual male of which quantitative data are given in section II p. 125 ff.) towards the other male are represented as Maltese crosses; the same male's activities towards females as black dots. The activities of the one *conchonius* male who in one observation period showed a considerable amount of courting the other male are represented as a normal cross (activities towards the other male) and black dots (the corresponding activities towards the females). No additional data are available to decide whether this male was in any other way comparable to homosexual *nigrofasciatus* males.

Figg. 49a-e are set up in the same way as 48a-e except that the time spent 'fighting' each of the other individuals is added to the time spent in pure attacking. 'Fighting' is a category of behaviour embracing threat, all phases of duelling, keeping-off (either male or female), mutual attempts at keeping-off and trying to get around a keeping-off male. All of these activities are mostly accompanied by threat; only in keeping-off the other female threat is often absent. If threat occurred simultaneously with attack this was — for the purpose of the graphs — scored as attack only.

Leaving aside the data from the homosexual males it is evident from the graphs that male *conchonius* and *nigrofasciatus* discriminate almost prefectly between males and females in that courting is virtually exclusively displayed at females whereas the great majority of attacking is directed at the other male. The latter point is even clearer when 'fighting' is added to pure attack. In both species a considerable amount of 'fighting' is shown towards the other male (see also sub (b)); in *conchonius* the amount of 'fighting' towards

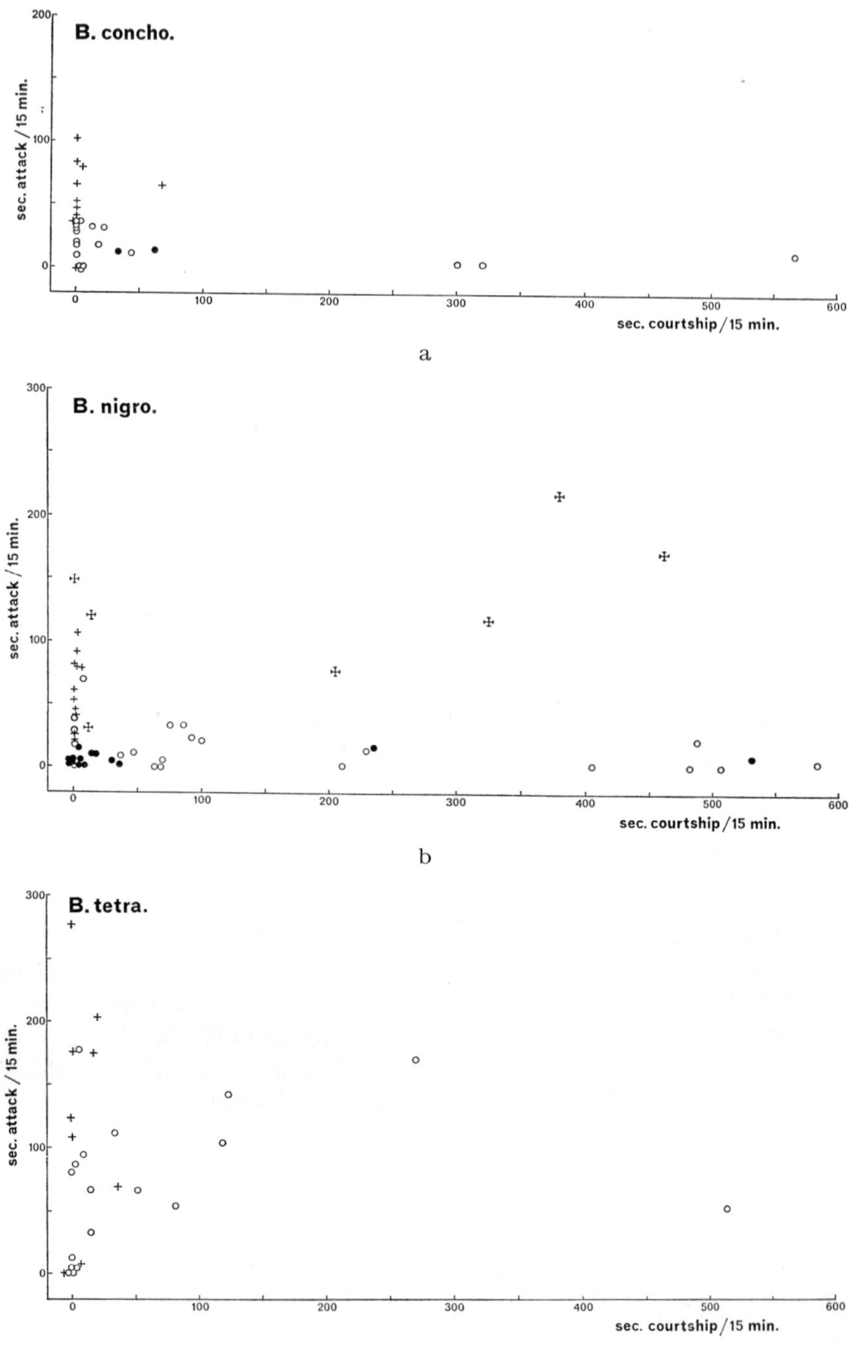

Fig. 48a-e. 5 species; courting and attacking scores of dominant males of 2 ♂♂ 2 ♀♀ groups; i = imports; further explanation in text.

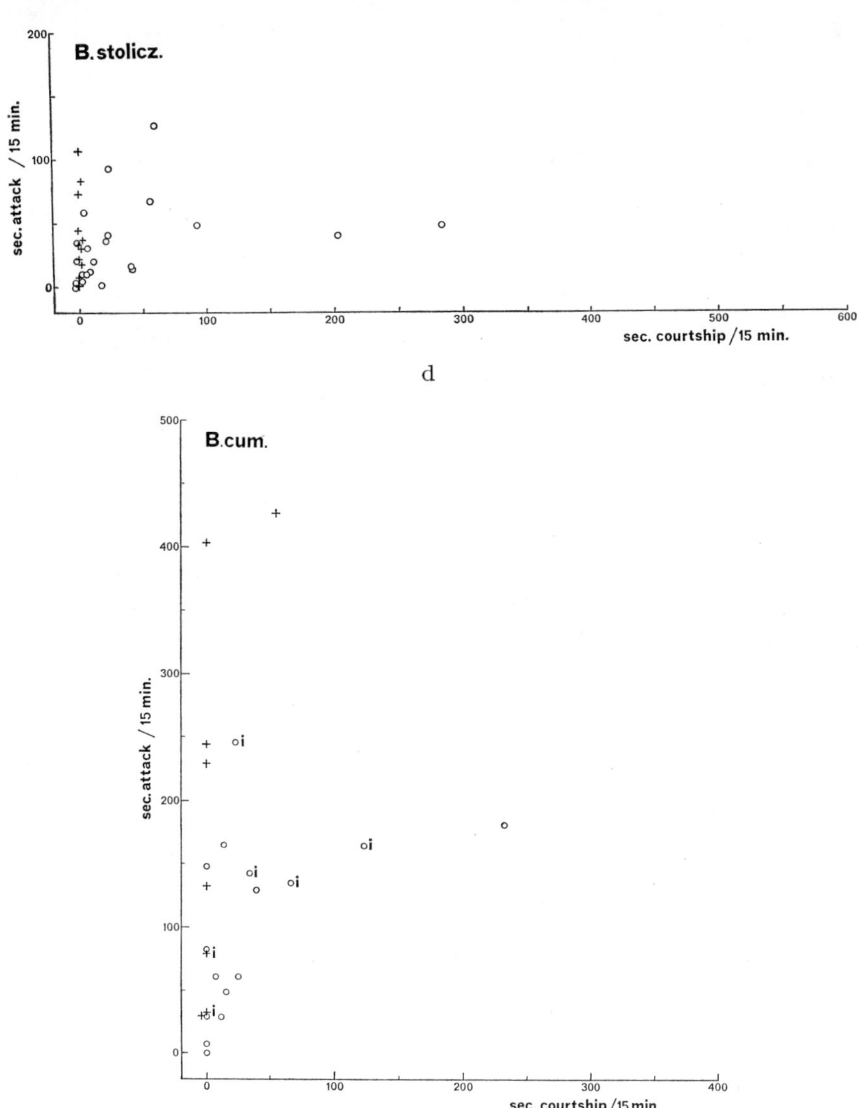

the females is negligible; in *nigrofasciatus* some amount of 'fighting' may be shown also to females. The last category, however, consists of keeping-off almost exclusively and so mainly represents aggression 'provoked' by the mere proximity of the non-courted female. (Of course keeping-off forms

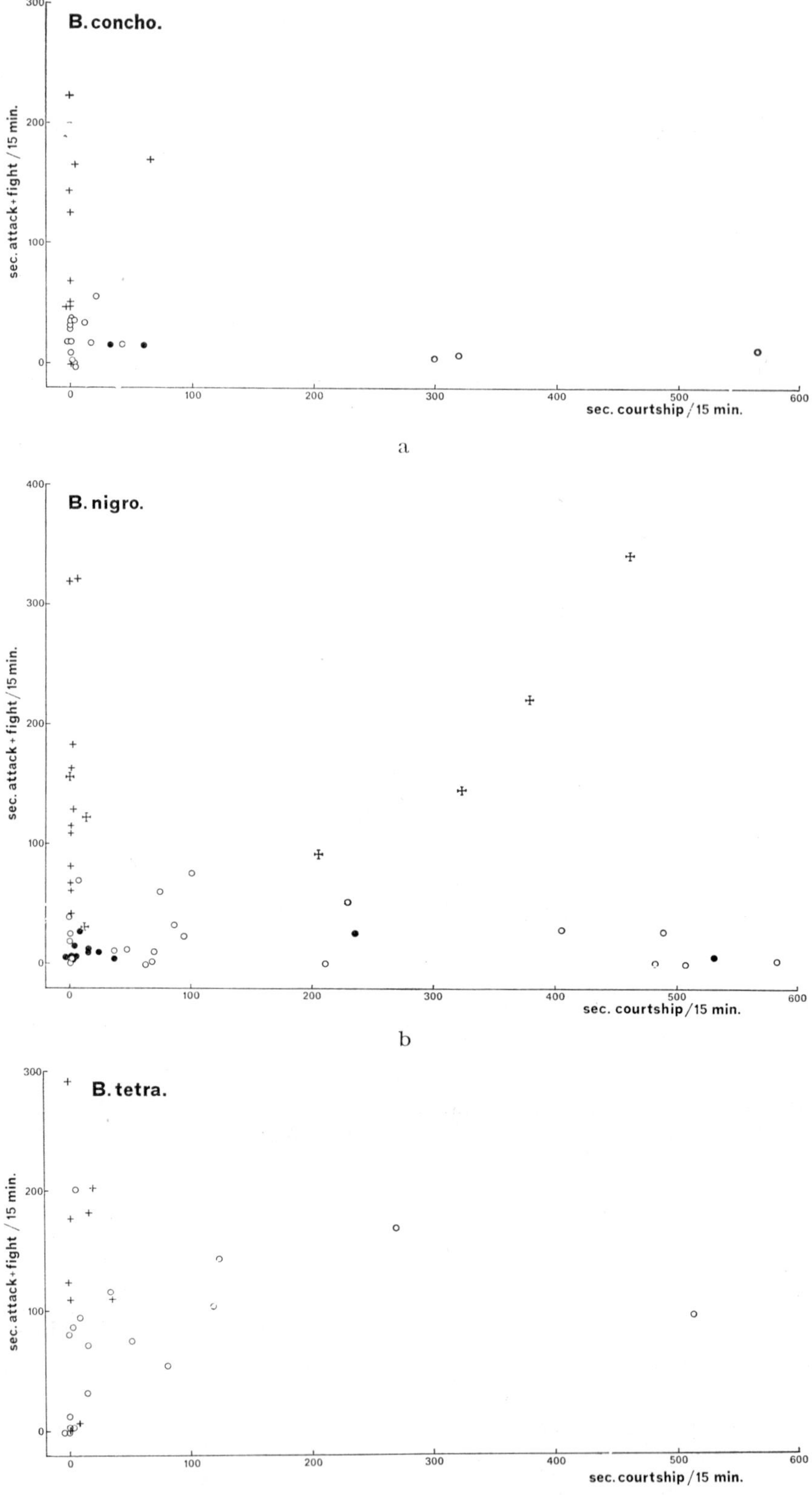

Fig. 49a-e. 5 species; courting and attacking + 'fighting' scores of dominant males of 2♂♂ 2♀♀ groups; explanation see text.

B. stolicz.
sec. attack + fight / 15 min.
sec. courtship / 15 min.
0
100
200
300
400
500

d

B. cum.
sec. attack + fight / 15 min.
sec. courtship / 15 min.
0
100
200
300
400
500

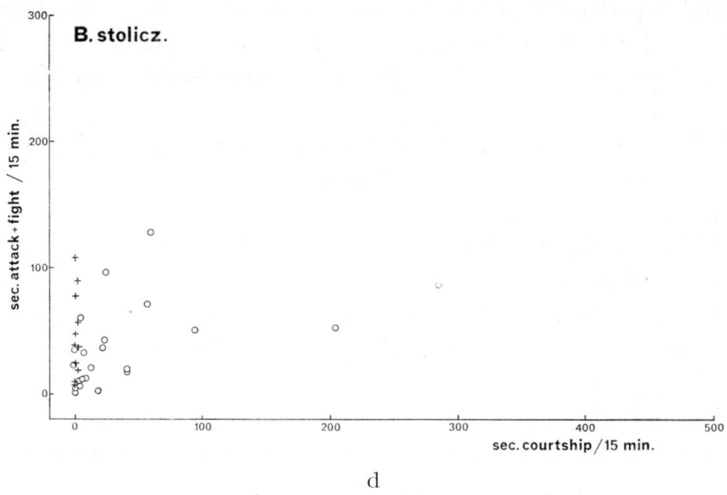

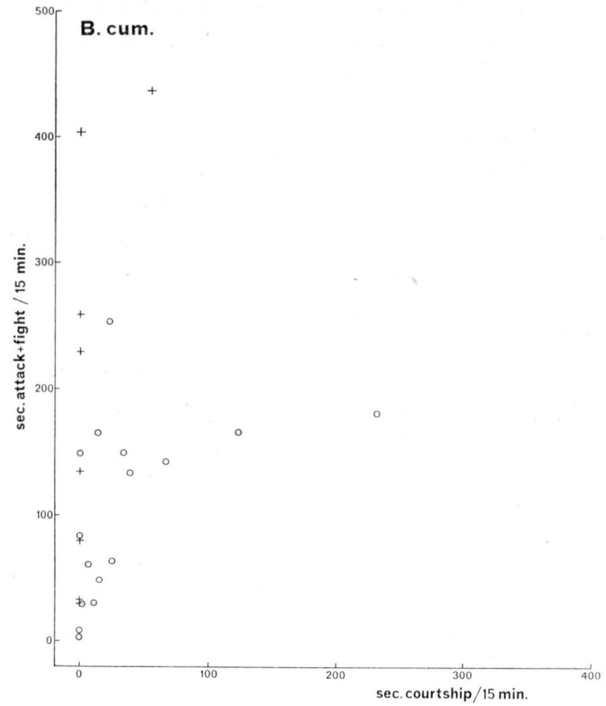

e

a considerable part of 'fighting' the other male too, but here also many other types of behaviour are involved). Further it must be emphasized that keeping-off, when displayed at a female, is by definition directed at the non-courted female.

The difference between the two species in the amount of keeping-off shown towards females is probably not due to a stronger attraction of the 'other' females to the courting pair in *nigrofasciatus* but rather to a greater 'irritability' of *nigrofasciatus* males as to the proximity of other females during courtship (see also sub (d)).

The graphs also show that in both species the amounts of attacking and of attacking plus 'fighing' directed at females decrease as the courting activity of the male increases.

The data for the homosexual male *nigrofasciatus* indicate that, if the other male is courted at all, the amounts of aggressive and courting behaviour are positively correlated. At comparable amounts of courting, the courted male is attacked much more than are courted females. On the other hand it is clear that the homosexual male does not attack females more than do normal males. If there is a difference it is to the other side.

The corresponding graphs for *tetrazona, stoliczkanus* and *cumingi* yield a very different picture. Like in the first two species, the other male is courted very rarely only. (In contrast to *nigrofasciatus* and perhaps *conchonius* there are no specifically homosexual males on record in these three species). However, females are attacked much more than in the first two species, almost as much as males. In all three species there are observation periods in which at least one of the females is attacked more than the other male (*tetrazona* both ♀♀ 1×; *stoliczkanus* 1♀ 5×, both ♀♀ 2×, *cumingi* 1♀ 1×, both ♀♀ 2×). This never happened in *conchonius* and only once (1♀) in *nigrofasciatus*. The fact that in the latter three species females are attacked more, relative to the other male — as compared with the first two species — is no doubt partly due to the tendency of the inferior males in these species to spend at least part of their time hidden in a corner of the tank or between vegetation (see also sub (b), p. 183). (This tendency is greater in the inferior males than in the females. Hiding did not occur in the inferior males of the first two species). What *is* clear is that females of the latter 3 species are — at comparable amounts of courting — attacked much more than are those of the first two species.

The same difference between the 2 groups of species holds if 'fighting' is added to the values of pure attack (figg. 49^{a-e}). It may be that, according to these parameters, *stoliczkanus* females — at comparable levels of court-

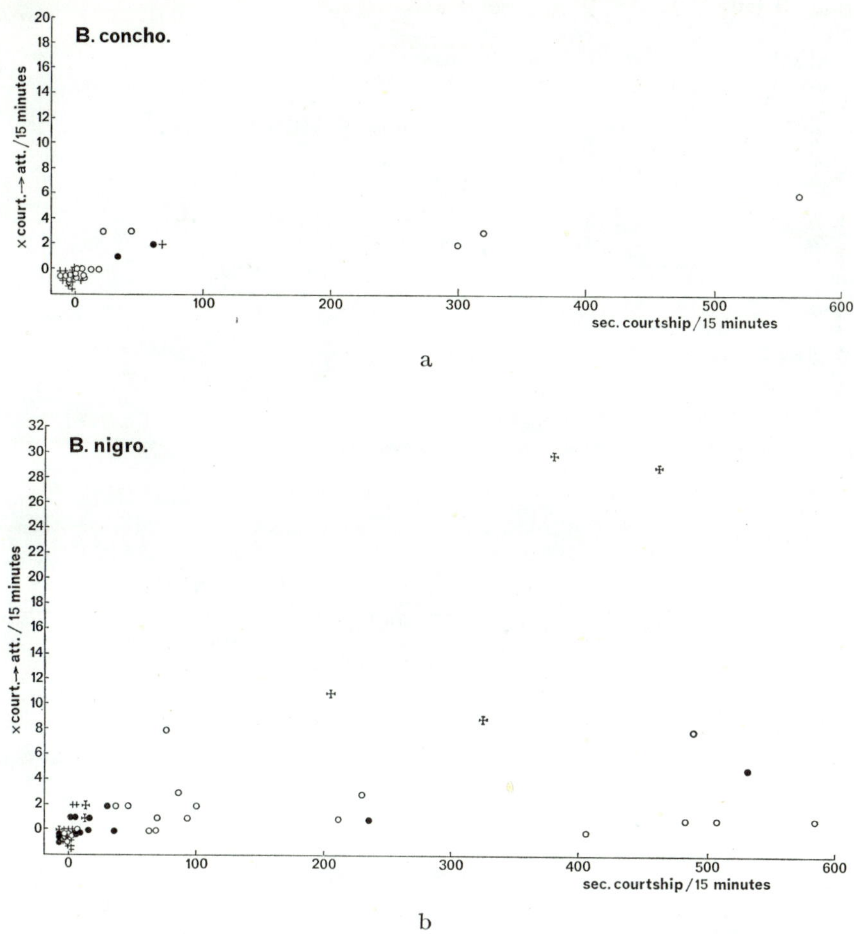

Fig. 50a-e. 5 species; dominant males of 2♂♂ 2♀♀ groups; frequencies of courting being immediately followed by attack at the courted female; explanation see text.

ship — are not treated aggressively more than are *nigrofasciatus* females: the data on *stoliczkanus* are too few to be sure of this. Even so, however, it is obvious that (1) the aggressive behaviour of *stoliczkanus* males towards females consists mainly of pure attack, whereas the aggressive behaviour *nigrofasciatus* males show towards females contains much keeping-off. (2) In contrast to *nigrofasciatus* the 'fighting' category of *stoliczkanus* males towards females — which is usually small, see below — contains other types of 'fighting' besides keeping-off.

It is also evident from the graphs that in the latter three species — in

B. tetra.
×court.→att./15 minutes
sec. courtship/15 minutes

c

B. stolicz.
×court.→att./15 minutes
sec. courtship/15 minutes

d

B. cum.
×court.→att./15 minutes
sec. courtship/15 min.

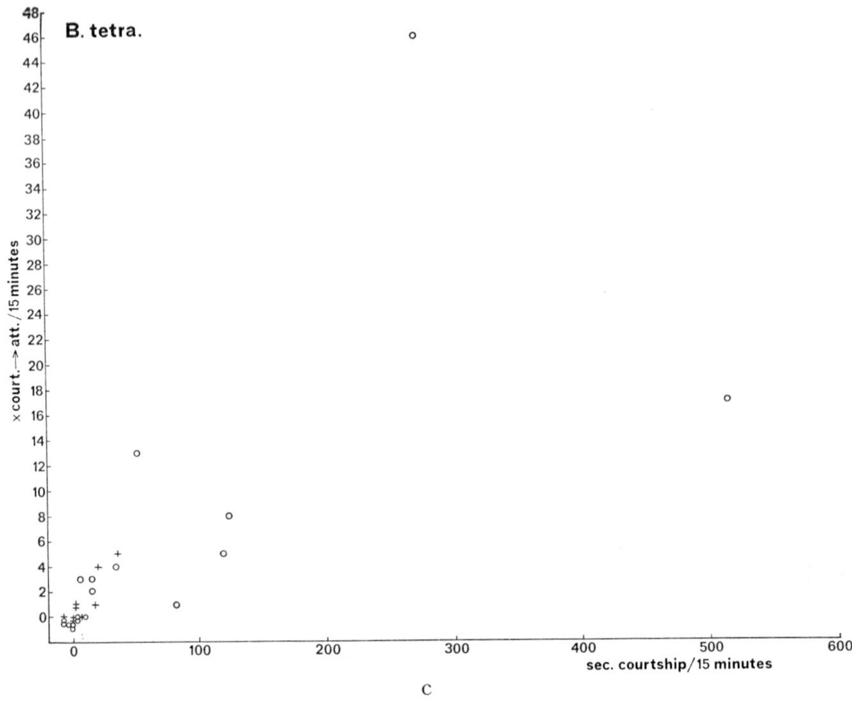

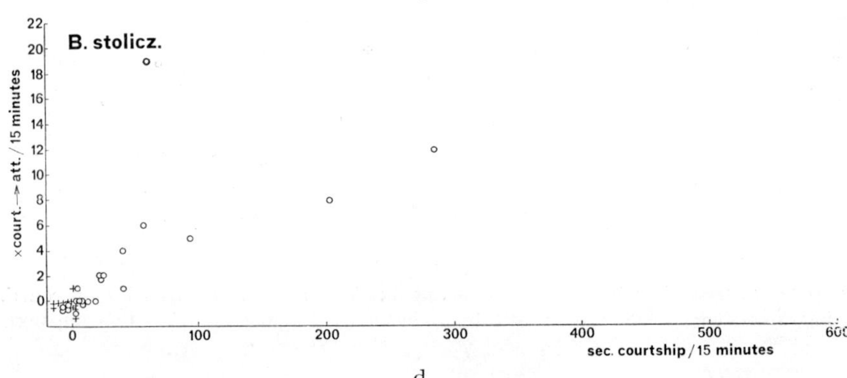

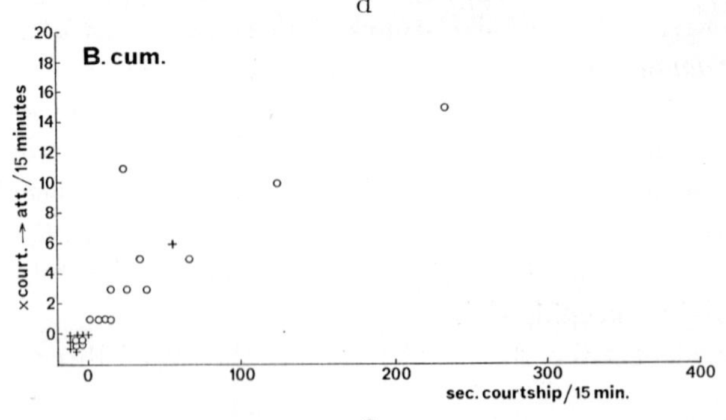

e

contrast to the first two species — the 'fighting' categories displayed towards the other male are small relative to pure attacking (see also sub (b)). Further, the graphs of the latter 3 species suggest that also in these species the amounts of attacking and of attacking plus 'fighting' decrease as the time spent courting increases.

To investigate more closely the relationship between courting and aggressive behaviour shown by a male towards the *same* female graphs 50^{a-e} have been set up. In these graphs the time spent courting (abscissa) is plotted against the number of times the courting male switched to attacking the same female without interruption (ordinate). Data are from the same (dominant) males as those of the former graphs; behaviour is again measured in 15′ periods; the legends of the graphs are the same as in the fore-going ones. Points for which courting is zero have been left out.

For *conchonius* and *nigrofasciatus* there is hardly if any positive correlation between the two measures, (though on a random basis the chances that courting is interrupted by attack should be expected to increase with increasing courting time). In the other 3 species, on the other hand, the graphs show a clear positive correlation. Besides, in these 3 species, the chances that courtship is interrupted by attack are much higher — at comparable amounts of courting — than in the first two species. The *nigrofasciatus* graph also shows that the homosexual male when courting the male is much more liable to attack the courted fish than when he is courting a female.

Just like courting may be interrupted by attack on the courted fish, courting may be immediately preceded by attack at the same animal. In all 5 species the frequencies of attack→courtship are slightly higher than those of courtship→attack more often (in more observation periods) than the other way round. If in figg. 50^{a-e} the frequencies of courtship following attack are plotted instead of those of attack following courting, the same general picture emerges.

It may be concluded that the differences between the species in aggressive behaviour towards the females remain the same even if only aggression towards the *courted* female is taken into account.

The differences found between the species in the attacking tendencies of courting males are probably related to the fact that it is much more easy to get high courting figures in a 2♂♂ 2♀♀ group of *conchonius* and *nigrofasciatus* than in any of the other 3 species [1]). The difficulty in obtaining

---

1) The latter difference is not well illustrated by the graphs as a result of the sampling technique used (see p. 170).

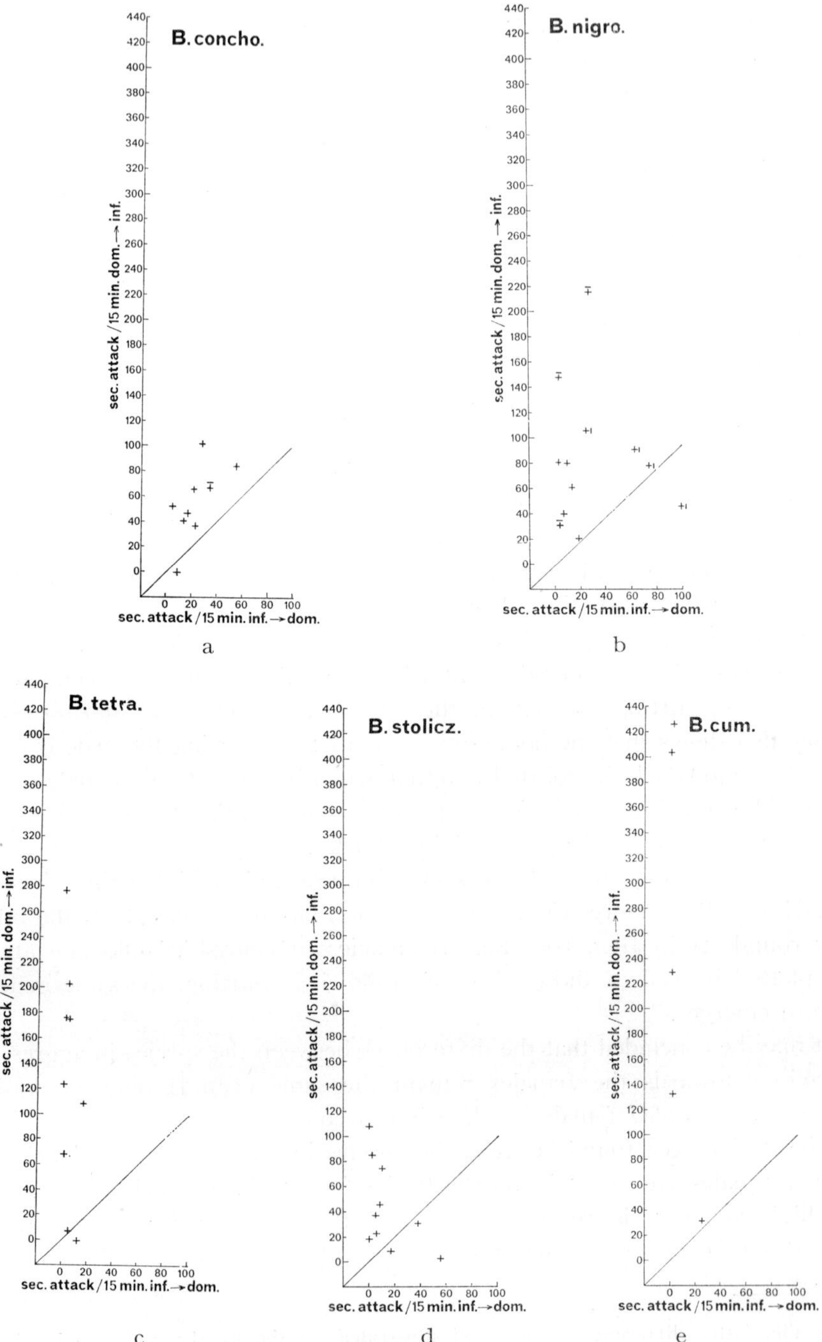

Fig. 51a-e. 5 species; mutual attacking scores of dominant and inferior males of 2♂♂ 2♀♀ groups; ∓ = dominant male 'homosexual'; +| = inferior male 'homosexual'.

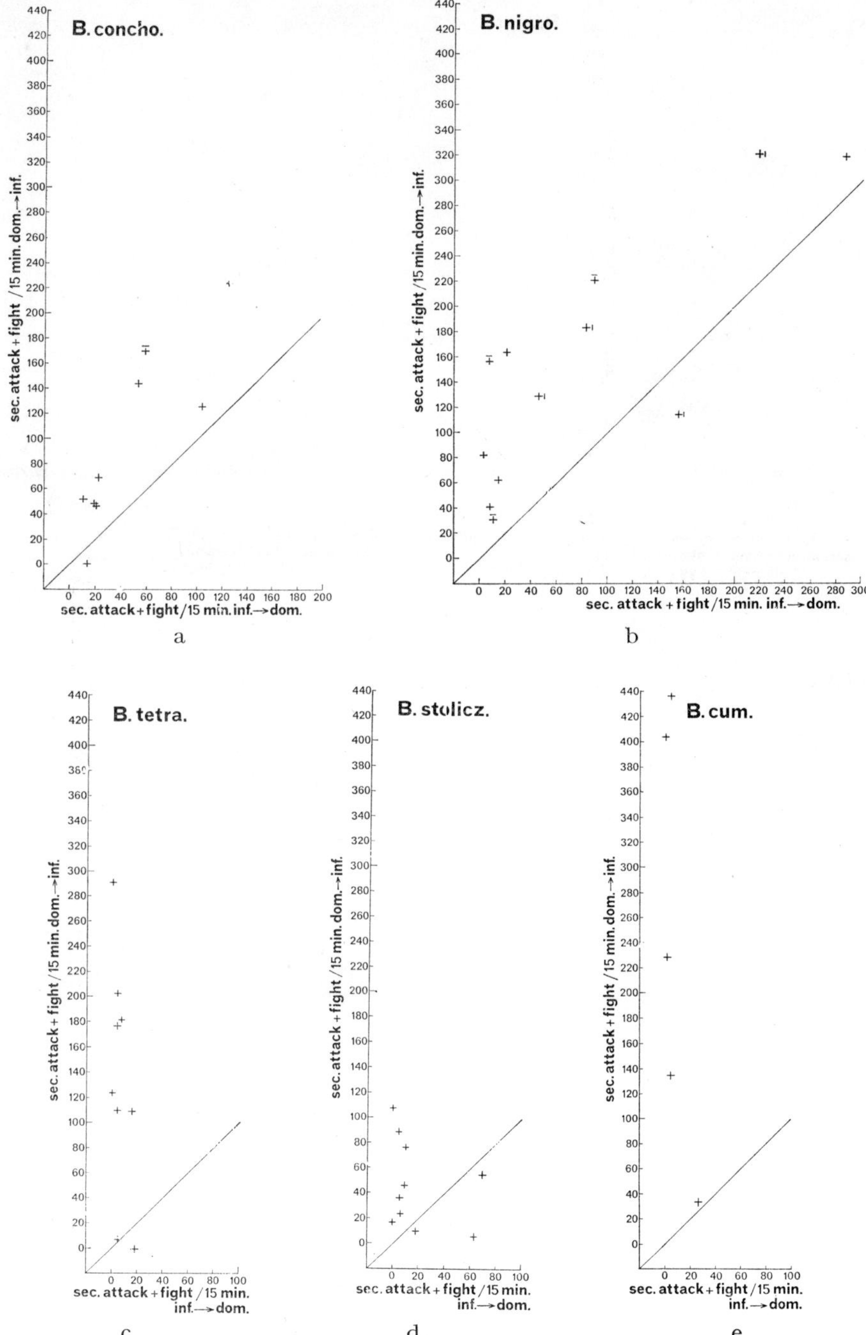

Fig. 52a-e. 5 species; mutual attacking + 'fighting' scores of dominant and inferior males of 2♂♂2♀♀ groups; ∓ = dominant ♂ 'homosexual'; +| = inferior ♂ 'homosexual'.

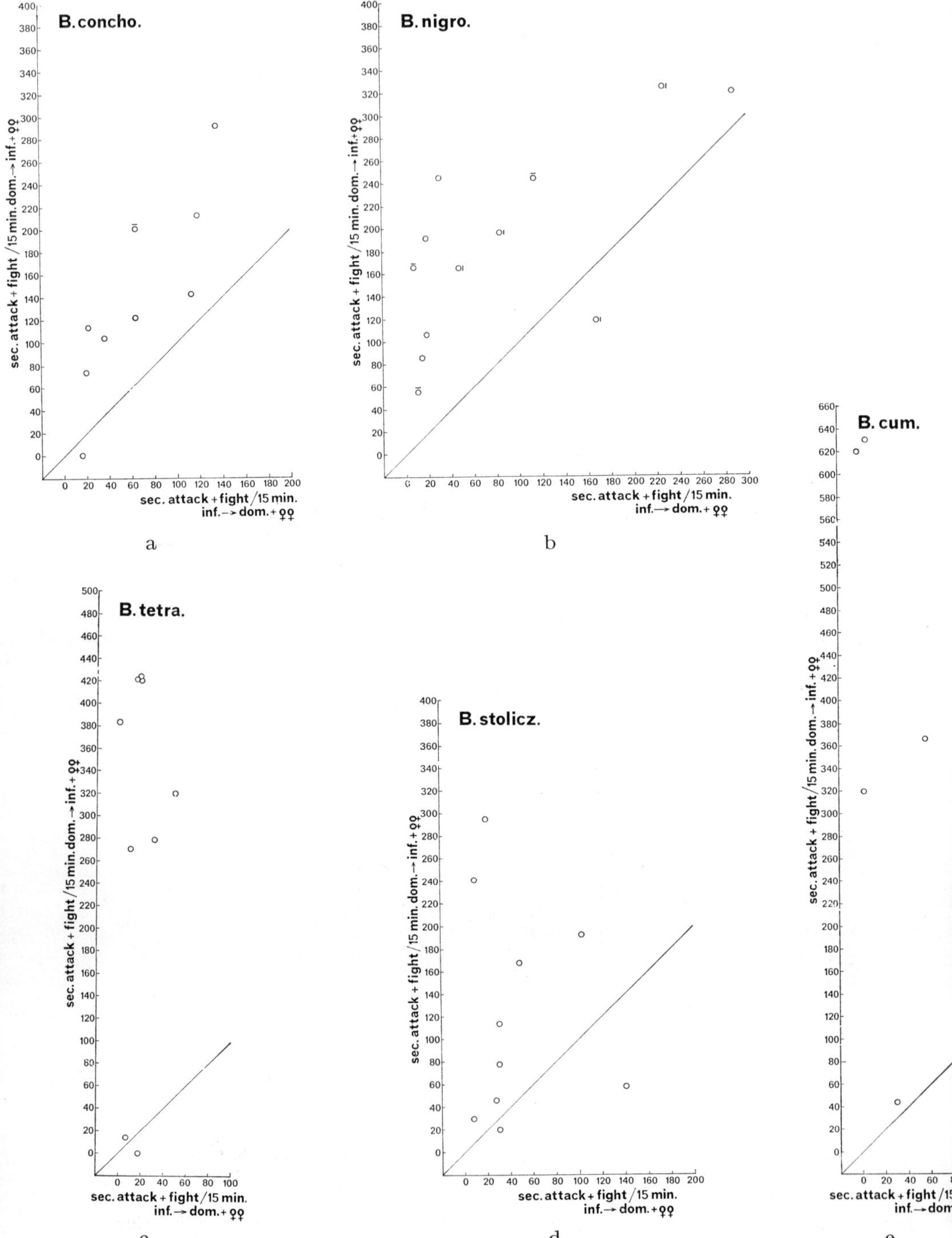

Fig. 53a-e. 5 species; total attacking + 'fighting' scores, at other 3 individuals, of dominant and inferior males of 2♂♂ 2♀♀ groups; $\overline{\text{o}}$ = dominant ♂ 'homosexual'; o| = inferior male 'homosexual'.

high courting scores in the latter 3 species is no doubt also related to the more irregular incidence of reproductive periods in these species such as was demonstrated in section II (6♂♂ 6♀♀ groups).

(b) The competitive relations between the males.

The agonistic activities of both males are illustrated in figg. 51^{a-e}, 52^{a-e} and 53^{a-e}. Figg. 51^{a-e} represent — for each of the 5 species — the number of seconds spent in attacking the other male by the dominant (ordinate) and by the inferior (abscissa) males respectively. (Data are from corresponding 15′ periods of the same observation series). Figg. 52^{a-e} and 53^{a-e} are set up in the same way and represent the number of seconds spent by the males in attacking plus 'fighting' one another and the number of seconds spent by the males in attacking plus 'fighting' the other three individuals. In the *nigrofasciatus* graphs it has been indicated (see legend) whether either the dominant or the inferior male was a homosexual male. (Again, only one homosexual individual is involved, figuring as the dominant male in one group and as the inferior in another). The one 'homosexual' *conchonius* male has been similarly marked.

From all three series of graphs and in all 5 species it is evident that (as might be expected) most often the dominant male spends more time in aggressive behaviour than does the inferior. (Most of the exceptions occur when aggressive activity of the dominant male is comparatively low). However, the species greatly differ among each other in the relative amounts of aggressive behaviour shown by the two males. From figg 51^{a-e} and 52^{a-e} it appears that in *conchonius* and *nigrofasciatus* the inferior male displays considerable amounts of attacking and 'fighting' relative to the activity of the dominant male. In the other 3 species the aggressive activities of the inferior male are much lower relative to those of the dominant male except in some periods characterized by comparatively low aggressive activity in the dominant. From the same graphs it follows that, whereas 'fighting' forms an important part of the aggressive activities of both males (as compared to pure attack) in *conchonius* and *nigrofasciatus*, 'fighting' as a category is almost negligible in the other 3 species.

When figg. 52^{a-e} are compared with figg. 53^{a-e} it follows that, whereas aggressive behaviour directed at the females is almost negligible in both males of the first 2 species, both males display considerable amounts of aggressive behaviour towards females in the latter 3 species. (For the dominant males this difference was already indicated in the graphs of the former sub-section (a)).

The data discussed so far all point to the conclusion that *conchonius* and *nigrofasciatus* inferior males are much less hampered in their aggressive activities than are the inferior males of the other 3 species. (It will be shown below that the same is true for courting). In fact, *stoliczkanus* inferiors spent at least part of their time hiding in a corner of the tank or between vegetation in 9 out of 11 observation series; in 2 of these cases the inferior hardly ever left his corner. In *tetrazona* the inferior male — as far as it was not chased around the tank by the dominant — kept hidden for part of his time in 1 case and for practically the whole time in another 6 cases out of a total of 9; in the remaining 2 cases all four individuals repeatedly fluttered against the front glass of the tank in a group (see also section IV). In *cumingi* the inferior male swam around more or less freely in only 1 out of 8 cases; in the other 7 it tried to hide in a corner of the tank or near to the surface or by pressing itself against the bottom — as long as it was not chased by the dominant. In all 3 species the few cases in which the inferior male moved around freely were invariably characterized by low aggressive activity of the dominant. In *stoliczkanus* the cases in which the inferior male stayed in his corner almost continuously, corresponded to the highest values found for aggressive activity on the part of the dominant. In all 3 species also the females tended to hide but not as strongly as did the inferior males. Hiding was never observed in either *conchonius* (10 cases) or *nigrofasciatus* (17 cases), either in inferior males or in females. It may be noted that *stoliczkanus* inferior males often hid even at comparatively low aggressive activity of the dominant males, whereas *nigrofasciatus* inferiors never hid even though aggressive activity of the dominants was often much higher than that displayed by *stoliczkanus* dominants. One must conclude that either *stoliczkanus* dominants are more efficient in driving their opponents from the main part of the tank or *stoliczkanus* inferiors are more readily intimidated.

The behaviour of *tetrazona, stoliczkanus* and *cumingi* inferior males — and to a lesser degree the behaviour of the females of these species — is suggestive of an animal finding itself within the territory of a rival and not being able to leave it. This would be in accordance with the strong territorial tendencies of the males of these species — as opposed to male *conchonius* and *nigrofasciatus* — during reproductive activity as was described in section II. In fact, especially *tetrazona* and *stoliczkanus* inferior males often seem to claim their tiny hiding places as miniature territories, often attacking both the dominant male and the females as they enter it and often being successful in driving them away. *Cumingi* inferior males didn't seem so much to restrict themselves to one particular hiding place but rather kept still where they were left after the latest chase of the dominant. This is probably due to the unrestrained nature of attacking behaviour in *cumingi* dominants (see below p. 190 and section IV (d) which are not apt to stop for an ensconced inferior.

Territorial behaviour was — on rare occasions — observed also in *conchonius* and *nigrofasciatus* in the sense that the dominant fish confined its attacks and most of his other movements to one half of the tank only, but this was seen in periods of low reproductive activity only. No tendency to hide was apparent in these cases, nor did the corresponding inferior male ever establish a counter-territory.

The differences between the species in the relationship between the two males become even clearer when the courting activities of the males are also considered. Though *conchonius* and *nigrofasciatus* inferior males in most cases courted less than their dominant rivals, they often were able to display considerable amounts of courtship themselves. Data are too few and the samples taken of different degrees of reproductive activity in the dominants are too unequal in different species to allow a complete comparison; therefore only some typical cases will be mentioned. One *conchonius* inferior courted the two females for 184″ in 15′ in the same series in which the dominant courted for 587″. In the other series on record in which the dominant displayed much courting (626″) the inferior male courted sporadically only, but this was apparently not due to intervention by the dominant but to low courting motivation in the inferior male itself. One *nigrofasciatus* inferior male courted for 337″ (dominant 571″) another for 66″ (dominant 637″); One homosexual *nigrofasciatus* inferior courted for 473″ (222″ of which were directed at the dominant male) whereas the dominant (a normal male) courted for 284″ only. In contrast, *tetrazona* inferior males never displayed more than 1.5″ of courting and *cumingi* inferiors never more than 2″. *Stoliczkanus* inferiors never courted more than 34″ and usually much less. In the one observation in which the dominant male showed as much as 492″ courtship, the inferior didn't perform more than 1″ of courting.

In *conchonius* and *nigrofasciatus* inferior males may be successful in mating. For *conchonius* only one case is on record in which the inferior male mated once during a period of 15′ (dominant 26x), but from other observations it is known that *conchonius* inferiors may often be much more successful. The highest number of matings by an inferior male *nigrofasciatus* on record is 8 *per* 15′; this number was scored twice in the observations discussed in this section. In contrast, no inferior male of the other 3 species was ever seen to mate during the observations.

Mating scores of dominant males also show some interesting differences between the species. The highest score on record for a *conchonius* dominant is 26 matings *per* 15′; for *nigrofasciatus* dominant males this was 24; the next was 16. (In each of these two observation series the inferior males scored 8). The highest mating scores for both *tetrazona* and *cumingi* dominants were 6. In spite of many efforts I was unsuccessful in recording any period in a *stoliczkanus* 2♂♂ 2♀♀ group in which the dominant male mated, even though *stoliczkanus* dominants have occasionally been seen to mate in 2♂♂ 2♀♀ groups.

The mating scores of the dominant males in the different species suggest that in *tetrazona, stoliczkanus* and *cumingi* it is not only the inferior males which suffer from the competitive situation but also the dominants. The data on mating scores are in accordance (except perhaps for *cumingi*, see section IV p. 277) with the differences between the species in the 'sensitivity' of the males to competitive situations (see section II).

(c) degree of 'emancipation' of the females in the 5 species.

The evidence the observations of the 2♂♂ 2♀♀ groups yield on this characteristic is rather patchy. This is mainly due to the variability of female behaviour within each species. Still, the data are suggestive and in accordance with those of section II.

First some of our — unquantified — experiences in establishing 2♂♂ 2♀♀ groups should be mentioned. In *conchonius* it never happened that a female became the dominant fish [1]). In *nigrofasciatus* one of the females may behave rather aggressively in some groups but she never becomes really dominant. The display of aggressive behaviour in a mixed group seems to be a property of relatively few individual females in this species; most females are very placid indeed in the presence of reproductively active males. Groups containing such a 'notoriously' aggressive female were not used for the observations because such females were especially apt to attack males which were courting them, thus interfering with their courtship.

In *stoliczkanus* and *cumingi* it happened quite often that a female became dominant, especially — but not necessarily — if she was slightly larger than the other three individuals. Females may become dominant in *tetrazona* groups too but this occurred definitely less frequently than in *stoliczkanus* or *cumingi*. They often do display considerable aggressive behaviour, however. Though perhaps individual females differ in the degree of aggressiveness in each of the latter 3 species, there is no clear-cut difference between 'notoriously' aggressive individuals and placid ones.

In figg. $54^{a\text{-}e}$ the number of seconds *per* 15′ spent by each female in

---

1) This might perhaps be done by selecting females much larger than the males, but I doubt whether even this will suffice to produce female-dominated groups with any considerable frequency.

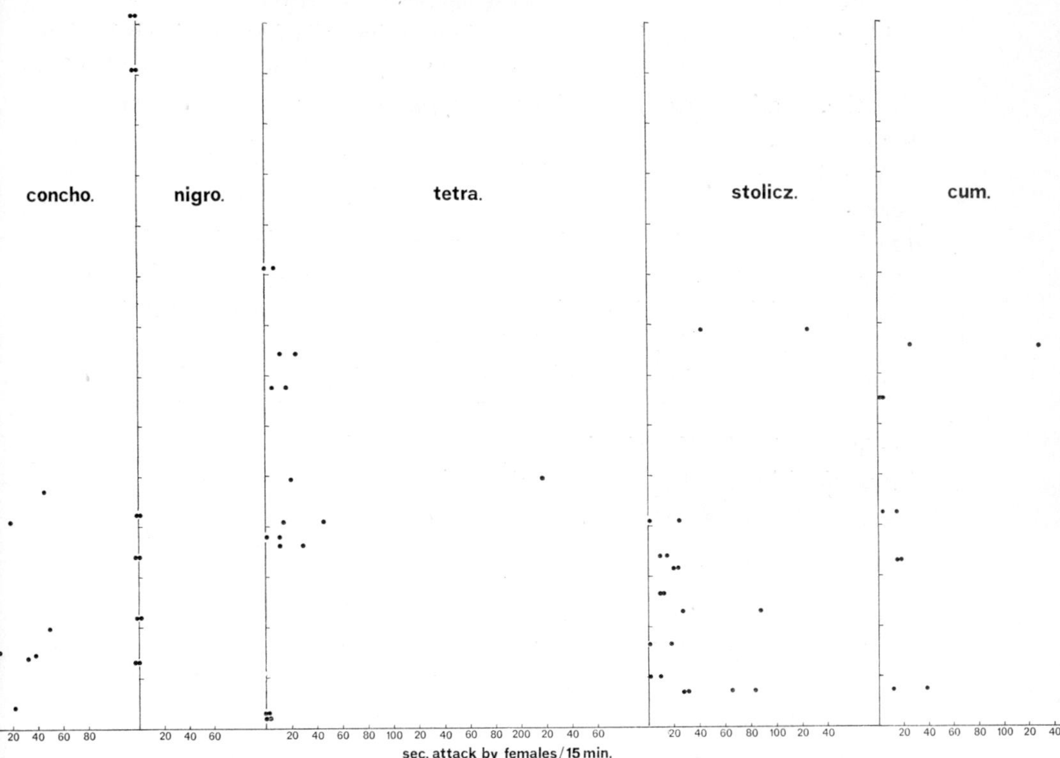

Fig. 54[a-e]. 5 species; 2♂♂ 2♀♀ groups; attacking scores of females (at other 3 individuals) relative to combined reproductive activities of the 2 males.

attacking plus 'fighting' the other three individuals (abscissa) are plotted against the combined reproductive activities of the two males in the corresponding observation periods (ordinate: number of seconds of attacking plus 'fighting' plus courting of both males added together). It follows from the graphs that in *conchonius* at least one of the two females quite often displays some aggressive behaviour. (A considerable part of this is directed at the males). The second female never showed more than 6″ of aggressive behaviour. Typically, the aggressive activity of both females seems to approach zero as the males become more active reproductively. (The one case in the graph in which the more aggressive female scored 45″ of aggressive behaviour at 469″ of combined activity of the males probably is a-typical because this female was observed last in a series in which reproductive activity was rapidly declining. Therefore, during the 15′ observation period of this female the reproductive activity of the two males must have been much lower than was measured during the males' own observation periods). In *nigrofasciatus* neither of the two females ever showed more than 1″ of aggressive behaviour, irrespective of the degree of activity in the males. (It may be reminded here that the, relatively rare, 'notoriously' aggressive females were excluded from the records).

The degree of aggressive activity of the females of the other 3 species is very variable from case to case. The following points, however, are clear. (1) In all 3 species there is at least one case in which a female displayed much more aggressive behaviour than was ever recorded for *conchonius* females. (2) Such high aggressive activities on the part of females do not seem to be restricted to periods of relatively low reproductive activity of the males, such as was found in *conchonius*. (3) Whereas in *conchonius* the less aggressive of the two females never scored more than 6″ of aggressive behaviour, the less aggressive females in *tetrazona* scored above 10″ in 4 out of 9 cases, in *cumingi* in 3 out of 5 cases; in *stoliczkanus* 5 out of 10 less aggressive females scored 20″ or more.

(d) aggressiveness in the males of the 5 species.

The sampling technique used in collecting the data from 2♂♂ 2♀♀ groups may well have biased the values found for aggressive behaviour. Therefore — on the basis of these observations — a comparison of the species as to the degrees of aggressiviness in the males can be *ad hoc* only. Still, the data are probably reasonably representative and in accordance with those of section II. More standardized data will be discussed in the next section.

(1) *B. conchonius* and *nigrofasciatus*.

It may be seen in figg. $51^{a,b}$ that the time spent attacking the other male — both in the dominant and the inferior males — is in general longer in *nigrofasciatus* than in *conchonius*. The same difference, though less pronounced, may be found in figg. $52^{a,b}$. In figg. $53^{a,b}$ the difference is even less clear, mainly due to one *conchonius* dominant male who scored as much as 292″ of attacking plus 'fighting' the other three individuals. (The same individual scored highest in graph 52^{a}; his record is characterized by extremely much 'fighting' the other male, intensive courtship to one female (567″) and a considerable amount of attacking (31″) and keeping-off (25″) the other female; the latter female was courted 21″ only).

We may conclude from this comparison that, though the 'fighting' scores of *conchonius* males often are not much lower than those of *nigrofasciatus* males, their pure attacking scores are. Though the samples of courtship durations of the two species are rather different, they do cover the same range. Therefore the sampling technique probably does not impair the conclusion stated above.

A second difference in aggressiveness between the two species may be found in the greater 'irritability' of *nigrofasciatus* males, as compared to *conchonius* males, with respect to the proximity of a second female during courtship (see p. 176 and figg. 49[a, b]). This difference between the two species relates to comparable amounts of courting; it is probably most pronounced in periods in which courtship doesn't exceed some 400″ *per* 15′ (*cf.* figg. 48[a, b], 49[a, b]).

Both classes of differences indicate stronger aggressive tendencies in *nigrofasciatus* males than in *conchonius* males. This is in accordance with the general difference in aggressiveness found for the two species in the 6♂♂ 6♀♀ groups (section II).

(2) *B. tetrazona* and *stoliczkanus*.

As may be seen from figg. 51[c, d], 52[c, d] and 53[c, d], *tetrazona* dominant males in general scored much higher values for aggressive behaviour towards the inferior male than did *stoliczkanus* dominants. This is true for all three scoring methods used. The opposite is true for the inferior males of the two species: *tetrazona* inferior males show much less aggressive behaviour than do *stoliczkanus* inferiors, even if aggressive behaviour towards females is taken into account. Whereas *stoliczkanus* inferiors with aggressive dominants are able to behave aggressively at least towards the females, *tetrazona* inferiors are much less able to do so. The difference in the aggressive activities of the inferior males of the two species is probably due to the much higher aggressiveness of *tetrazona* dominants as compared with *stoliczkanus* dominants.

It is inferred here that in these two species higher aggressive activity of the dominant male progressively reduces such activity in the inferior. More or less the same phenomenon seems to be present within at least the data of *stoliczkanus* (see fig. 51[d]). This is a quite different effect from what was found in *conchonius* and *nigrofasciatus* in which the aggressive activities of the two males seem to be more or less positively correlated (see figg. 51[a, b], 52[a, b], 53[a, b]). This difference between the two groups of species is no doubt related to the much stronger territorial tendencies of *tetrazona* and *stoliczkanus* and/or to the tendency of the inferior males of these species to hide and to abstain from nearly all aggressive and courting behaviour when confronted with aggressive dominants. The difference is also in accordance with the fact that much of the aggressive activity of *conchonius* and *nigrofasciatus* males consists of 'fighting' (often in direct competition for a female) whereas the aggressive activity of *tetrazona* and *stoliczkanus* males is mainly made up of pure attack only. Therefore we may conclude that in *tetrazona* and *stoliczkanus* the amounts of aggressive behaviour shown by the inferiors is very strongly influenced by the activities of the dominant male. The best indication for aggressiveness, therefore, is to be found in the behaviour of the dominant males. This leads to the conclusion that *tetrazona* males are in general more aggressive than are *stoliczkanus* males. This conclusion is in accordance with what was found in section II. The conclusion can hardly be impaired by the sampling technique, as the samples of courtship durations of the two species are very similar to one another (see also section IV, p. 204).

The same difference in aggressiveness between the two species is suggested by figg. 48^{c}, ^{d}. As far as the data go, *tetrazona* dominants — at comparable levels of courting activity — attack the females much more than do *stoliczkanus* dominants.

(3) *conchonius-nigrofasciatus vs tetrazona-stoliczkanus.*

It is virtually impossible — on the basis of the data presented so far — to compare degrees of aggressiveness between the two pairs of species. The reasons for this are the following.

Technically, the samples of courtship durations in the 2♂♂ 2♀♀ observations are rather different in the two pairs of species. As stated before, this might have induced a bias in the aggressive scores as well.

Structurally, the aggressive scorings for the two pairs of species are incomparable because of the difference in hiding tendencies of the inferior males, the differences in the ratio of 'fighting' to attack and the difference in the distribution of the dominant male's aggressive behaviour over the inferior male on the one hand and the females on the other. These differences would remain even if the samples of courtship durations would be made the same in the two pairs of species.

(4) *B. cumingi.*

Dominant males of the 2♂♂ 2♀♀ groups lived up to their reputation of unrestrained aggressiveness established in sections I and II of this chapter. They scored higher values for aggressive behaviour than any of the other 4 species, irrespective of which scoring method was applied (see graphs 51^{e}, 52^{e}, 53^{e}). Besides, almost all of their aggressive behaviour, both towards the inferior male and towards the females, consisted of pure attack almost exclusively; 'fighting' as a category was quite negligible. Not only the inferior male but also the females often were not able even to take refuge in any kind of cover but hung passively near the water surface as long as they were not attacked. Only rarely an inferior male managed to display some aggressive behaviour at least towards a female. *Cumingi* is the only one of the 5 species in which it is not 'safe' to keep a small group together for more than 1 or 2 days, even if it is an ambisexual group. Both the inferior male and the females may be irremediably damaged if the association is maintained beyond this limit. Serious damage, especially in inferior males, may occur in *tetrazona* small groups too but this is much less common in ambisexual groups of this species (see also section IV sub (c)).

SECTION IV. BEHAVIOUR OF THE 5 SPECIES IN 2♂♂- AND 3♂♂ GROUPS

The data presented in this section were collected mainly in order to assess differences between the 5 species in the degree and kind of aggressiveness of the males. Data on such differences have been discussed in the fore-going sections but these data were either not sufficiently quantitative (section II) or not sufficiently standardized (section III). The observations discussed in this section were made according to a standardized time schedule and involve quantitative behaviour records of all individuals of each group.

Procedure.

All groups consisted of conspecific males only; during the observation periods they were kept in 60 × 35 × 35 cm tanks with asbestos-cement sides and bottom and a glass front. The bottom of all tanks was covered with a layer of sand. The number of males in a group was either 2 or 3; the tanks were either completely bare or supplied with some scattered vegetation both growing from the bottom and floating at the surface. Thus four different situations were created:

2♂♂ in a planted tank	(2♂♂ pl.)
2♂♂ in a bare tank	(2♂♂ b.)
3♂♂ in a planted tank	(3♂♂ pl.)
3♂♂ in a bare tank	(3♂♂ b.).

Of each of the 5 species at least 2 groups of each of these four types were studied. The observations of these groups were all made under the standardized artificial light (a.l.) regime described on p. XI. In addition, of each species, one or several other groups of the same types were studied. The latter groups were almost exclusively kept under natural light (n.l.) conditions. Table 7 summarizes the groups used, the light conditions under which they were kept and — as far as the natural-light groups are concerned — the dates at which the observations were made.

Each single group was composed of individuals from one (mostly 125 cm) stock tank in which they had lived together with several conspecifics of both sexes (and often other species) for at least several weeks. The individuals of those groups which were observed under the artificial light regime had been under the same regime while in the stock tanks for at least several weeks. The individuals of a group were selected so as to be of approximately equal size. No other principles of selection were applied except for general healthiness and sex (see also section II, p. 163).

Of the groups studied under the artificial light regime, for each of the 5 species, every two groups of the same type (2♂♂ pl., 2♂♂ b., 3♂♂ pl., or 3♂♂ b.) were established simultaneously [1]) using animals from the same stock tank; so one can be sure that these pairs of groups were made up of different individuals. For the rest, some individuals may have been a member of more than one of the experimental groups but this must have been a rare occurrence because different groups were often composed from different stock tanks and the number of males in each stock tank was usually large.

All groups were started around 17.00 on the day preceding the first day of observation. The fishes were taken from the stock tank and introduced simultaneously into the experimental tank. There they were fed shortly afterwards. Observations were made at fixed times on the first and second day after introduction. On both days they were again fed around 17.00. For the artificial light experiments observations were made beginning at 09.00, 11.30 and 14.30 on both days; for the natural light experiments observations began at 08.00, 11.00 and 14.00.

As was mentioned above, in the artificial light experiments two similar groups were established simultaneously. Of one of these (called group A hereafter) at the times indicated above the 2 or 3 individual males were observed one after the other, each for a period of 15 minutes. Following this the fishes of the other group (called group B hereafter) were observed, each for a period of 5 minutes. Group A remained group A throughout the experiment; at all observation sessions the observations of group A preceded those of group B. Which of the two groups was chosen to be group A was decided in a more or less random fashion on the morning of the first observation day before the fishes had been seen.

In the natural light experiments only one group was studied at a time. At the times indicated above each of the individuals was observed for a

---

1) The few exceptions to this rule are marked with an asterisk in table 7.

TABLE 7

Survey of 2♂♂ and 3♂♂ groups used. Explanation see text

Group	Light	Date
B. conchonius		
3♂♂ pl.	a.l.	
3♂♂ pl.	a.l.	
3♂♂ b.	a.l.	
3♂♂ b.	a.l.	
3♂♂ pl.	n.l.	11-12/6
3♂♂ b.	n.l.	5-6 /7
2♂♂ pl.	a.l.	
2♂♂ pl.	a.l.	
2♂♂ b.	a.l.	
2♂♂ b.	a.l.	
B. nigrofasciatus		
3♂♂ pl.	a.l.	
3♂♂ pl.	a.l.	
3♂♂ b.	a.l.	
3♂♂ b.	a.l.	
3♂♂ b.	n.l.	5-6/6
2♂♂ pl.	a.l.	
2♂♂ pl.	a.l.	
2♂♂ b.	a.l.	
2♂♂ b.	a.l.	
B. tetrazona		
3♂♂ pl.	a.l.	
3♂♂ pl.	a.l.	
3♂♂ b.	a.l.	
3♂♂ b.	a.l.	
3♂♂ b.*	a.l.	
2♂♂ pl.	a.l.	
2♂♂ pl.	a.l.	
2♂♂ b.	a.l.	
2♂♂ b.	a.l.	
2♂♂ pl.	n.l.	14-15/6
2♂♂ b.	n.l.	30-31/5
B. stoliczkanus		
3♂♂ pl.	a.l.	
3♂♂ pl.	a.l.	
3♂♂ b.	a.l.	
3♂♂ b.	a.l.	
3♂♂ pl.	n.l.	19-20/6
2♂♂ pl.	a.l.	
2♂♂ pl.	a.l.	
2♂♂ b.	a.l.	
2♂♂ b.	a.l.	
2♂♂ pl.	n.l.	22/5
2♂♂ b.	n.l.	3-4/7
B. cumingi		
3♂♂ pl.*	a.l.	
3♂♂ pl.*	a.l.	
3♂♂ b.	a.l.	
3♂♂ b.	a.l.	
2♂♂ pl.	a.l.	
2♂♂ pl.	a.l.	
2♂♂ b.	a.l.	
2♂♂ b.	a.l.	
2♂♂ b.	n.l.	16-17/5

a.l. = artificial light regime. n.l. = natural light.

period of 15 minutes in the same way as was done in the artificial light experiments.

In both the artificial light and natural light experiments the order in which the individuals of a certain group were observed was changed in a regular way from one observation session to the next; for instance:

2♂♂:			
	day I	09.00	AB
		11.30	BA
		14.30	AB
	day II	09.00	BA
		11.30	AB
		14.30	BA

3♂♂:			
	day I	09.00	ABC
		11.30	BCA
		14.30	CAB
	day II	09.00	CBA
		11.30	BAC
		14.30	ACB

The observations were spoken into a tape recorder and analyzed later. The incidence and duration of every single behaviour element was recorded. Simultaneously, at irregular intervals during the observations the general situation of the group was described taking into account such activities as schooling, fluttering, territorial behaviour or hiding.

Not all groups were observed for the entire period of two days. A few experiments had to be broken off after the first day because one of the fishes had jumped out of the tank at the end of the first or at the beginning of the second day, because one or more fishes got ill during the experiment or because an inferior fish had been too seriously damaged by the attacks of the dominant. A few of the planned observation sessions are partly or completely missing due to failure of the recorder. Tables 8, 9 and 14 summarize the observations available from all groups of each of the 5 species (see below).

The observation times were chosen with the intention to get a fair sample of the intensities and kinds of activity of the fishes over the whole day. The observations discussed in section II of this chapter have revealed that the species differ greatly in their daily activity rhythms. Therefore, observations made at any one specific time of the day only can hardly be expected to be a sound basis for comparison between the species. This remains true even though the rhythms of activity displayed under the very special conditions of the 2♂♂- and 3♂♂ experiments were not in all respects the same as those shown by the 6♂♂ 6♀♀ groups. Therefore observation times had to be scattered over the day in a way to yield representative data for all 5 species.

It may be questioned whether the observation times were *sufficiently* scattered to give a fair picture of the fishes' total activity. First, it must be emphasized that no observations were made later than 15.30. However, many unstandardized observations of small groups (both ambisexual and all-male) made both under artificial and natural light conditions indicate that after 18.00 (1) the reproductive activity of *conchonius* and *nigrofasciatus* is negligible and that (2) though especially *tetrazona* and *cumingi* males may display considerable aggressiveness during the evening, such activity was as a rule much lower and never higher than before 18.00 both in the latter two species and in *stoliczkanus*. Therefore the observation samples taken are probably reasonably representative even in the absence of evening observations.

Second, it is certain that at least *conchonius* maintains its early activity rhythm even in small all-male groups. This would in no way invalidate the sampling of the artificial light experiments in which observations were started soon after the lights had been switched on. However, as the natural light experiments were made during the summer, several hours of daylight preceded the first observations of these experiments. Obviously this seriously impaired the sampling of at least the *conchonius* groups. It is possible that the samples of some of the other species were similarly impaired. Therefore, for part of the comparisons made in this section, the natural light experiments were not used.

Results.

The data of the 2♂♂- and 3♂♂ groups will be analyzed in two steps. First some aspects of the general situation in the groups will be discussed. Special attention will be given to the presence or absence of dominant-inferior relationships and to the presence and strength of hiding tendencies in the inferior male(s). Second, the behaviour of the individual fishes will be analyzed in more detail; attention will be focussed on agonistic interactions. Possible causal connections between the situation in the group and the detailed behaviour of the group members will be discussed.

Comparisons between the species will be made according to the following scheme. First the data of *conchonius* are compared with those of *nigrofasciatus* and the data of *stoliczkanus* are compared with those of *tetrazona*. These comparisons are made in order to test the hypotheses brought forward in the preceding sections that *nigrofasciatus* males are more aggressive than *conchonius* males and *tetrazona* males more aggressive than *stoliczkanus* males. It may be reminded that, apart from these differences in aggressiveness, the general structure of the reproductive behaviour was found to be very similar in *conchonius* and *nigrofasciatus* and likewise in *stoliczkanus* and *tetrazona*.

Second, the data of *conchonius* and *nigrofasciatus* on the one hand and those of *stoliczkanus* and *tetrazona* on the other will be compared. It will be seen from this comparison that the general structural differences in behaviour found to exist between these two pairs of species are also to some extent reflected in the behaviour of the 2♂♂- and 3♂♂ groups. This again precludes a simple quantitative comparison between the two pairs of species.

Finally, the data of *cumingi* will be discussed and compared with those of the other 4 species.

(a) *B. conchonius* and *nigrofasciatus*.

Situational data.

The situations obtaining during the observations are shown diagrammatically in table 8. Each 5 minute observation period is represented as a triangle (for the 3♂♂ experiments) or as a circle (for the 2♂♂ experiments). For this purpose the 15 minute observation periods were broken down into three 5 minute periods each. As in each observation session each individual was observed for either 5 minutes or 15 minutes, each observation session contains either three or nine 5 minute periods for the 3♂♂ experiments and either two or six 5 minute periods for the 2♂♂ experiments.

Each large vertical column represents the observation of one group of fishes. The successive observation sessions are indicated at the left side of the table. For the 15 minute observations, within each observation session each individual's record is represented as a small vertical column of three symbols; these are to be read from top to bottom. For the 5 minute observations the individuals' records are given in one horizontal row. For each observation session the order from left to right of the small columns or single symbols corresponds to the order in which the individuals were observed. If necessary it is indicated which of the small columns or single symbols represents the dominant male's record.

The following situations were distinguished and represented by the following symbols. Open triangles or circles with a point in the centre symbolize periods in which none of the males was clearly dominant and in which all individuals spent much of their time schooling and fluttering against the front glass of the tank. In all other situations one of the males was clearly dominant. Plain open triangles and circles represent situations in which the inferior male(s) swam around freely (except for the time they were chased by the dominant) and showed no hiding tendencies. Completely black symbols indicate that the inferior(s) were hiding in the corners of the tank or (as was mostly the case in both species, see tables 11, p. 233) near to the surface, again as long as they were not chased by the dominant. Half-black symbols mean that the inferior(s) showed at least some tendency to hide but also moved around freely for part of the time. In one case, in the 3♂♂ b.A experiment of *nigrofasciatus*, the half-black triangles mean that during the observation period of the dominant one inferior was hiding continuously while the other moved about freely. In all other cases half-black triangles for the dominant mean that both inferiors were hiding for part of the time.

In most cases the same individual remained dominant throughout the experiment. Two kinds of exception did occur. In the *conchonius* 3♂♂ b.A experiment, on day II, two of the males (including the dominant of day I) took turns in being dominant, changing roles every few minutes. The third male was continuously inferior to the current dominant. The observations of the first two males are represented as open triangles crossed by a small

TABLE 8

B.concho.

sessions ↓ / groups →	3♂♂ pl.A	3♂♂ pl.B	3♂♂ b.A	3♂♂ b.B	3♂♂ pl.n.l.	3♂♂ b.n.l.
I9	△△△ △△△ △△△	△△△	△△△ △△△ △△△	△△△	△△△ △△△ △△△	△△△ △△△ △△△
I11	△△△ △△△ △△△	△△△	△△△ △△△ △△△	△△△	△△△ △△△ △△△	△△△ △△△ △△△
I14	△△△ △△△ △△△	△△△	△△△ △△△ △△△	△△△	△△△ △△△ △△△	△△△ △△△ △△△
II9	△△△ △△△ △△△	△△△	△̶△△̶ △̶△△̶ △̶△△̶	△△△	△△△ △△△ △△△	△△△ △△△ △△△
II11	△△△ △△△ △△△	◭△△	△△̶△̶ △△̶△̶ △△̶△̶	△△△	△△△ △△△ △△△	△△△ △△△ △△△
II14	△△△ △△△ △△△	△△△	△△△ △△△ △△△	△△△	△△△ △△△ △△△	△△△ △△△ △△△

a

B.nigro.

sessions ↓ / groups →	3♂♂ pl.A	3♂♂ pl.B	3♂♂ b.A	3♂♂ b.B	3♂♂ b.n.l.
I9	◬◬◬ ◬◬◬ ◬◬◬	△△△	◬◬◬ ◬◬◬ ◬◬◬	▲▲▲	◬◬◬ ◬◬◬ ◬◬◬
I11	◬◬◬ ◬◬◬ ◬◬◬	△△△	◭◭◭ ◭◭◭ ◭◭◭	▲▲▲	◬◬◬ ◬◬◬ ◬◬◬
I14	◬◬◬ ◬◬◬ ◬◬◬	△△△	△▲◭ △▲◭ △▲◭	▲▲▲	◬◬◬ ◬◬◬ ◬◬◬
II9	◬◬◬ ◬◬◬ ◬◬◬	△△△	▲▲▲ ▲▲▲ ▲▲▲	▲▲▲	◬◬◬ ◬◬◬ ◬◬◬
II11	◬◬◬ ◬◬◬ ◬◬◬	△△△	▲▲▲ ▲▲▲ ▲▲▲	▲▲▲	◬◬◬ ◬◬◬ ◬◬◬
II14	◬◬◬ ◬◬◬ ◬◬◬	△△△	▲▲▲ ▲▲▲ ▲▲▲	▲▲▲	◬◬◬ ◬◬◬ ◬◬◬

b

B.concho.

sessions ↓ / groups →	2♂♂ pl.A	2♂♂ pl.B	2♂♂ b.A	2♂♂ b.B
I9	◐● ◐◐ ◐◐	○○	○○ ○○ ○○	○○
I11	●● ●● ●●	○○	○○ ○○ ○○	○○
I14	●● ●● ●●	◐◐	○○ ○○ ○○	○○
II9	●○ ●○ ◐	●◐	○○ ○○ ○○	○○
II11	●● ●● ●●	●●	◐○ ◐○ ◐○	○○
II14	●● ●● ●●	●●	○○ ○○ ○○	○○

c

B.nigro.

sessions ↓ / groups →	2♂♂ pl.A	2♂♂ pl.B	2♂♂ b.A	2♂♂ b.B
I9	●● ●● ●●	○○	●● ●● ●●	◐◐
I11	●● ●● ●●	●●	●● ●● ●●	●●
I14	●● ●● ●●	●●	●● ●● ●●	●●
II9	●● ●● ●●	●●	●● ●● ●●	●●
II11	●● ●● ●●	●●	●● ●● ●●	●◐
II14	●● ●● ●●	●●	●● ●● ●●	◐◐

d

horizontal line. Throughout these periods the two current inferiors remained free to move around. During the last observation session the original dominant was in the dominant role again.

Besides these relatively quick and transient changes in dominance, more permanent dominance reversals may take place. One such reversal occurred in the *nigrofasciatus* 3♂♂ b.A experiment at some time between the last observation session of day I and the first session of day II. In the *nigrofasciatus* 3♂♂ pl.A experiment one of the males gradually became more aggressive and seemed to become dominant in the course of the second day but, shortly after the last observation session, he was easily beaten by one of the other males which suddenly started to threaten and to attack the others and became dominant.

It may be seen from the table that in all six 3♂♂ *conchonius* groups there was a dominant male from the beginning; the inferiors never showed any hiding tendency except for one inferior during one 5 minute period only. [1]) In contrast, in the *nigrofasciatus* 3♂♂ groups, in 2 out of the 5 groups no clear dominance was established during the experiment and all males kept schooling and fluttering for most of the time. (As mentioned above, in one of these groups a male became fully dominant shortly after the last observation session of the second day). The same situation obtained during the first observation session of the 3♂♂ b.A experiment. In only 1 of the remaining 3 groups with clear dominance both inferiors kept moving freely throughout the experiment. In the other two groups in all observations at least one and mostly both inferiors hid for at least part of the time; in most cases both inferiors hid continuously.

In the 2♂♂ groups dominance was established in all groups of both species before the beginning of the observations. In *conchonius* hiding by the inferior occurred in 3 out of the 4 groups but the frequency of hiding was negligible in one of these cases and hiding was absent during the first two observation sessions of another. Even in the one group in which the inferior showed hiding in nearly all observations there were some periods in which he didn't do so continuously. In contrast, in *nigrofasciatus* hiding by the inferior was very common and continuous in almost all observations.

We may conclude that (1) it takes more time for a *nigrofasciatus* male, at least in a 3♂♂ group, to establish full dominance than it does for a *conchonius* male (2) once dominance has been established *nigrofasciatus* inferiors hide more than do *conchonius* inferiors both in 2♂♂ and in 3♂♂ groups.

1) Though this male was reasonably active during most observations, there were some doubts as to its general healthiness during the second day of the experiment.

The second conclusion is also supported by many unstandardized observations of small all-male groups; no such data are available to support or to contradict the first conclusion. Both conclusions contrast sharply with my experiences with small ambisexual groups (2♂♂ 2♀♀ mostly) in similar tanks. In such groups in both species dominance is usually found to be established early in the morning after introduction and intensive reproductive behaviour is then shown. The same tendency was found in the 6♂♂ 6♀♀ groups of both species in large tanks (see section II pp. 108, 118). In both species inferior males in small ambisexual groups rarely if ever hide (see also section III, p. 184).

The data of table 8 also suggest that (3) in both species hiding by the inferior(s) is more common in 2♂♂ than in 3♂♂ groups (4) in *nigrofasciatus* it takes more time for a male to become fully dominant in a 3♂♂ group than it does in a 2♂♂ group.

Discussion.

Conclusion (2) may be reformulated so as to say that the individual distance between males in an all-male group is greater in *nigrofasciatus* than in *conchonius*. This difference might be explained by the assumption that *nigrofasciatus* males, especially when dominant, are more aggressive than are *conchonius* males in the same role. The stronger hiding tendencies in the inferiors of the former species would then be a result of the more intensive aggressive behaviour of the dominant. This interpretation would be in accordance with the findings of the fore-going sections, especially because the behaviour of a dominant male in a small all-male group is certainly much more similar to the behaviour of males in large and small ambisexual groups than is the behaviour of the inferiors. However, the difference in individual distance found would not *prove* the assumption because much the same results might have been found if the aggressive behaviour were not different in the two species but *nigrofasciatus* males, especially when inferior, would be more liable to flee from the attacks of a conspecific male. These two possibilities will be referred to hereafter as hypotheses (1) and (2). As a third possibility, both assumptions might be true. In order to be able to decide between these possibilities we will have to look more closely at the details of the agonistic interactions between the males. Before doing so, however, the following relevant remarks must be made.

First, it must be stressed that the differences we are going to look for are differences in *motivation;* therefore in analysing the data we must try to eliminate as much as possible such differences as existed in the stimulus situation for the two species.

1. One difficulty in interpreting the quantitative data of this section is that in the 2♂♂ and 3♂♂ groups the behaviour of the dominant and of the inferior fishes will inevitably be to some extent each other's mirror image. Since an inferior flees almost exclusively when it is attacked, the amount of time spent fleeing by the inferior(s) is primarily determined by the time spent attacking by the dominant. No data are available on the absolute speed of fleeing by the inferior such as might allow a comparison between the species. On the other hand it seems likely that the time spent attacking by the dominant is to some extent influenced by the inferior's readiness to flee when attacked and by its speed of fleeing. Chasing, which makes up the majority of the dominant's attacking behaviour, can only occur if the inferior flees in response to initial attack. A chase often comes to an end because the chased fish stops fleeing. It seems likely that fleeing in an attacked fish often releases prolonged attack in the other fish. Fast fleeing releases more (and probably stronger) attack than does slow fleeing (see DUNHAM *et al.*, 1968, p. 20-21). Again, no data are available on differences between the species in the absolute speed of attacking by the dominants. The behaviour records of the dominants do involve discrimination of attack intensities insofar that full attack was distinguished from distance attack, inhibited attack and courting posture (which may be considered as a weak intensity of attacking, see Appendix 2). As was stated on p. 74 distance attack and inhibited attack probably result from the attacker slowing down rather than from the fleeing fish speeding up. However, in the absence of data on the absolute speeds of either the attacking or the fleeing fish, it is impossible to say whether differences between the species in the occurrence of low intensity attack is due to slower fleeing by the chased fish or to 'spontaneous' slowing down of the attacker.

It is obvious that the mutual mirroring of the behaviours of dominant and inferior fishes will greatly interfere with the possibility to choose between the above hypotheses on the basis of the quantitative data available. The best way to get around this difficulty would be to study directly the interactions between 2 fishes instead of observing one fish at a time or to use models which can be moved at various constant speeds. The first method was applied by DUNHAM *et al.* (1968) in studying agonistic interactions of one species: *B. stoliczkanus*, but no comparable data are present for the other species. In the absence of such data we can try to find a provisional solution in the following ways.

First, unquantified impressions of behaviour intensities which are difficult to measure quantitatively may supplement the quantitative data. For instance it is possible that *nigrofasciatus* inferiors do sometimes flee more rapidly than do *conchonius* inferiors but often they do not. Nevertheless *nigrofasciatus* dominants as a rule attack and bite much more vigorously than do *conchonius* dominants even if the attacked fish doesn't flee fast. Second, hypothesis (1) may also be tested not by comparing the behaviour of dominants of the two species but by comparing the inferiors. The fact that *nigrofasciatus* inferiors are as a rule more inferior than are *conchonius* inferiors would lead one to expect that, if *nigrofasciatus* males were *not* more aggressive, *nigrofasciatus* inferiors would exhibit less or at best as much aggressive behaviour as *conchonius* inferiors. If at least some *nigrofasciatus* inferiors would be found to show more aggressive behaviour than do *conchonius* inferiors under comparable circumstances, this would be a strong argument in favour of hypothesis (1). Conversely, if *nigrofasciatus* dominants would, under comparable circumstances, display more fleeing than do *conchonius* dominants, this would be an argument for the other hypothesis. It is clear that in this way neither of the hypotheses may be disproven. If for instance *nigrofasciatus* inferiors attack less than do *conchonius* inferiors this wouldn't disprove hypothesis (1) because it could equally well be due to the *nigrofasciatus* inferiors being more liable to flee and/or to their dominants being more aggressive.

The existence of what may be called a secondary mirror effect might be suspected; *i.e.* that flight by the dominant is largely caused by attacks by the inferior and/or that attack by the inferior is largely caused by flight by the dominant. If this would be so, and if both results would be found: higher attacking scores in *nigrofasciatus* than in *conchonius* inferiors *and* higher fleeing scores in *nigrofasciatus* than in *conchonius* dominants, again either or both hypotheses might be true.

However, two sets of arguments may be brought to bear upon this possibility. First from unquantified observations it follows that, though inferior males may attack in response to flight by a dominant and though dominants may flee in response to an inferior's attack, (1) by far most flights by dominants do *not* follow attack by an inferior (2) by far most attacks by inferiors do *not* occur in response to flight by the dominant and (3) by far most flights by dominants are *not* followed by attack by an inferior. Second, a secondary mirror effect is relevant to the interpretations only if both results (see above) are found. Anticipating upon the following analyses it may be stated that only arguments in favour of hypothesis (1) and none for the other hypothesis will be found. This outcome in its turn may serve as an argument against the very existence of a secondary mirror effect (see also p. 283).

Both strategies will be employed in analysing the data.

2. A second complication in comparing the data of the two species arises from the fact that hiding by the inferior apparently greatly reduces the aggressive activities of the dominant. Dominants are much more liable to attack an inferior fish which is swimming around freely than an inferior which is hiding. 1) Thus, even if *nigrofasciatus* inferiors hide more because

1) This is probably not due to a hiding inferior being actually hidden from the view

nigrofasciatus dominants are more aggressive (either by attacking more vigorously or by having spent more time attacking previous to hiding or both) one might still find lower aggressive scores in *nigrofasciatus* dominants, at least as far as the time spent attacking is concerned. Differences in the *intensity* of attacking would again be difficult to interprete due to the mirroring effect discussed above.

On the other hand, a hiding inferior may be considered as a more constant stimulus for the dominant than a moving one. At least differences in speed of fleeing are absent as long as the inferior keeps hiding and thus at least part of the mirroring effect is eliminated. Again, to take full advantage of this situation, the detailed interactions between the two fishes should be studied. This hasn't been done in the present observations. However, the observations as they were made might suggest another provisional argument in favour of hypothesis (1). If the hypothesis weren't true one would expect that the stronger hiding tendency of *nigrofasciatus* inferiors would result in lower aggressive scores (as expressed in time spent attacking) by *nigrofasciatus* dominants, even if only those observation periods of the two species are compared in which the inferior(s) hide continuously. If, on the other hand, it would be found that *nigrofasciatus* dominants in this situation spend more time attacking [1]) than do *conchonius* males, this could serve as an argument for hypothesis (1). It should be stressed, though, that this argument looses at least part of its strength through the following complication. Further, it will again be impossible to disprove hypothesis (1) in this way.

3. A third complication arises from the fact that the males of the two species differ in their colour patterns, mainly in the amount of black pigmentation. Thus, even if the mirroring effect is left out of consideration, the males of the two species are not presented with the same stimuli in the present observations. Anticipating upon some preliminary data to be discussed in the next chapter (p. 292) we might speculate on the influence the colour patterns might have had on the outcome of the present experiments. Experiments with a *nigrofasciatus* male which was induced to fight models with various combinations of black, grey and white vertical zones suggest that both the black body markings and the all-over black pattern found in *nigrofasciatus* act as a threat pattern on conspecific males. For the all-over

of the dominant; the vegetation in the tanks was, too sparse for that. Moreover, from the behaviour of the dominant it was often obvious that it 'knew perfectly well' in which corner an inferior was hiding.

1) The time spent attacking is not necessarily the best measure of the dominant's success in driving the inferior out of its hiding place. A better measure might be derived from the number of times the dominant succeeds in chasing the inferior from hiding position. This number, however, cannot be calculated from the available data.

black pattern this suggestion is indirectly supported by the observation that this pattern starts to appear whenever a male starts threatening and grows more intensive as a male spends more time threatening within a given period of time (*cf.* section II, p. 119).

From a comparison of the conditions under which the model experiments were done with the situations in the experimental groups described in the present chapter (see chapter IV, p. 292) it seems probable that under normal circumstances (*i.e.* in the large ambisexual groups and also in part of the 2♂♂ 2♀♀ groups) the black patterns indeed act as threat patterns and elicit escape in conspecific males in the same way as do threat postures (see Dunham *et al.,* 1968). If this is true it is also very likely that in the present experiments the dominant male's black patterns have a similar effect on the behaviour of the inferior(s). Since *nigrofasciatus* males have more black patterns than have *conchonius* males this effect alone could account for at least part of the difference in individual distance found between the two species in the present experiments; *a fortiori* if the inferior's marking patterns would have a similar effect upon the dominant.

On the other hand, if in spite of the threatening effect of the black markings inferior *nigrofasciatus* males would still be found to be more aggressive than *conchonius* inferiors, the argument in favour of hypothesis (1) would be considerably strengthened. At the same time, the presence of a stronger threat pattern in *nigrofasciatus* inferiors — as compared with *conchonius* inferiors — would seem to weaken any argument for hypothesis (2) even if *nigrofasciatus* dominants would be found to show higher fleeing tendencies than *conchonius* dominants. However, the situation is probably not even as simple as that. Dunham *et al.* (1968, p. 24) have described how in *B. stoliczkanus* the threat posture, which elicits mainly escape in 'natural' situations, may instead elicit attack if it is performed by an inferior male in an established 2♂♂ group. This phenomenon is typical not only for *B. stoliczkanus* but for the other 4 species as well. Especially in the 3 supposedly more aggressive species: *B. nigrofasciatus, B. tetrazona* and *B. cumingi* it seems to be the rule rather than an incidental occurrence, both in 2♂♂ and in 3♂♂ groups. It is quite possible that threat patterns have a similar dual effect as have threat postures. *A fortiori*: if the assumption of the threat posture (*i.e.* presenting a maximal area to the opponent *plus* optimal display of the conspicuous markings) by a cornered inferior elicits attack rather than retreat in a dominant male, it is highly unlikely that the same inferior's marking patterns alone will keep the dominant from attacking, let alone induce him to flee.

We may conclude from this that, if *nigrofasciatus* dominants could be shown to flee more readily than *conchonius* dominants, this would stand out as a strong argument for hypothesis (2) in spite of the differences between the species in colour markings. On the other hand, if threat patterns — like threat postures — could elicit attack instead of escape when displayed by an inferior fish, this would weaken any argument in favour of hypothesis (1) which could be derived from the aggressive behaviour of the dominant fishes even if only those periods are compared in which the inferior fish hides continuously. However this may be, I think that the latter differences are best taken at their face value as long as no definite data are available to assess whether threat patterns may actually elicit attack or not.

To conclude these remarks we may ask whether the complications discussed here may in any way impair the conclusions reached in the fore-going two sections. I think this influence is negligible at least as far as the differences in degree of aggressiveness between *conchonius* and *nigrofasciatus* and between *stoliczkanus* and *tetrazona* are concerned. Neither pronounced inferiority nor hiding did occur in the 6♂♂ 6♀♀ groups of any of the species; so the mirroring effect must, if present, have been negligible in these situations. Nor can hiding have influenced the aggressive scores of the males in this situation. The same is true for the 2♂♂ 2♀♀ experiments of *conchonius* and *nigrofasciatus*. Only in the 2♂♂ 2♀♀ experiments of *stoliczkanus, tetrazona* and *cumingi* pronounced inferiority of one male did occur and this male often showed hiding (see section III). However, in these experiments it was found that *tetrazona* dominants were in general much more aggressive than *stoliczkanus* dominants, also towards their respective inferior males, even though *tetrazona* inferiors hid more consistently than did *stoliczkanus* inferiors (see section III, p. 184).

Likewise, it is improbable that the differences in colour patterns may impair the conclusions; both in the *conchonius-nigrofasciatus* and in the *stoliczkanus-tetrazona* pair it was the species carrying most black patterns which was found to be the most aggressive. Only in the 2♂♂ 2♀♀ experiments of *stoliczkanus* and *tetrazona,* because of the strong dominance of one of the males, the possible dual effect of the black marking patterns discussed above might have played a rôle. However, as stated before, the possibility of an aggression-provoking effect of the black marking patterns can better be left out of consideration as long as there are no direct data to test it.

A different question is whether the qualitative differences in colour patterns (sexual dichromatism, definition of contrast) may have been responsible for the qualitative differences in behaviour (attacking the courted female, territoriality) found to exist between *conchonius-nigrofasciatus, stoliczkanus-tetrazona* and *cumingi.* Lacking direct data on the effects of such qualitative differences of colour patterns it is impossible to exclude this possibility, though I don't think it to be very probable either. This point will be further discussed in the next chapter.

Summarizing from the above discussion: in order to decide between the two hypotheses on the basis of the quantitative data, in the following paragraphs we will look at 1. aggressive scores of dominants with hiding inferiors, 2. aggressive scores of inferiors and 3. fleeing scores of dominants. Some qualitative data will be added.

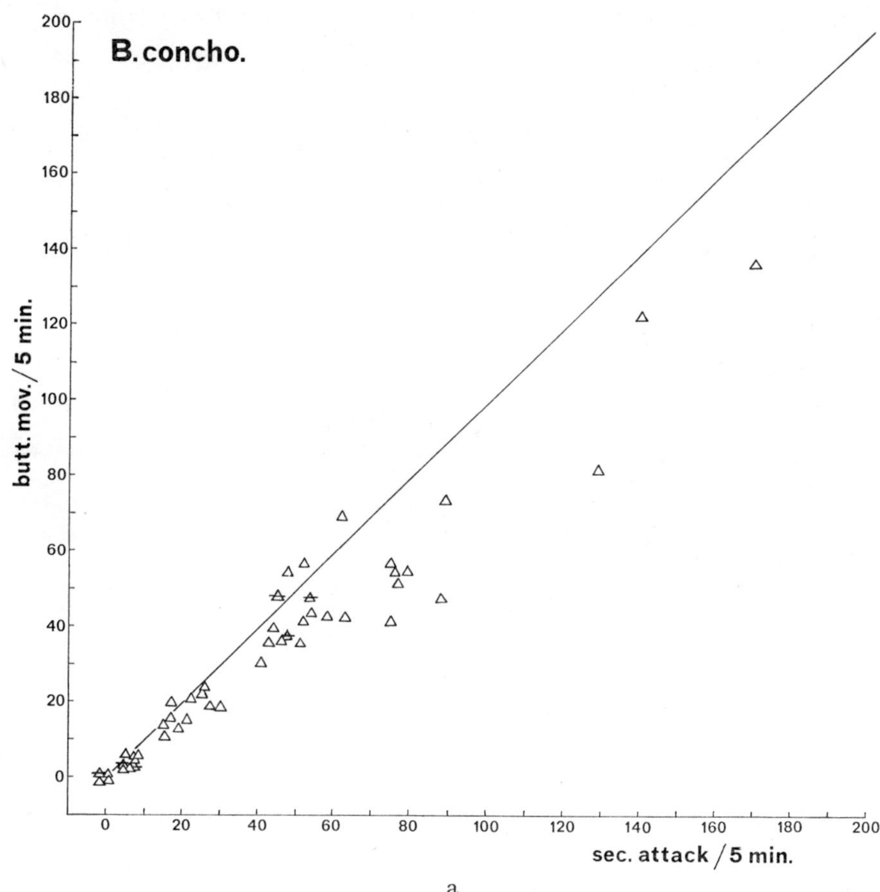

a

Fig. 55[a-e]. *B. conchonius, B. nigrofasciatus*; aggressive scorings of dominant males in 3 ♂♂ (triangles) and 2 ♂♂ (circles) groups; 45° diagonal for reference; a, b, d, e: a.l. series; c: n.l. series *conchonius*; periods with established dominance only; further explanation see text.

Possible arguments for hypothesis (1).

Behaviour of the dominants. Figg. 55[a-e] represent — for the 2♂♂ and 3♂♂ experiments separately — the time spent in full attack and the number of butting movements (bites, butts, intention butts, taps, intention taps all added together) by the dominants of the two species in all 5′ periods. For the 3♂♂ experiments the activities of the dominant directed at both inferiors have been added together [1]).

1) The symbols used in these and the following graphs are the same as used in table 8.

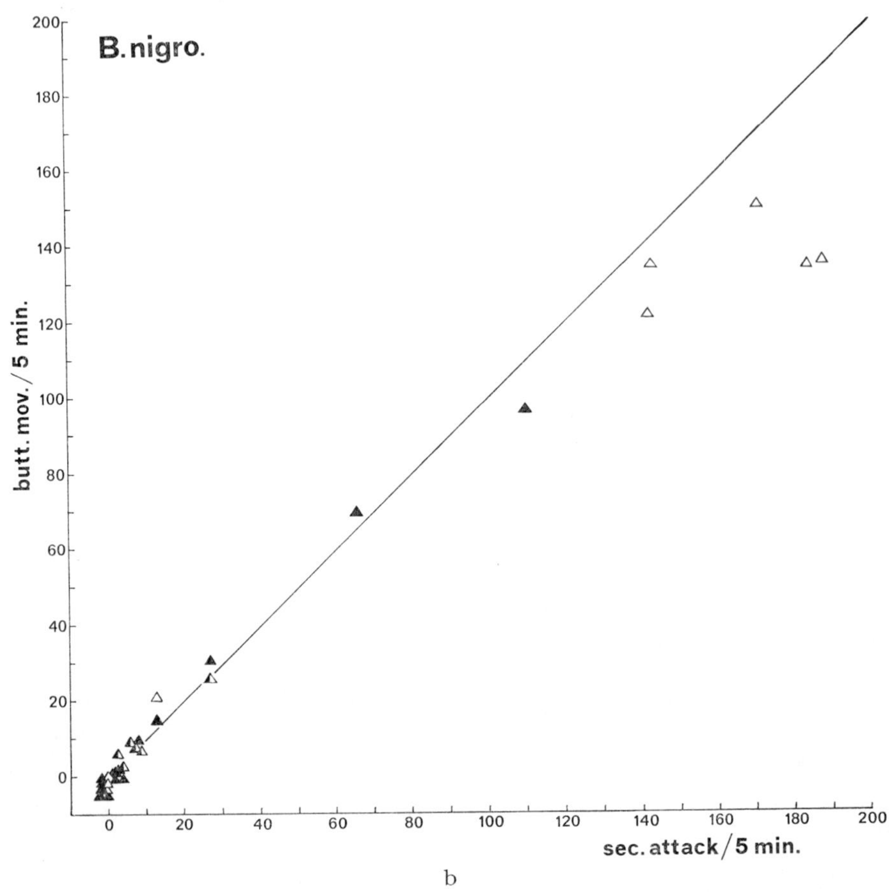

Legends fig. 55 see p. 205

The data from the two *conchonius* 3♂♂ n.l. experiments are given separately (fig. 55^c); the data from the two *nigrofasciatus* 3♂♂ experiments in which no full dominance was established (1 a.l. and 1 n.l. series) have been omitted from the graphs.

If we compare only the data from the artificial light experiments and only those periods in which dominance was clearly established, the following trends are apparent: (a) for both species there is a close positive correlation between the two measures; the frequency of butting movements (per seconds attack) is practically independent from the absolute amounts of

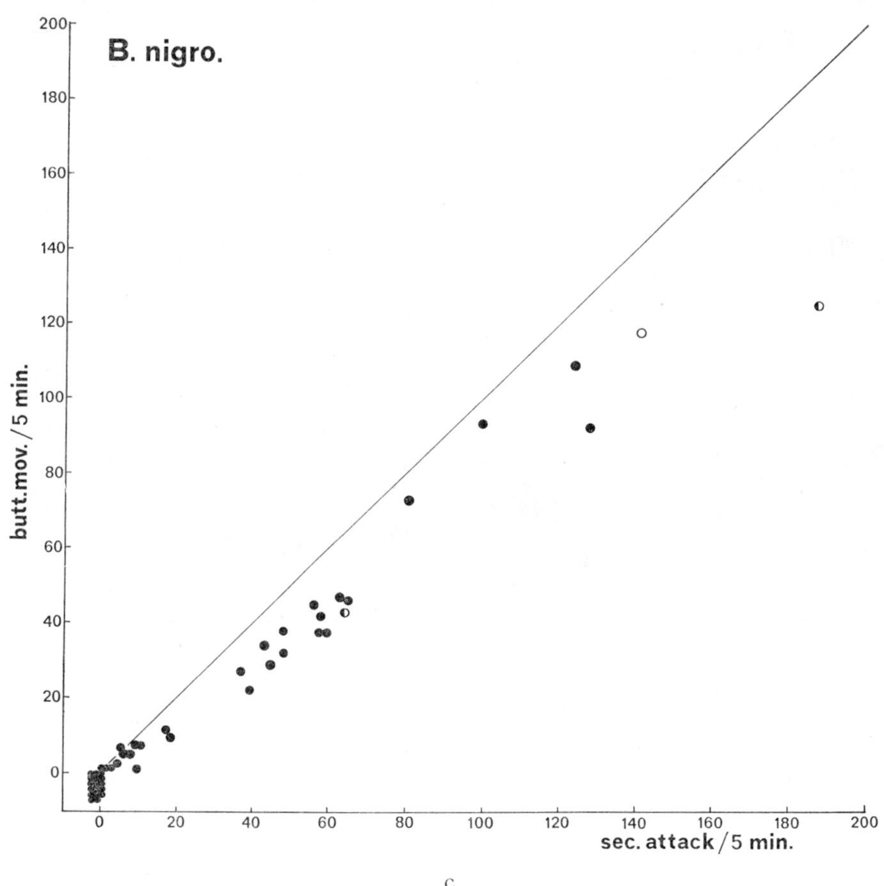

c

Legends fig. 55 see p. 205

both. (b) there is no obvious difference between the species in the frequency of butting movements. (c) there is no obvious difference between the 2♂♂ and 3♂♂ dominants in the frequency of butting movements. (d) in both measures *nigrofasciatus* dominants tend to score lower than do *conchonius* dominants. This is most clear in the 3♂♂ experiments. That this difference between the species is due to the much stronger hiding tendencies of *nigrofasciatus* inferiors is suggested by the fact that (e) within each species the aggressive scores of the dominants with hiding inferiors are on the average considerably lower than those of dominants with free inferiors. This is

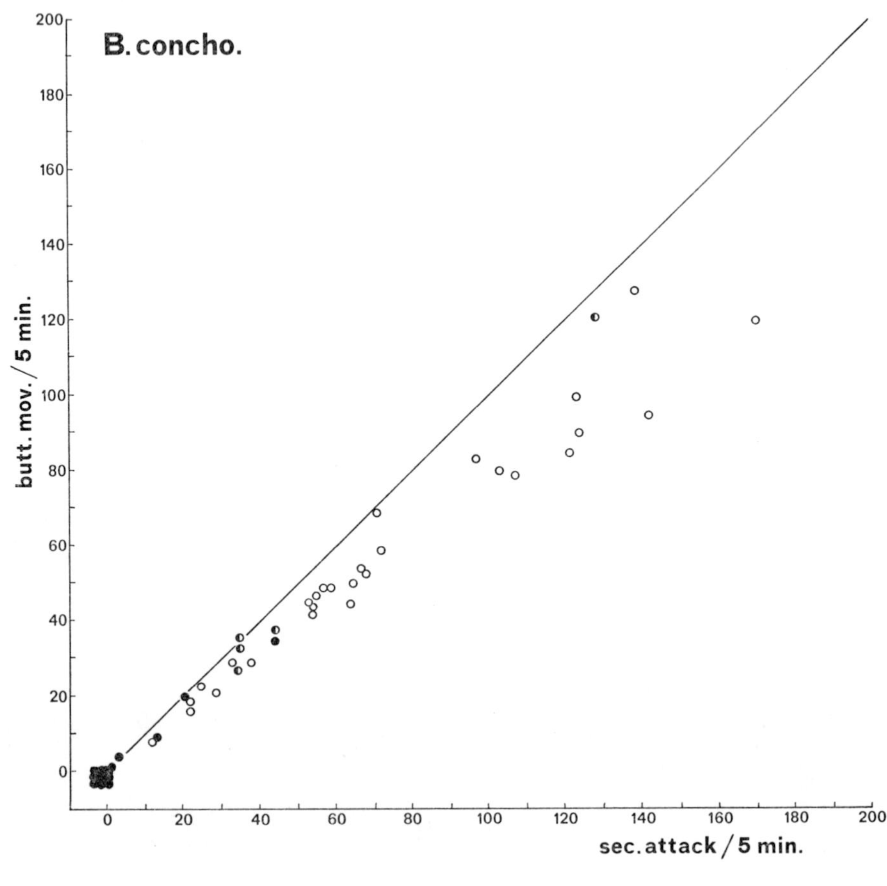

d

Legends fig. 55 see p. 205.

especially clear in the 2♂♂ experiments of *conchonius* and in the 3♂♂ experiments of *nigrofasciatus*. (The *conchonius* 3♂♂ experiments and the *nigrofasciatus* 2♂♂ experiments do not allow such a comparison because *conchonius* 3♂♂ inferiors practically never hide and *nigrofasciatus* 2♂♂ inferiors are rarely free). (f) if only the periods are compared in which the inferior(s) hide continuously (the comparison is possible for the 2♂♂ groups only) the aggressive scores of the *nigrofasciatus* dominants tend to be higher than those of *conchonius* dominants: in comparing graphs 55[d] and 55[e] we find that *conchonius* dominants scored more than 50″ attack per 5′

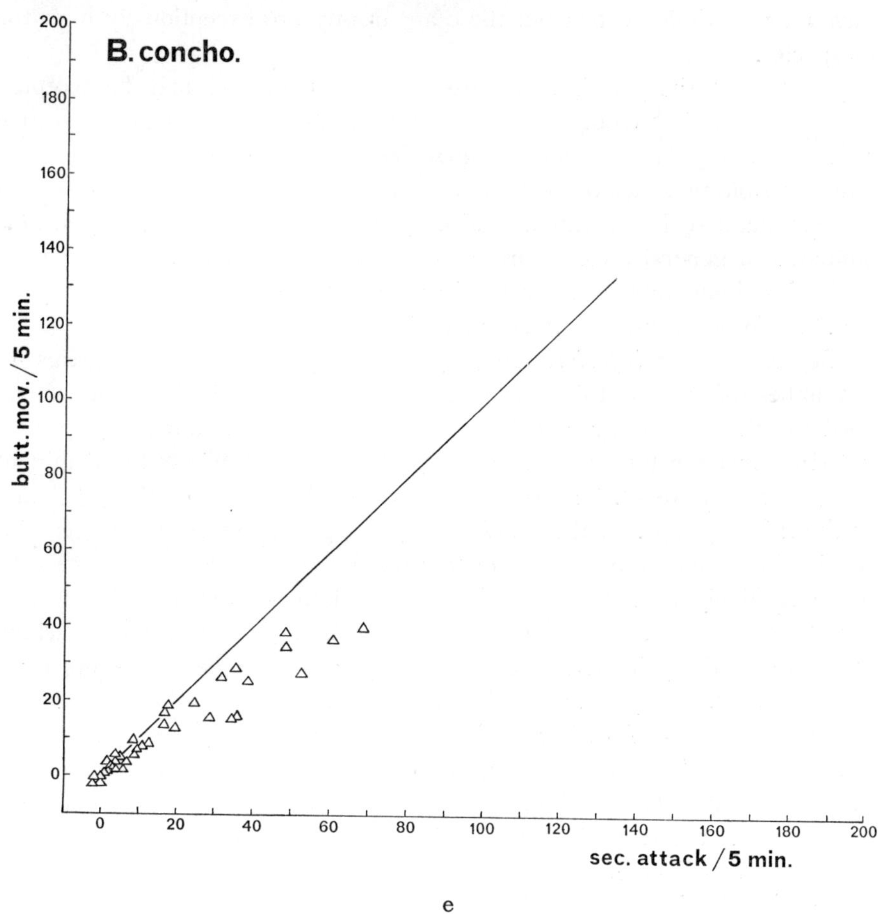

e

Legends fig. 55 see p. 205.

in none of the 14 periods in which the inferior hid continuously; in 9 out of the 14 cases the dominant scored zero for both measures. On the other hand, *nigrofasciatus* dominants with hiding inferiors scored more than 50″ attack per 5′ in 12 out of 45 cases; they scored zero for both measures in only 17 out of the 45 cases. Similar outcomes are obtained if butting scores are compared or if those periods in which the inferior hid for part of the time only are included. (g) in the one *nigrofasciatus* 3♂♂ experiment in which the inferiors remained free, the dominant scored very high values for both aggressive measures in 5 out of 6 cases. Comparable values do occur in *conchonius* 3♂♂ dominants but these apparently are exceptionally high for this species.

According to the principles discussed above, point (f) may be accepted as an argument in favour of hypothesis (1). Point (g), though suggestive of the same hypothesis cannot be qualified as an argument for it because of the possible presence of the mirroring effect discussed above.

The unquantified observation — already stated above — that *nigrofasciatus* dominants in general attack more vigorously and bite more fiercely than do *conchonius* dominants, even if the inferior fish doesn't flee fast, may be taken as another argument in favour of hypothesis (1).

Behaviour of the inferiors. Figg. 56[a-d] show — for the two species — the number of attacks (full attacks, distance attacks and inhibited attacks all added together) made by each inferior fish on the dominant per 5′ period, plotted against the total number of seconds spent fleeing for the dominant in the same periods [1]). The results are rather different for the 3♂♂ and for the 2♂♂ groups. In the *conchonius* 3♂♂ groups inferiors may exhibit considerable numbers of attacks on the dominant, but scores of more than 10 are relatively rare and practically confined to periods in which fleeing does not exceed 50″. In contrast, in *nigrofasciatus* 3♂♂ inferiors attacking scores are relatively higher, especially if only open triangles (inferiors free) are considered. In spite of the lower number of observation periods the highest scores found for *nigrofasciatus* exceed those of *conchonius* and these high scores go with much higher values for fleeing. It should be mentioned that all *nigrofasciatus* attacking scores higher than 10 (except the one half-black triangle) are from the two inferiors of the same group (5 points from each of the two fishes).

The *nigrofasciatus* graph further suggests that aggressive scores of inferiors which hide continuously are low, both relative to non-hiding inferiors of the same species and to (non-hiding) *conchonius* inferiors. Many more data would be needed to assess whether the aggressive scores of free inferiors of the two species differ at low levels of fleeing.

The apparent positive correlation between the two measures in the *nigrofasciatus* graph is no doubt partly due to the circadian synchronization of the activities of all males of a group (also in *conchonius* there often is a positive correlation between the two measures within single experimental series) and probably partly to a common correlation of the two measures with hiding.

---

1) In these analyses in general frequencies are used as parameters for behaviour patterns which are on the average short, for others total durations (in seconds per 5 minutes) are used.

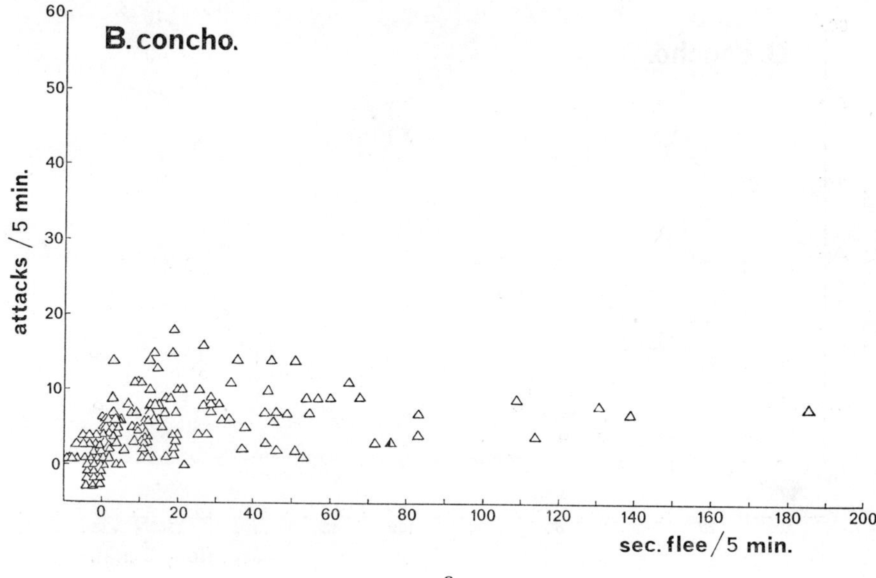

a

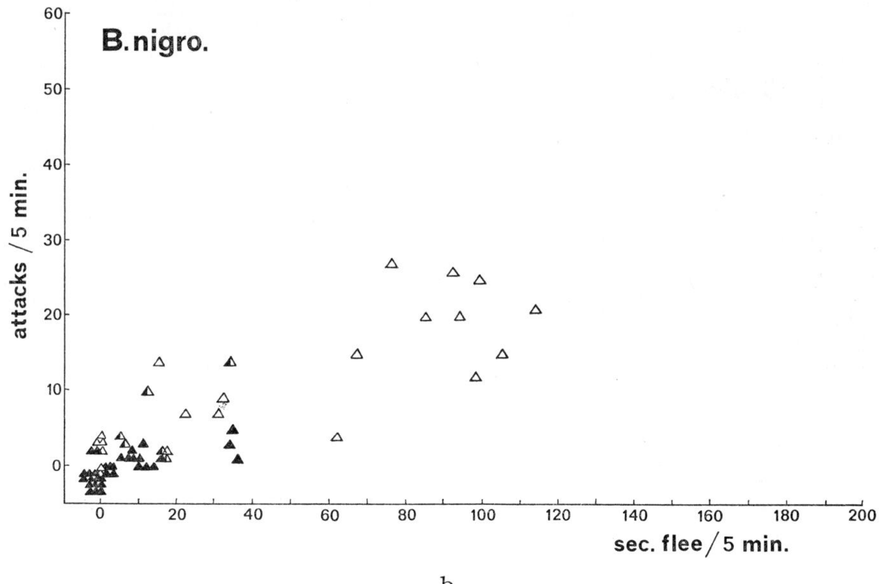

b

Fig. 56[a-d]. *B. conchonius, B. nigrofasciatus*; fleeing and attack scores of inferior males in 3♂♂ (triangles) and 2♂♂ (circles) groups; n.l. and a.l. series combined; periods with established dominance only.

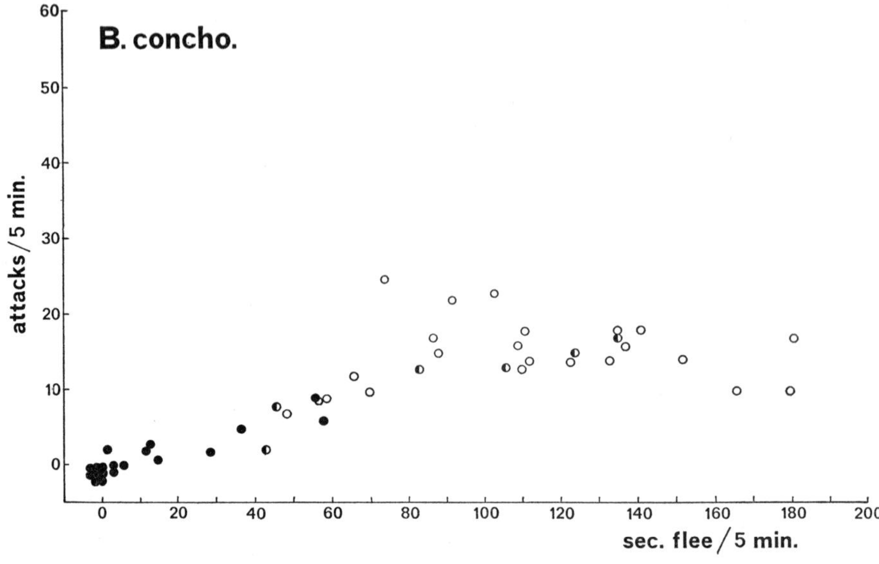

c

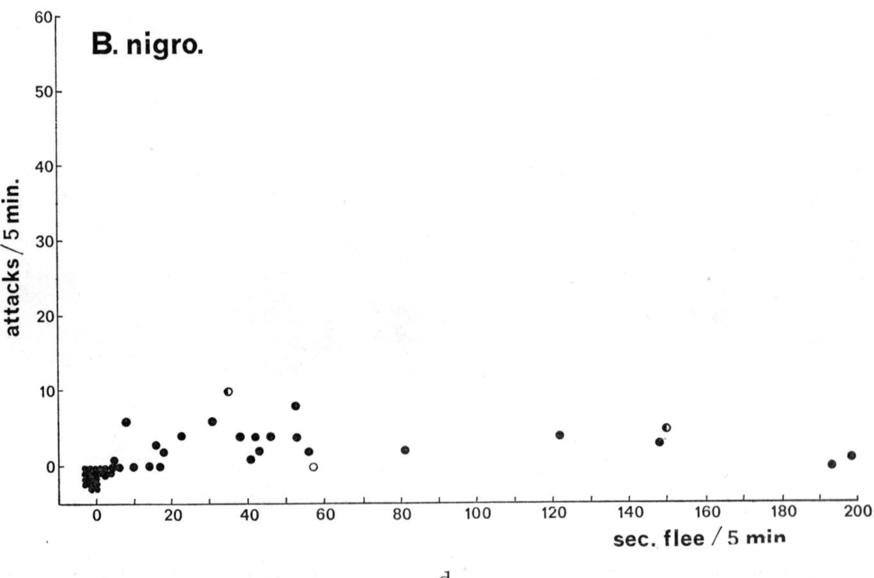

d

Legends fig. 56 see p. 211.

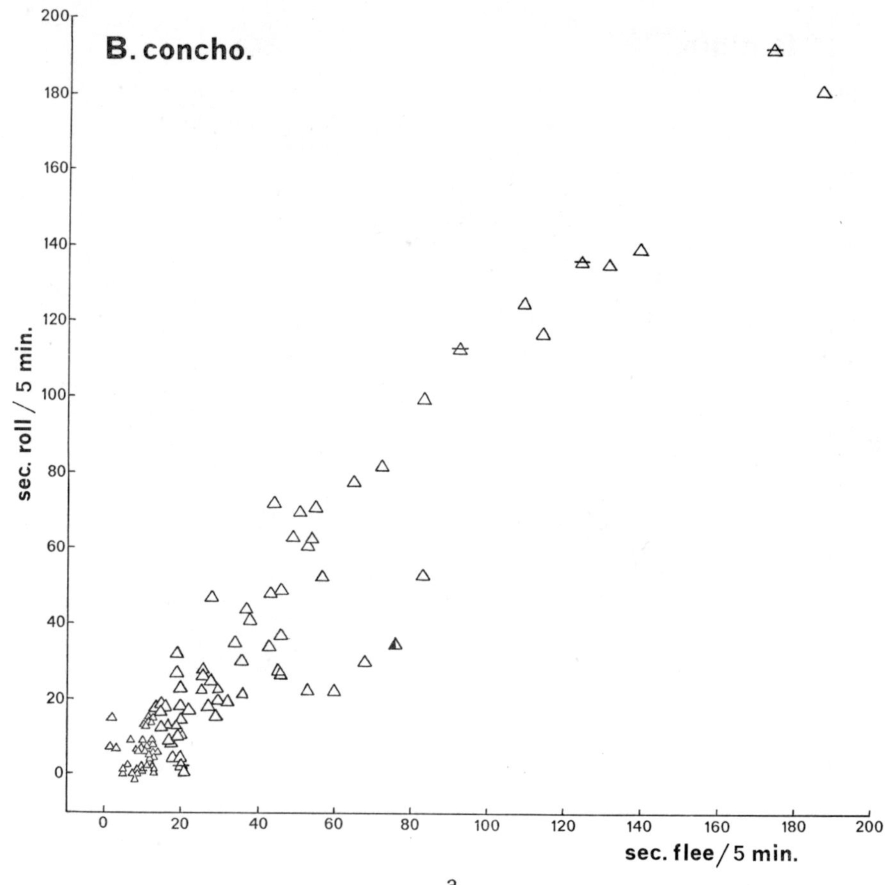

a

Fig. 57a-d. *B. conchonius, B. nigrofasciatus*; fleeing and (total) roll scores of inferior males in 3 ♂♂ (triangles) and 2 ♂♂ (circles) groups; n.l. and a.l. series combined; periods with established dominance only.

The 2♂♂ experiments yield a quite different picture. Attacking scores of free *conchonius* 2♂♂ inferiors are quite high, considerably higher even than those of 3♂♂ inferiors of the same species. (It must be noted that the attacking scores of the 3♂♂ inferiors would be higher if the attacks on the other inferior were included but this wouldn't make up for the difference with the 2♂♂ inferiors). The apparent positive correlation between attacking and fleeing in the 2♂♂ *conchonius* graph is probably largely due to a common correlation of the two measures with the hiding tendency.

In contrast, attacking scores of *nigrofasciatus* 2♂♂ inferiors are always rather low, at least compared to the free *conchonius* inferiors showing equal

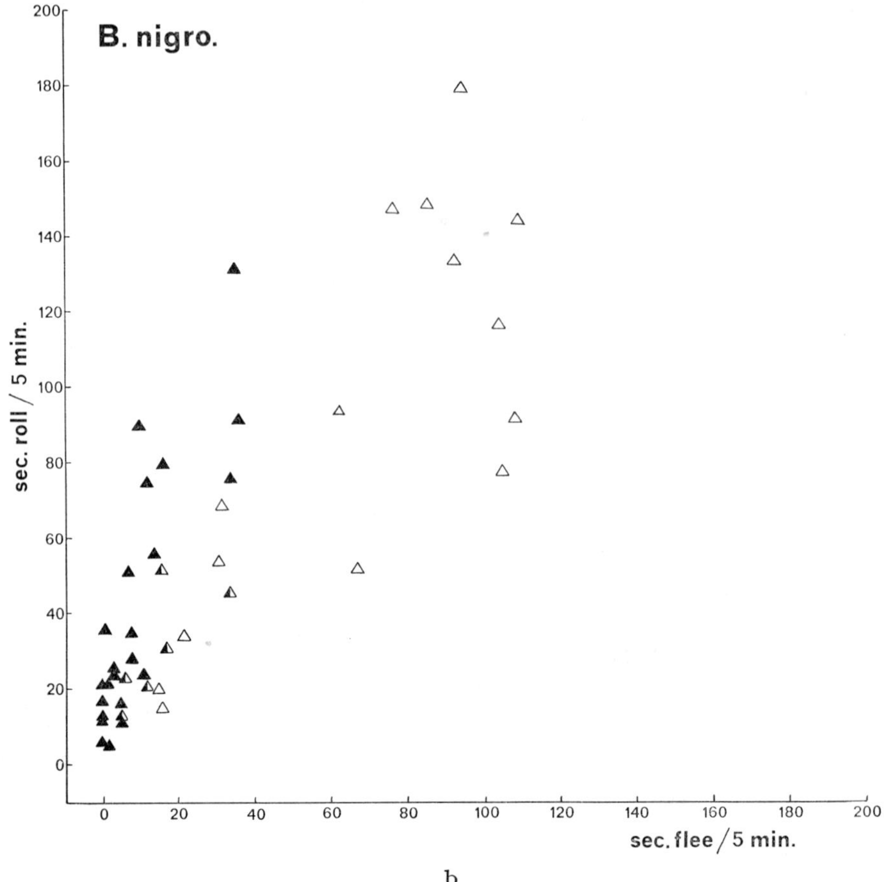

b

Legends fig. 57 see p. 213.

amounts of fleeing. That this difference between the species is probably due to the more complete submission of the *nigrofasciatus* inferiors is suggested by the fact that *nigrofasciatus* 2♂♂ inferiors hide much more than do *conchonius* 2♂♂ inferiors and by the fact that attacking scores of *nigrofasciatus* 2♂♂ inferiors are especially low at high values for fleeing. The attacking scores of continuously hiding inferiors of both species are low and do not seem to be obviously different in the two species.

According to the principles discussed above, the fact that in the 3♂♂ situation free *nigrofasciatus* inferiors may attack more than do free *conchonius* inferiors, may be taken as a further argument in favour of hypo-

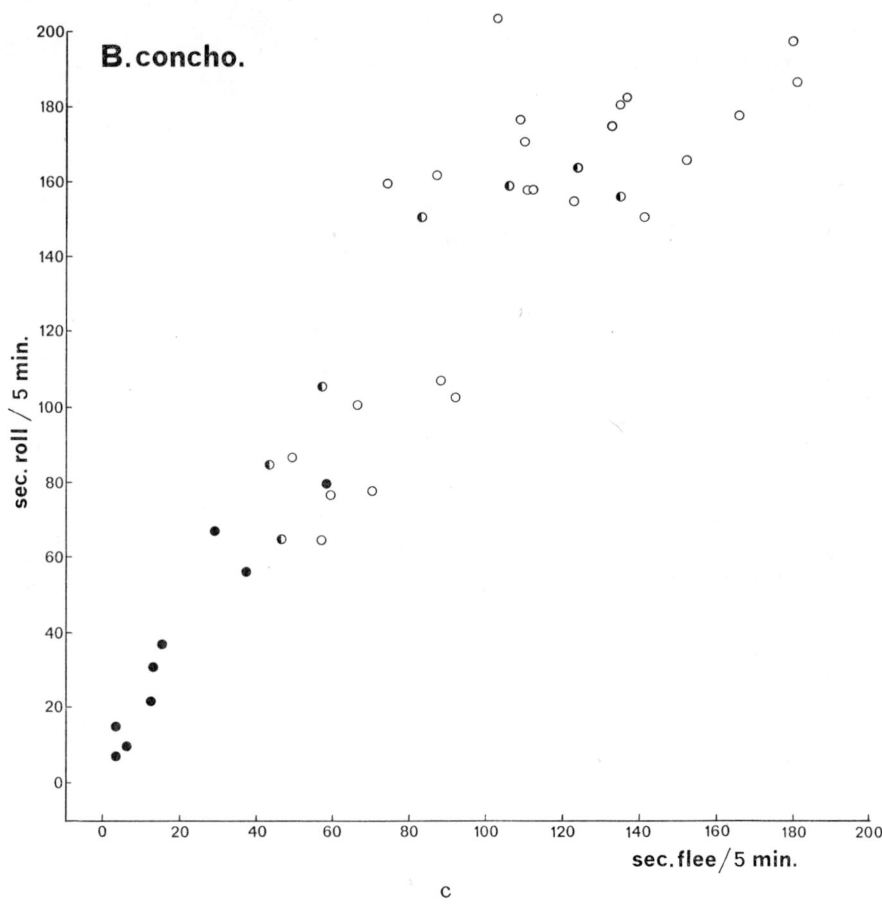

c

Legends fig. 57 see p. 213.

thesis (1). None of the other features of inferior behaviour presented so far either supports or contradicts the hypothesis.

Another, more indirect, argument may be derived from the amounts of ventral and dorsal rolling displayed by the inferiors in response to the attacks of the dominant. It was argued in chapter II (p. 94) that both types of rolling are partly aggressively motivated, dorsal rolling more so than ventral rolling. Therefore, the strength of the aggressive motivation of inferiors may be measured, besides by counting overt attacks, by assessing the amounts and kind of rolling shown, especially relative to the amount of fleeing. Figg. 57[a-d] show — for the two species — the time spent rolling

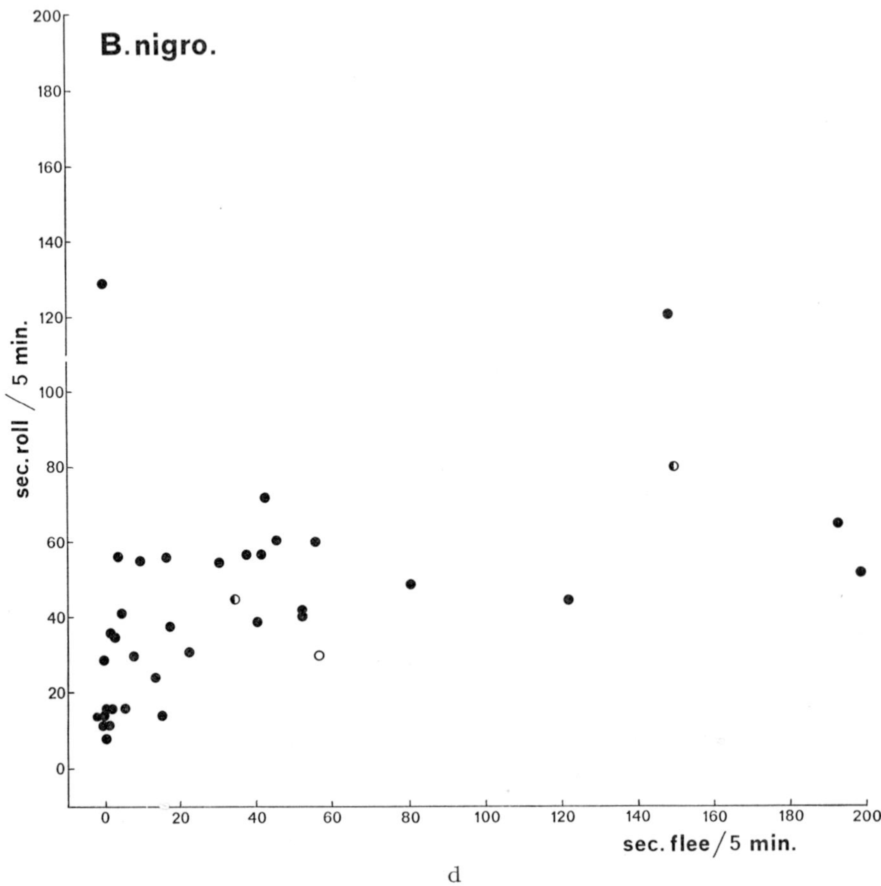

d

Legends fig. 57 see p. 213.

(ventral *plus* dorsal) and fleeing (for the dominant) by the inferiors of the 3♂♂ and the 2♂♂ groups for the dominants. The outcomes are again somewhat different for the 3♂♂ and for the 2♂♂ inferiors. For the 3♂♂ groups the graphs show that *nigrofasciatus* inferiors tend to exhibit more rolling than do *conchonius* inferiors at comparable levels of fleeing. The difference is most obvious for low levels of fleeing (or for hiding inferiors; these two categories cannot very well be distinguished in the present data). However, also in the higher categories of fleeing, most of the free *nigrofasciatus* inferiors roll more than do *conchonius* inferiors — though in a few periods the free *nigrofasciatus* inferiors do display relatively little rolling.

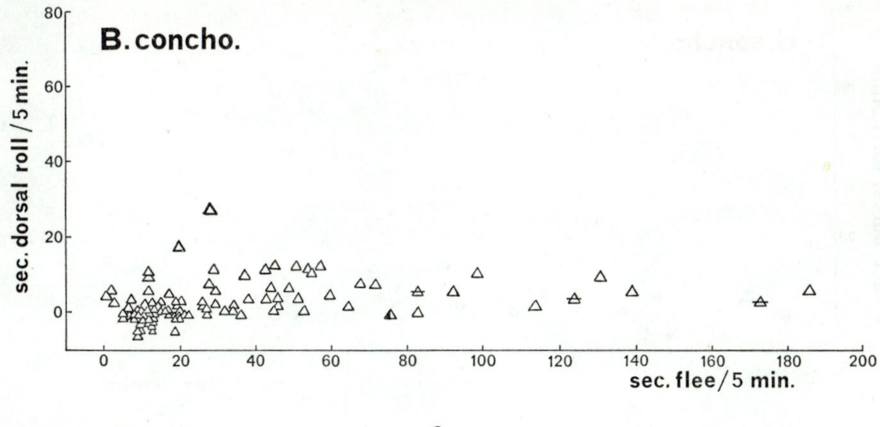

a

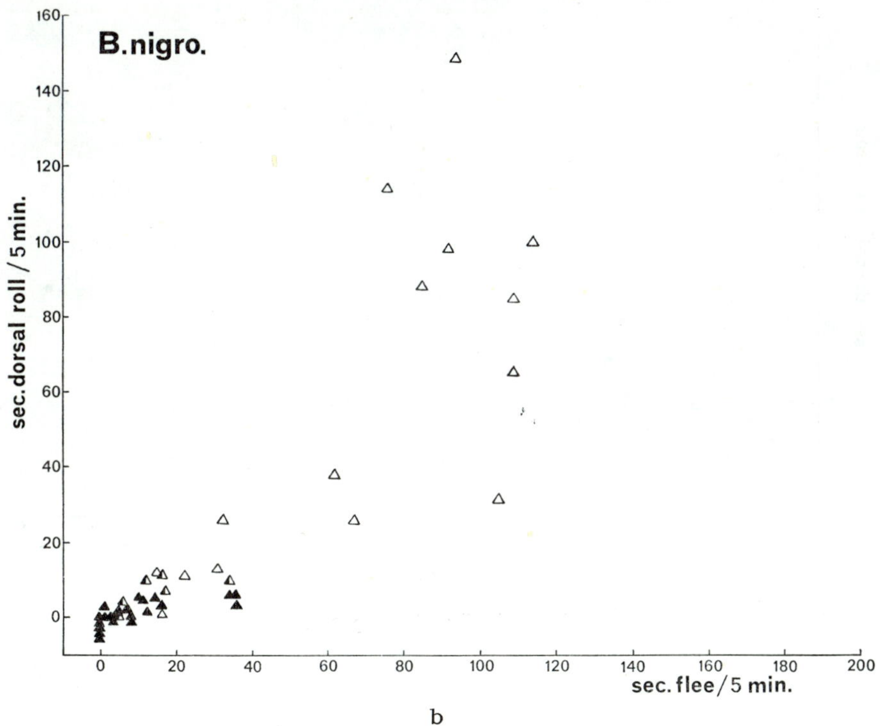

b

Fig. 58a-d. *B. conchonius, B. nigrofasciatus*; fleeing and dorsal roll scores of inferior males in 3 ♂♂ (triangles) and 2 ♂♂ (circles) groups; n.l. and a.l. series combined; periods with established dominance only.

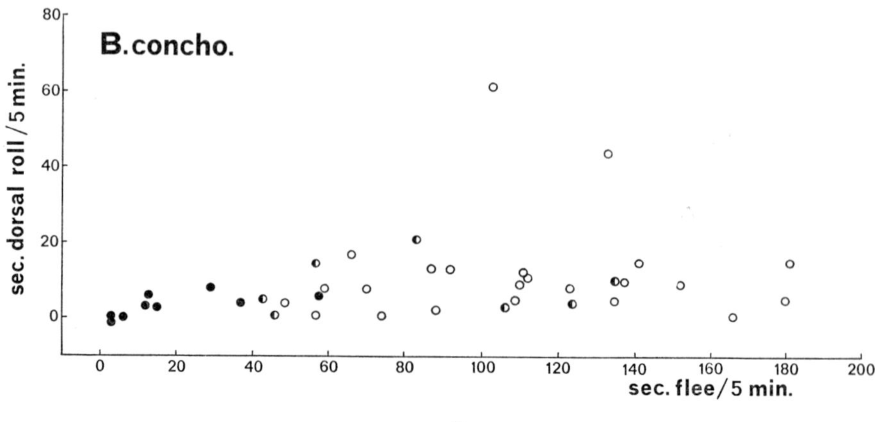

c

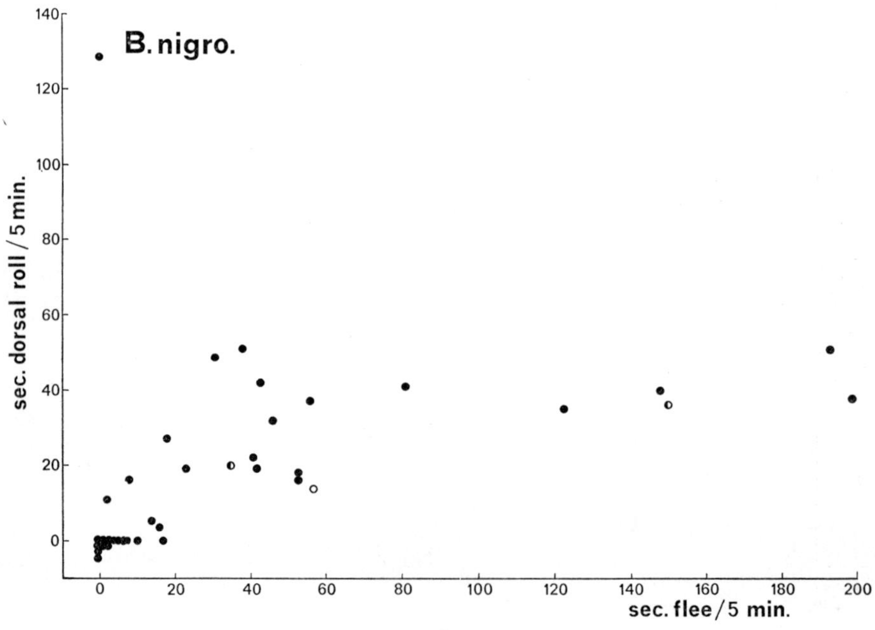

d

Legends fig. 58 see p. 217.

In the 2♂♂ groups *conchonius* inferiors display a considerable amount of rolling, even more — relative to the amount of fleeing — than do 3♂♂ inferiors of the same species. In contrast, the *nigrofasciatus* inferiors show much less rolling at high levels of fleeing. This difference is probably due to *nigrofasciatus* inferiors being more completely submitted: almost all *nigrofasciatus* inferiors in this category of fleeing hide continuously, whereas most of the *conchonius* inferiors in the same category of fleeing do not.

On the other hand, at low levels of fleeing, *nigrofasciatus* inferiors reach much higher values for rolling than do *conchonius* inferiors. This is so in spite of the fact that in this category of fleeing the *nigrofasciatus* inferiors hide at least as much as do *conchonius* inferiors.

If rolling is taken as an indication of aggressive motivation, we may conclude that, even in a 2♂♂ situation, inferior *nigrofasciatus* are more strongly aggressively motivated than are *conchonius* inferiors, provided that they do not flee too much. The same holds for all categories of fleeing (though again especially for the lower categories) in the 3♂♂ inferiors.

Even more pregnant are the differences in the amount of dorsal rolling alone shown by the inferiors of the two species. Figg. $58^{a\text{-}d}$ represent the amounts of dorsal roll shown by the inferiors, toward the dominant, plotted against the amounts of fleeing for the dominant, again per 5′. The graphs show that dorsal roll is much more common in *nigrofasciatus* than in *conchonius* inferiors in almost all situations, even in the 2♂♂ situation in which the inferior males are most completely submitted; (*i.e.* in a situation in which pure fleeing and hiding are the predominant activities, overt attacking and even the sum of dorsal and ventral rolling being very low).

The frequent occurrence of the more aggressive variant of rolling in the inferiors of the species in which inferiors tend to be more completely submitted is another strong argument in favour of hypothesis (1).

It may be noted that there is a remarkable parallellism between the graphs showing the frequencies of overt attack and those showing the amounts of rolling. For instance *cf* figg. 56^a and 56^b with 58^a and 58^b and figg. 56^c and 56^d with 57^c and 57^d.

If, as is suggested in the fore-going paragraphs, aggressive motivation may be expressed not only in overt attack but also in ambivalent behaviour patterns, one more activity of the inferiors should be taken into consideration: threat. Figg. $59^{a\text{-}d}$ show the frequencies of threat by the inferior fishes towards the dominant per 5′, plotted against the amount of fleeing. It follows from the graphs that in all situations threat is less common in *nigrofasciatus* than in *conchonius* inferiors. Apparently, in inferior *nigrofasciatus* the occurrence of threat is almost completely suppressed, probably as a result

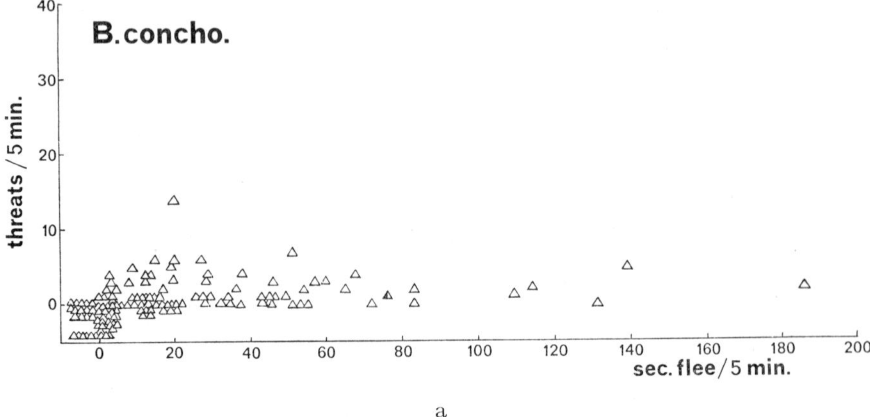

a

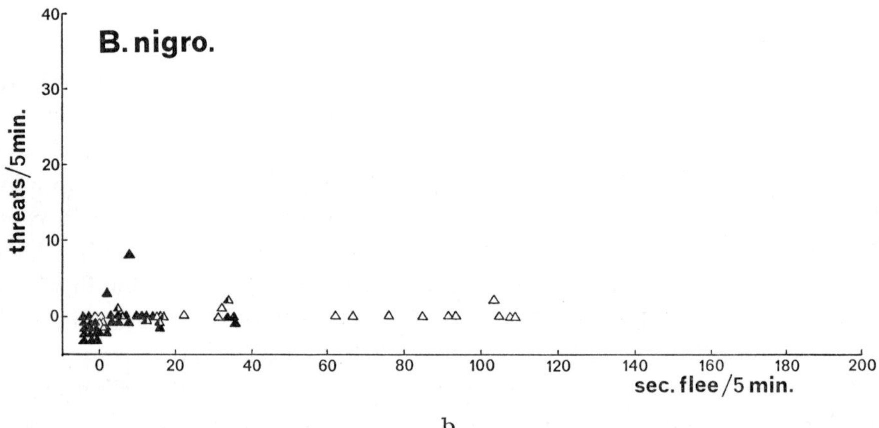

b

Fig. 59a-d. *B. conchonius, B. nigrofasciatus*; fleeing and threat (frequency) scores of inferior males in 3♂♂ (triangles) and 2♂♂ (circles) groups; n.l. and a.l. series combined; periods with established dominance only.

of their more complete submission as compared with *conchonius* inferiors. The question why the supposedly higher aggressive tendencies of *nigrofasciatus* inferiors — as compared with conchonius inferiors — are expressed in the form of rolling and (sometimes) in overt attack and not in threat will be discussed in a future paper (Kortmulder, in prep.). The phenomenon is probably related to the relative independence of threat from the aggressive and fleeing motivations (see chapter II, sub (b) and (e), and p. 260 of this chapter).

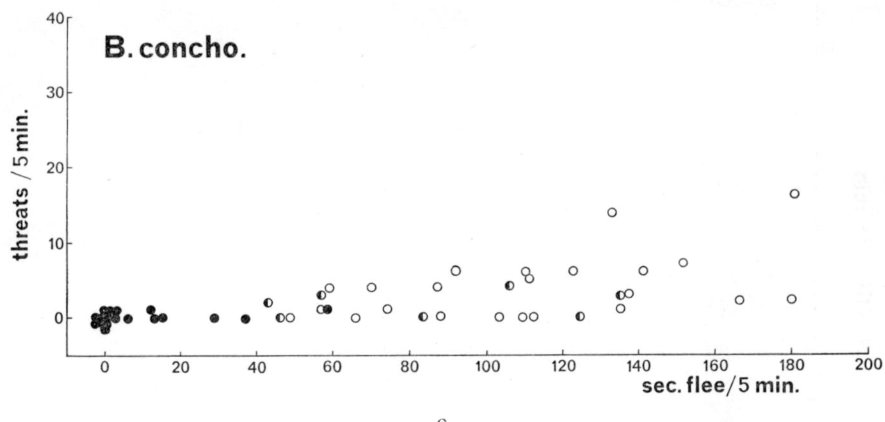

c

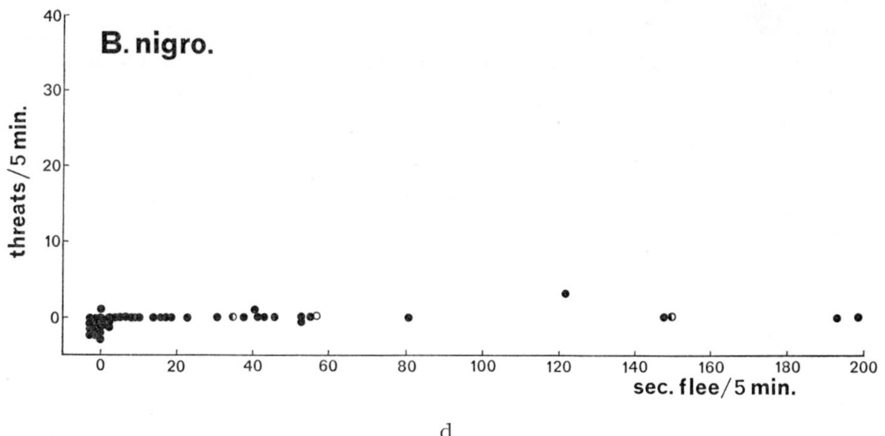

d

Possible arguments for hypothesis (2).

Behaviour of the dominants. Figg. 60[a-d] show — for the two species — the number of times the dominant flees for the inferior(s) plotted against the time spent in (full) attack 1). Neither in the 3♂♂ nor in the 2♂♂ situation there is any obvious difference between the species in the frequencies of fleeing. No argument in favour of hypothesis (2) may be found here.

Just like the aggressive motivation of the inferiors could be supposed to

1) Again, for the 3♂♂ situations, the activities of the dominant towards both inferiors have been added together.

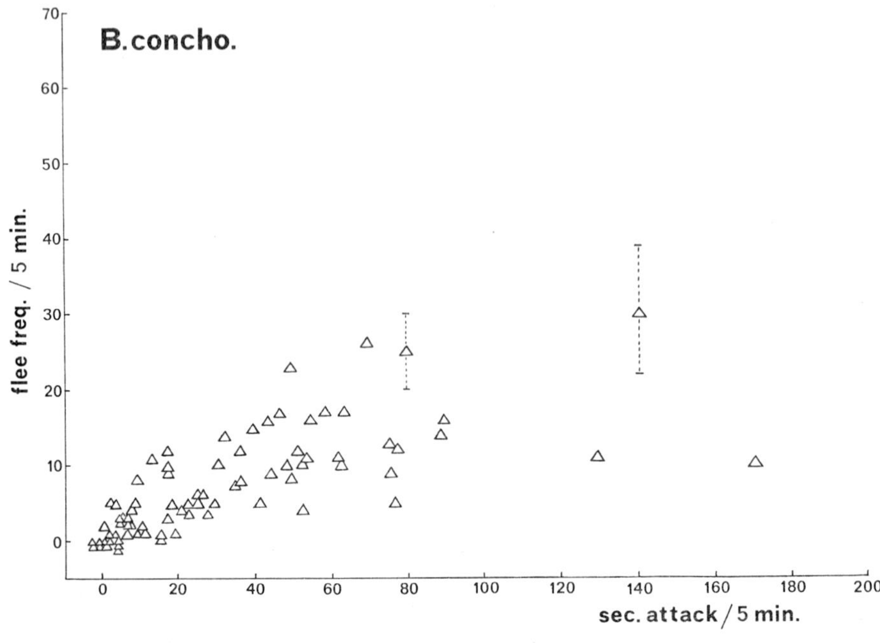

a

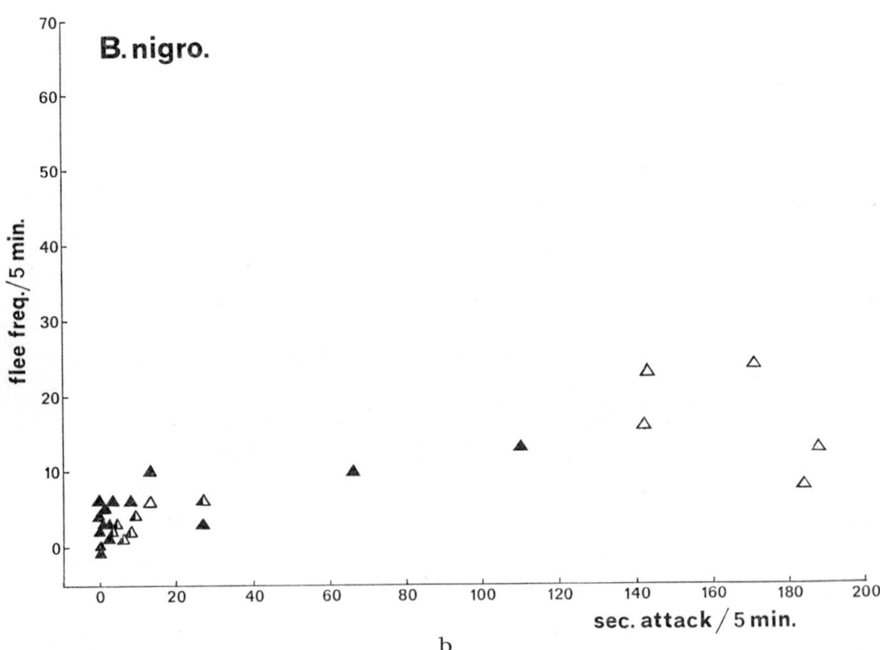

b

Fig. 60a-d. *B. conchonius, B. nigrofasciatus*; attacking and flee (frequency) scores of dominant males in 3♂♂ and 2♂♂ groups; n.l. and a.l. series combined; periods with established dominance only.

B. concho.
flee freq./5 min.
sec. attack / 5 min.
70
60
50
40
30
20
10
0
0
20
40
60
80
100
120
140
160
180
200

c

B. nigro.
flee freq./5 min.
sec. attack / 5 min.
70
60
50
40
30
20
10
0
0
20
40
60
80
100
120
140
160
180
200

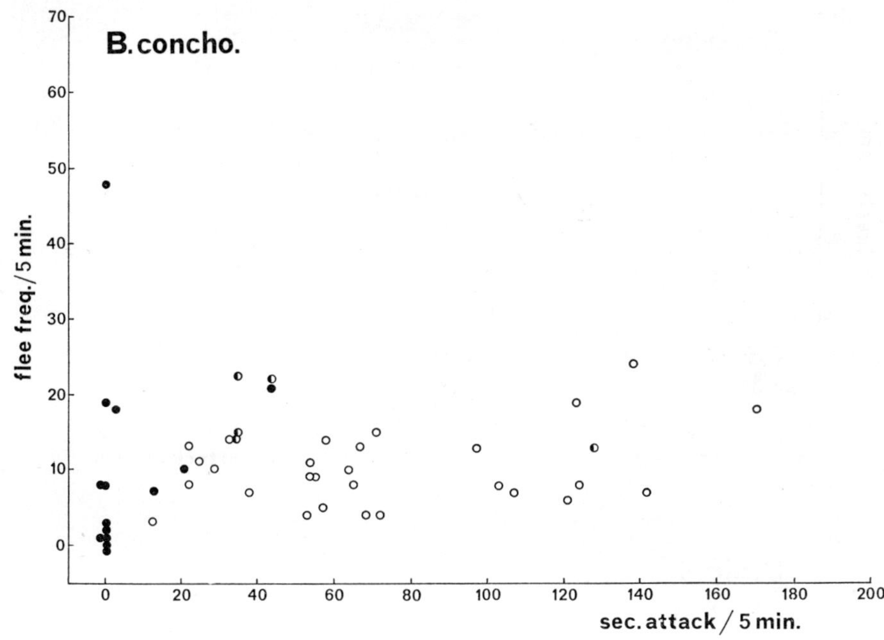

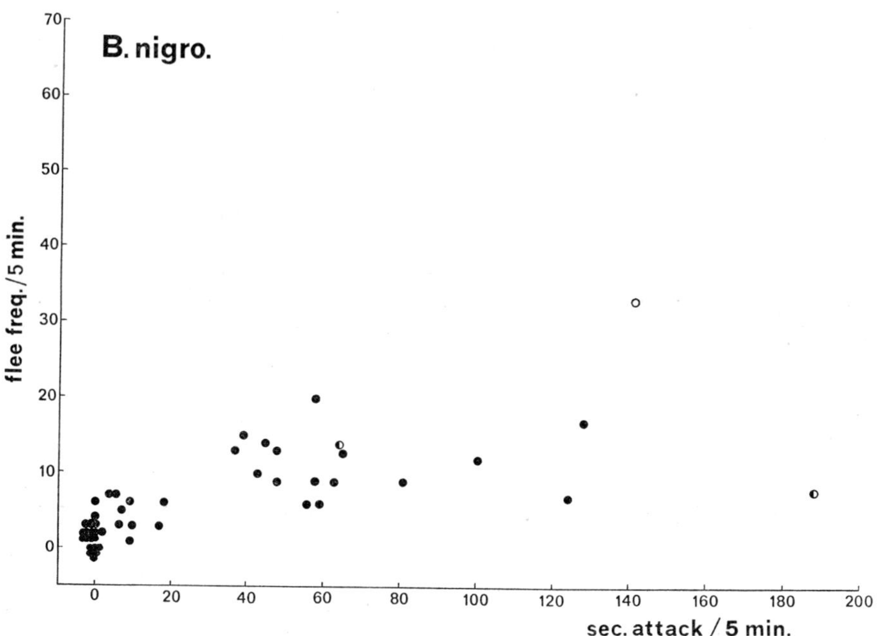

d

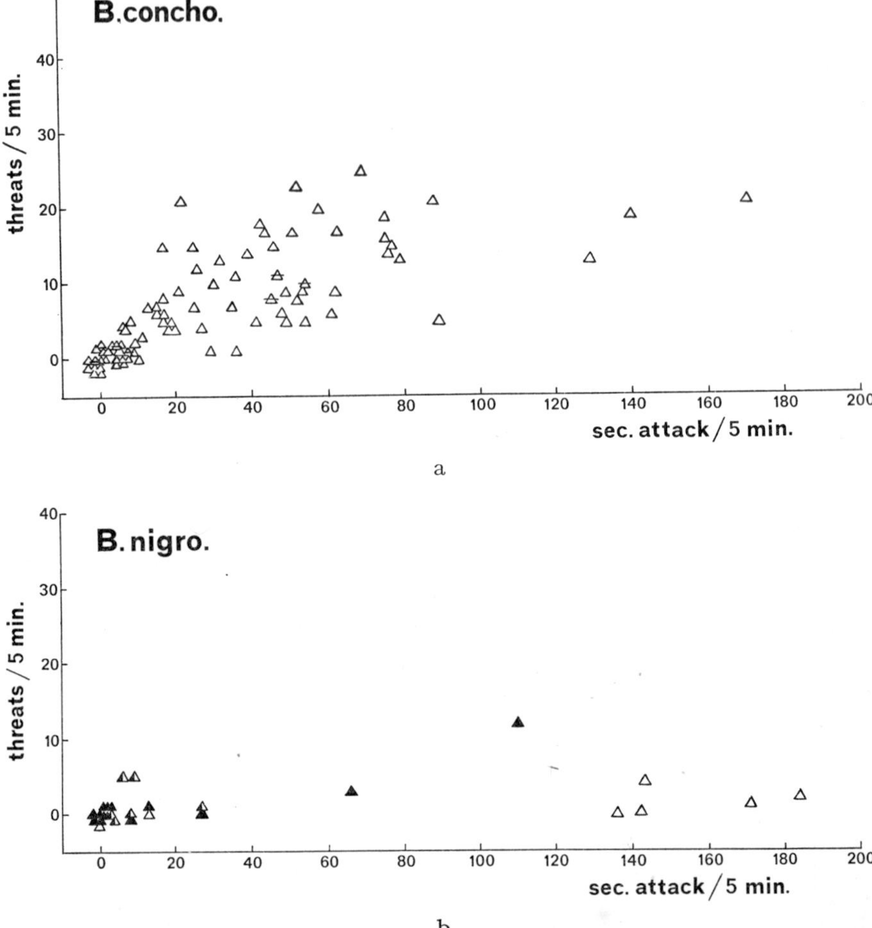

Fig. 61[a-d]. *B. conchonius, B. nigrofasciatus*; attacking and threat (frequency) scores of dominant males in 3♂♂ and 2♂♂ groups; n.l. and a.l. series combined; periods with established dominance only.

be expressed in ambivalent behaviour patterns as well as in overt attack, it might be surmised that the fleeing motivation of the dominants might be expressed in threat as well as in overt fleeing (rolling is virtually absent in dominants). Figg. 61[a-d] and 62[a-d] show the frequencies and the total durations of threat per 5′ plotted against the time spent attacking for the dominants of the two species. For the 3♂♂ experiments it is found that the *nigrofasciatus* dominants display considerably less threat — both in frequency and in total duration — than do *conchonius* dominants at com-

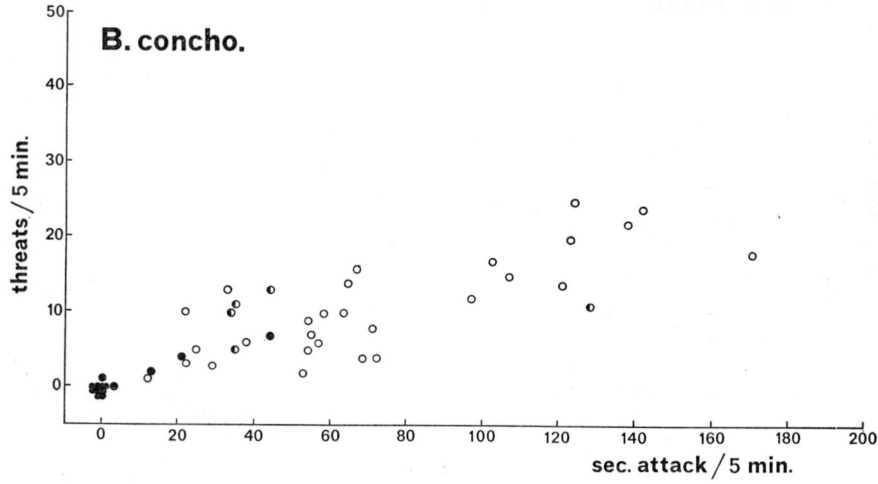

c

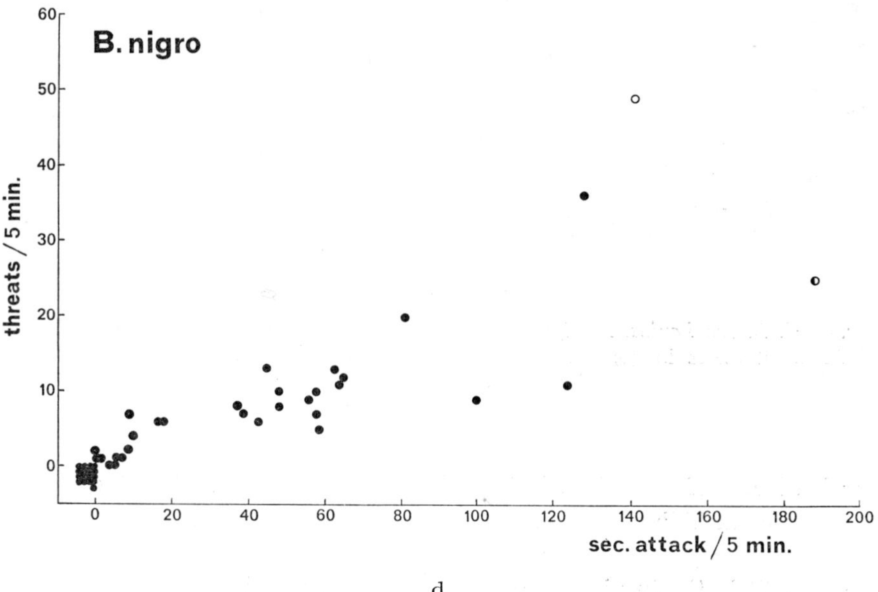

d

Legends fig. 61 see p. 224.

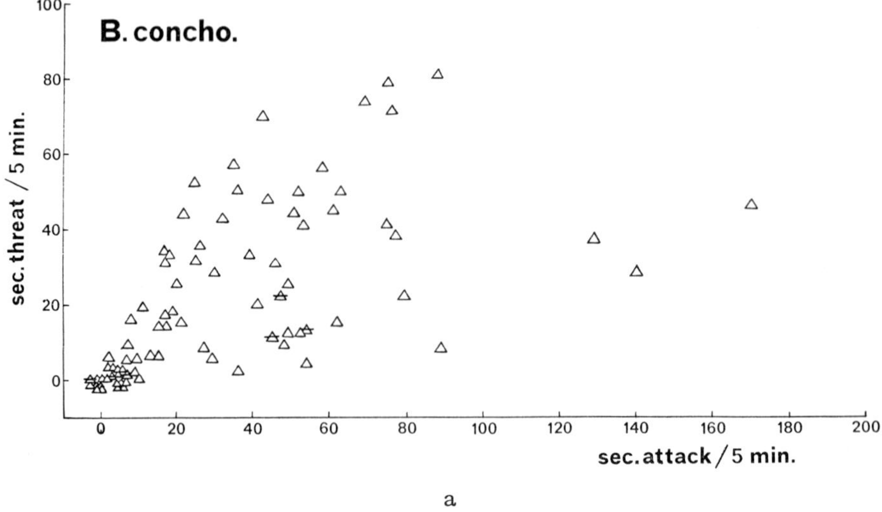

a

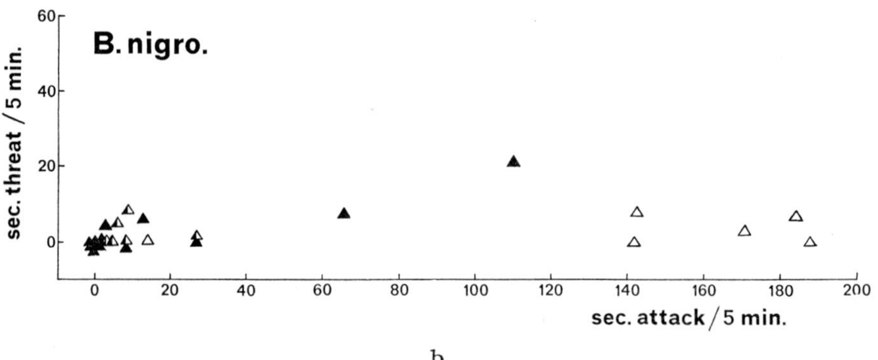

b

Fig. 62a-d. *B. conchonius, B. nigrofasciatus*; attacking and threat (total duration) scores of dominant males in 3♂♂ and 2♂♂ groups; n.l. and a.l. series combined; periods with established dominance only.

parable levels of attacking. This difference cannot be (entirely) due to the differences in hiding tendencies of the inferior fishes, because threat in the dominant *nigrofasciatus* is rare even in the periods in which the inferiors are free. This finding is rather surprising because it is the opposite of what might be expected on the basis of hypothesis (2) and because in the 2♂♂ experiments no obvious differences are found between the species either in the frequency or in the total duration of threat by the dominant. More data

B. concho.
sec. threat / 5 min.
sec. attack / 5 min.
160
140
120
100
80
60
40
20
0
0 20 40 60 80 100 120 140 160 180 200

c

B. nigro.
sec. threat / 5 min.
sec. attack / 5 min.
100
80
60
40
20
0
0 20 40 60 80 100 120 140 160 180 200

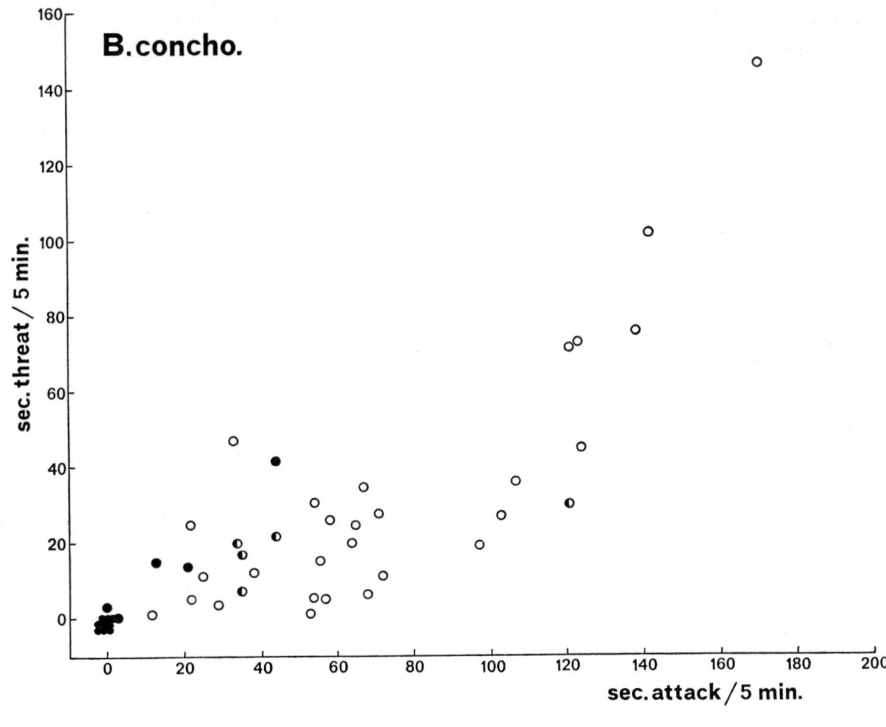

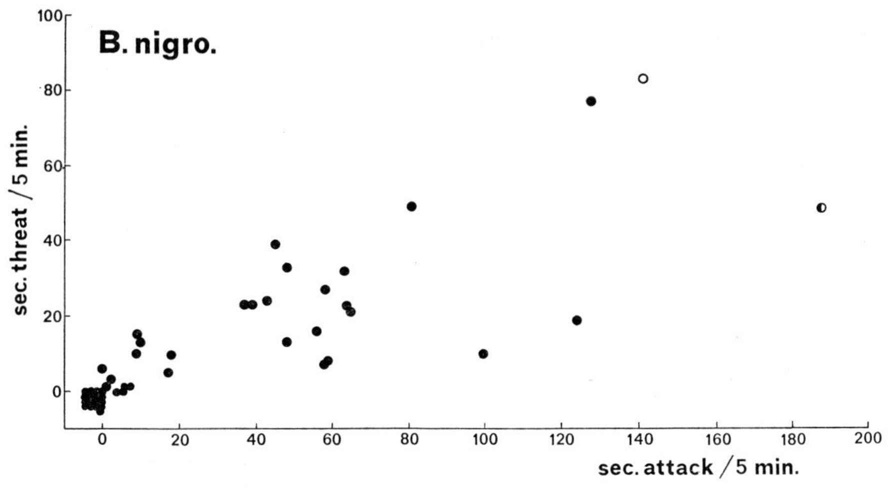

d

will be needed to interprete this finding. Anyway, it does't provide any evidence in favour of hypothesis (2).

Another difference between the species found was that, at least in the 3♂♂ groups, it takes a *nigrofasciatus* male more time than a *conchonius* male to attain full dominance. This finding might indicate that — at least in the initial stages of interaction — *nigrofasciatus* males are more afraid of each other than are *conchonius* males. However, since such greater bashfulness on the part of *nigrofasciatus* dominants-to-be may easily be interpreted as a result of the stronger threat patterns of their inferiors-to-be, the facts do not prove a difference in fear *motivation* between the species.

Behaviour of the inferiors. Yet another phenomenon suggestive of hypothesis (2) is that, when full dominance is in process of being established in a *nigrofasciatus* group, the individuals which are becoming inferior may sometimes start to move to the upper levels of the tank and tend to hide even before the fish which is becoming dominant has shown any considerable amount of attack or threat. This observation contrasts strikingly with the situation in *conchonius* groups in which inferior fishes often do not hide at all even if they have been attacked by the dominant for a long time. However, also this difference between the species may easily be interpreted as a result of the difference in threat patterns. The quick submission on the part of the inferior fishes in *nigrofasciatus* might be quite satisfactorily interpreted as a result of the stronger threat pattern of the fish which has 'just made up its mind' to become dominant [1]).

We may summarize this sub-section by saying that several arguments have been found in favour of hypothesis (1) whereas no arguments could be found to support hypothesis (2). According to the principles discussed above this doesn't mean that hypothesis (2) isn't true, but for the time being it is better abolished for lack of evidence. The greater aggressiveness of *nigrofasciatus* males and the stronger threat patterns in the same species, which in some respects counterbalance the effects of the first characteristic, seem to be sufficient to explain all differences between the species found so far.

Some additional data on the behaviour in the 2♂♂ and 3♂♂ groups are given in appendix 2. These data do fit the general picture of the differences between the species but do not discriminate between the 2 hypotheses.

---

1) An alternative, or complementary, explanation might be that the fishes which are becoming inferior know by experience how aggressive an active conspecific male may be and therefore start to withdraw as soon as one becomes active, see also sub (d), p. 282.

(b) *B. stoliczkanus* and *tetrazona.*

Situational data.

The situations obtaining during the observations are shown in table 9. The table is arranged in the same way as table 8; the same symbols are used. The *stoliczkanus* 3♂♂ pl.B. experiment was broken off after the first day because of disease of the dominant fish. The *tetrazona* 3♂♂ pl.A experiment had to be stopped after the first day because the male which had become dominant after the last observation session of day I had jumped out of the tank early in the morning of day II. The males of the corresponding 3♂♂ pl.B group, which had been observed for 5′ each in the observation sessions of day I, were observed for 15′ each in the observation sessions of day II. Also in the *tetrazona* 3♂♂ b.B experiment one of the males jumped out of the tank, probably late in the evening of the first day. It was replaced by another male (slightly smaller than the other two) early in the morning of day II. The other two males were still schooling and fluttering at that time. The *stoliczkanus* 2♂♂ pl.n.l. group was observed for one day only.

No transient changes in dominance, such as were found in one of the *conchonius* experiments, took place during any of the observation sessions. One case of permanent dominance reversal obtained in the *stoliczkanus* 2♂♂ pl.B group some time between the observations of days I and II.

It may be seen from the table that in *stoliczkanus* full dominance was established from the beginning of the observations in all groups, both 2♂♂ and 3♂♂. In contrast, in the *tetrazona* 3♂♂ groups dominance was not established on the first day in 4 out of 5 cases; in the 5th group (3♂♂ pl.A) one male became dominant late in the afternoon, when the last observation session was over. As was mentioned above the observations of this group could not be continued on day II. In only 2 of the 4 remaining groups dominance was established in the course of day II; in the other 2 groups all males kept schooling and fluttering also throughout day II. In the *tetrazona* 2♂♂ groups dominance was established prior to the first observation of day I in only 2 out of 6 cases. In 2 out of the 4 remaining groups dominance was established in the course of the first day (during the first observation session in the one case and between the first and second observation sessions in the other). In another group dominance was established between the first and second observation sessions of day II; in only one group both males kept schooling and fluttering also throughout day II. In the 3♂♂ groups established dominance was found in only 27 of the 153 5′periods (18%), in the 2♂♂ groups in 137 of the 168 periods (82%). This indicates that the establish-

TABLE 9

B. stolicz.

sessions ↓ groups →

3 ♂♂ pl. A | 3 ♂♂ pl. B | 3 ♂♂ b. A | 3 ♂♂ b. B | 3 ♂♂ pl. n. l.

I9, I11, I14, II9, II11, II14

a

B. tetra.

sessions ↓ groups →

3 ♂♂ pl. A | 3 ♂♂ pl. B | 3 ♂♂ b. A | 3 ♂♂ b. B | 3 ♂♂ b. a. l.

I9, I11, I14, II9, II11, II14

b

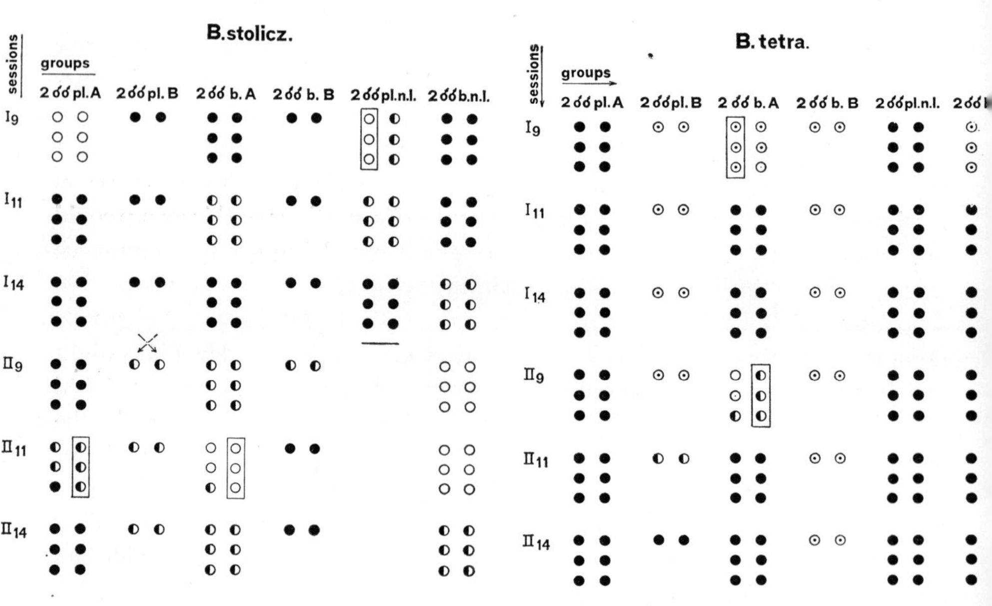

c

d

ment of dominance was less strongly inhibited in the 2♂♂ than in the 3♂♂ groups. In both kinds of groups it was more strongly inhibited than in the corresponding *stoliczkanus* groups. (Similar results are obtained if observation sessions are chosen as units instead of 5′periods).

Hiding was more or less common in inferiors of both species, both in 2♂♂ and 3♂♂ groups. There were, however, distinct quantitative differences between the species and between the two kinds of groups. In the *stoliczkanus* 3♂♂ groups the inferiors hid continuously in 56 of the 189 5′periods (30%), part-time hiding was found in 63 periods (33%) and the inferiors were free in 70 periods (37%). For the *tetrazona* 3♂♂ groups only 27 5′periods with established dominance are available to compare with the *stoliczkanus* figures. In 24 of these 27 periods (89%) both inferiors hid continuously and they were free in no periods. The only 3 cases in which hiding was not continuous (3♂♂ b.A, day II, 11.30) fell between 30 and 45 minutes after dominance had (quite suddenly) become established. Up from 45 minutes after the establishment of dominance hiding of both inferiors was continuous. These figures may suggest that hiding tendencies are stronger in *tetrazona* than in *stoliczkanus* 3♂♂ inferiors. In the *stoliczkanus* 2♂♂ groups the inferior hid continuously in 71 of the 150 5′periods (47%), hiding was part-time in 53 (35%) and the inferior was free in 26 periods (17%). These figures indicate that hiding tendencies are stronger in 2♂♂ than in 3♂♂ inferiors of this species. In the *tetrazona* 2♂♂ groups the inferior hid continuously in 128 (= 93%) of the 137 5′periods in which dominance had been established; it was completely free in only 3 periods (2%) and these periods (including the 2 of 2♂♂ b.A, day II, 09.00) all fell within 15 minutes after the males had still been schooling and fluttering. These figures do not indicate a difference between the hiding tendencies of 2♂♂ and 3♂♂ inferiors of this species (both approach 100%). The figures do, however, indiciate that hiding tendencies of *tetrazona* 2♂♂ inferiors are much stronger than those of *stoliczkanus* 2♂♂ inferiors.

We may summarize the above conclusions as follows. (1) it takes more time for a *tetrazona* male, both in a 2♂♂ and in a 3♂♂ group, to attain full dominance than it does for a *stoliczkanus* male (2) once dominance has been established *tetrazona* inferiors, at least in 2♂♂ groups and probably also in 3♂♂ groups, hide more than do *stoliczkanus* inferiors (3) in *stoliczkanus* hiding by the inferior(s) is more common in 2♂♂ than in 3♂♂ groups; in *tetrazona* hiding by the inferior(s) is very strong in both kinds of groups (4) in *tetrazona* it takes more time for a male to become dominant in a 3♂♂ than in a 2♂♂ group.

All four conclusions are supported by many unstandardized observations of small all-male groups. The first two conclusions are also in accordance with my experiences with small ambisexual groups (see also section III, p. 184) and with the data on the 6♂♂ 6♀♀ groups in large tanks (see section II, p. 134 and 149). The four conclusions are — apart from some noteworthy differences in accent — essentially parallel to those formulated for *conchonius* and *nigrofasciatus* on p. 198 of this section.

TABLE 10

Highest values for height of hiding inferior(s);
(explanation see text)

a. *stoliczkanus* 3♂♂ groups

groups: sessions	3♂♂ pl. A		3♂♂ pl. B		3♂♂ b. A		3♂♂ b. B		3♂♂ b. n.l.	
I_9	0	0	1/2	1/2	—	—	—	—	3/4	3/4
I_{11}	1/3	2/3	1/2	0	—	—	—	—	1/2	1/2
I_{14}	1/2	1/3	1/3	0	2/3	9/10	—	—	1/2	1/2
II_9	1/3	1/2			—	1/2	—	—	—	0
II_{11}	1/2	1/2			2/3	1/3	—	—	—	—
II_{14}	1/2	3/4			1/2	3/4	—	—	—	—

b. *tetrazona* 3♂♂ groups

groups: sessions	3♂♂ pl. A		3♂♂ pl. B		3♂♂ b. A		3♂♂ b. B		3♂♂ pl. n.l.	
I_9	—	—	—	—	—	—	—	—	—	—
I_{11}	—	—	—	—	—	—	—	—	—	—
I_{14}	—	—	—	—	—	—	—	—	—	—
II_9			—	—	—	—	—	—	—	—
II_{11}			—	—	9/10	9/10	—	—	—	—
II_{14}			4/5	9/10	9/10	9/10	—	—	—	—

c. *stoliczkanus* 2♂♂ groups

groups: sessions	2♂♂ pl. A	2♂♂ pl. B	2♂♂ b. A	2♂♂ b. B	2♂♂ pl. n.l.	2♂♂ b. n.l.
I_9	—	1/2	0	3/4	0	9/10
I_{11}	3/5	1/2	1/2	9/10	0	2/3
I_{14}	2/3	1/2	3/4	9/10	1/2	0
II_9	1/2	1/3	1/4	5/6		—
II_{11}	9/10	1/2	0	9/10		—
II_{14}	3/4	0	0	3/4		1/4

d. *tetrazona* 2♂♂ groups

groups: sessions	2♂♂ pl. A	2♂♂ pl. B	2♂♂ b. A	2♂♂ b. B	2♂♂ pl. n.l.	2♂♂ b. n.l.
I_9	1/3	—	—	—	0	—
I_{11}	9/10	—	9/10	—	3/4	9/10
I_{14}	9/10	—	9/10	—	5/6	9/10
II_9	9/10	—	1/3	—	1/2	9/10
II_{11}	9/10	9/10	9/10	—	9/10	9/10
II_{14}	9/10	9/10	9/10	—	1/2	9/10

3♂♂ b. B 3♂♂ pl. n.l.

TABLE 11

Highest values for height of hiding inferior(s);
(explanation see text)

a. *conchonius* 3♂♂ groups

1 inferior hiding in 1 observation period (see table 8); height = 0

b. *nigrofasciatus* 3♂♂ groups

groups:	3♂♂ pl. A		3♂♂ pl. B		3♂♂ b. A		3♂♂ b.B		3♂♂ b. n.l.	
sessions										
I$_9$	—	—	—	—	—	—	5/6	3/4	—	—
I$_{11}$	—	—	—	—	3/4	?	9/10	9/10	—	—
I$_{14}$	—	—	—	—	3/4	—	9/10	9/10	—	—
II$_9$	—	—	—	—	1/5	1/4	9/10	9/10	—	—
II$_{11}$	—	—	—	—	9/10	9/10	9/10	9/10	—	—
II$_{14}$	—	—	—	—	9/10	9/10	9/10	9/10	—	—

c. *conchonius* 2♂♂ groups

groups:	2♂♂ pl. A	2♂♂ pl. B	2♂♂ b. A	2♀♀ b.B
sessions				
I$_9$	9/10	—	—	—
I$_{11}$	3/4	—	—	—
I$_{14}$	1/2	1/2	—	—
II$_9$	9/10	4/5	—	—
II$_{11}$	9/10	5/6	0	—
II$_{14}$	9/10	1/2	—	—

d. *nigrofasciatus* 2♂♂ groups

groups:	2♀♀ pl. A	2♂♂ pl. B	2♂♂ b. A	2♂♂ b.B
sessions				
I$_9$	3/4	—	9/10	9/10
I$_{11}$	5/6	0	9/10	9/10
I$_{14}$	5/6	1/3	9/10	9/10
II$_9$	2/3	1/2	9/10	1/2
II$_{11}$	9/10	9/10	9/10	9/10
II$_{14}$	9/10	9/10	9/10	0

Conclusion (2) may be elaborated by introducing one more parameter. The height at which a hiding inferior stands in a corner may vary a lot, both within and between observation periods. During the observations heights were estimated as fractions of the total depth of the water. 0 means that the hiding fish is very near to the bottom, sometimes actually resting on it; 9/10 means that the fish touches or almost touches the water surface. Tables $10^{a, b}$ and $10^{c, d}$ summarize — for the 3♂♂ and 2♂♂ experiments respectively — the maximal height attained by each of the inferior fishes in each of the observation sessions. It is evident from these tables that *tetrazona* inferiors, when hiding, tend to hide at higher levels than do *stoliczkanus* inferiors. This is true for both 3♂♂ groups (as far as the number of data allows comparison) and 2♂♂ groups. For the sake of comparison the corresponding figures for *conchonius* and *nigrofasciatus* inferiors are given in table 11[a, b] and 11[c, d]. It appears from these tables that both *conchonius* and *nigrofasciatus* inferiors, when hiding, mostly hide in the upper half of the tank. No obvious difference in height is present in these two species.

This means that the reformulation of conclusion (2) in terms of individual distance, which was given for *conchonius* and *nigrofasciatus* on p. 199, is valid *a fortiori* for *stoliczkanus* and *tetrazona*. The same hypotheses which were set up to account for the difference in individual distance of *conchonius* and *nigrofasciatus* males may be set up for *stoliczkanus* and *tetrazona* males. Hypothesis (1) will now say that *tetrazona* males, especially when in the dominant role, are more aggressive than *stoliczkanus* males; hypothesis (2) will say that *tetrazona* males, at least when inferior, are more liable to flee than are *stoliczkanus* males.

Again, in order to decide between the two hypotheses we will have to look at the details of the agonistic interactions between the males. As to the evaluation of these data the same theoretical considerations which were developed for *conchonius* and *nigrofasciatus* (p. 199 ff.) may be repeated here — substituting *stoliczkanus* for *conchonius* and *tetrazona* for *nigrofasciatus* — with one exception. Because *stoliczkanus* and *tetrazona* inferiors differ not only in the amount of time spent hiding but also in the height at which they hide, the difference between these two species in individual distance is more pronounced than in the other two species [1]). Therefore, if *tetrazona* dominants would be found to attack hiding inferiors more than do *stoliczkanus* dominants, this argument in favour of hypothesis (1) would carry more weight than the analogous argument applied to the other two species.

Possible arguments for hypothesis (1).

Behaviour of the dominants. Figg. 63[a-g] represent the time spent in full attack and the number of butting movements by the dominants of the two species in all 5′ periods. For the 3♂♂ experiments the activities of the dominant directed at both inferiors have again been added together. The data from the artificial light and natural light experiments are given separately; the data from all 5′ periods without established dominance have been left out.

In comparing the data of the two species we will mainly compare a.l. with a.l. and n.l. with n.l. experiments. The following trends are apparent from

---

1) It may be reminded that the dominant male as a rule stays low.

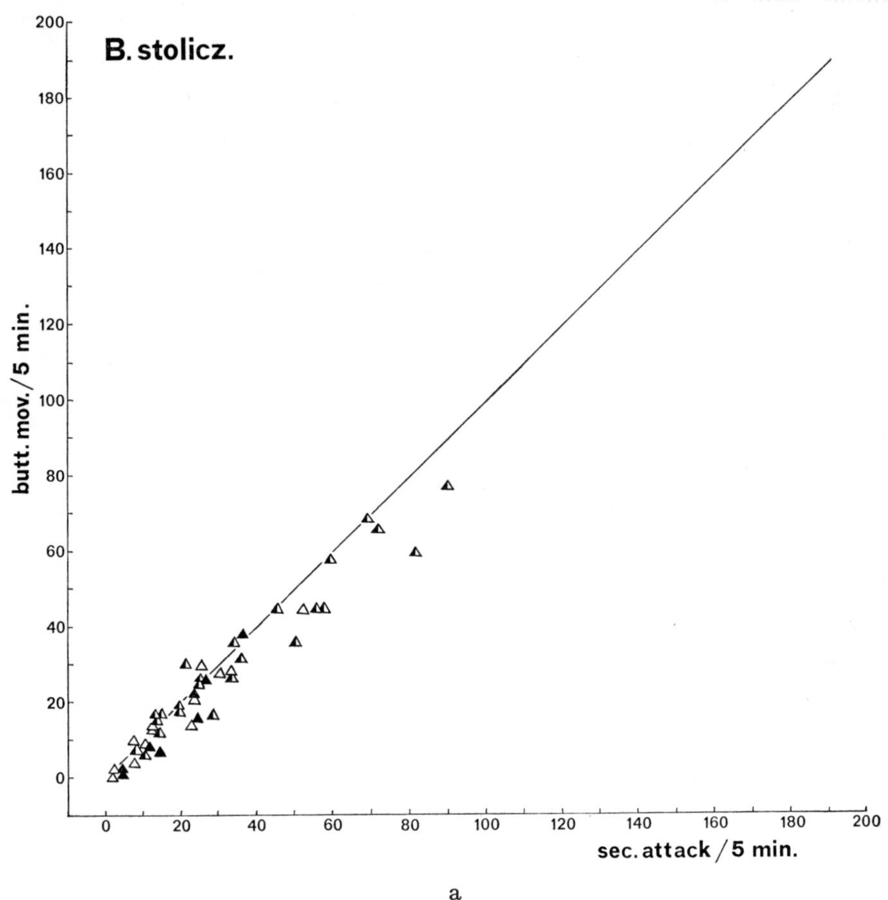

a

Fig. 63a-g. *B. stoliczkanus, B. tetrazona*; aggressive scorings of dominant males in 3♂♂ (triangles) and 2♂♂ (circles) groups; 45° diagonal for reference; a, b, d, e: a.l. series; c, f, g: n.l. series; periods with established dominance only.

the graphs: (a) for both species there is a distinct positive correlation between the two measures, though perhaps they are not as closely correlated as in *conchonius* and *nigrofasciatus*. The latter point is probably due to differences between individual dominants rather than to variations in butting frequency within any single experiment; for instance in graph 63d (*stoliczkanus* 2♂♂ a.l.), as far as the periods are concerned in which either of the two measures exceeds 10, all points above the 45° line are from 2 experiments (2♂♂ pl. A & B) and those below the line are from the 2 others (2♂♂ b. A & B). In most cases the frequency of butting movements (per seconds

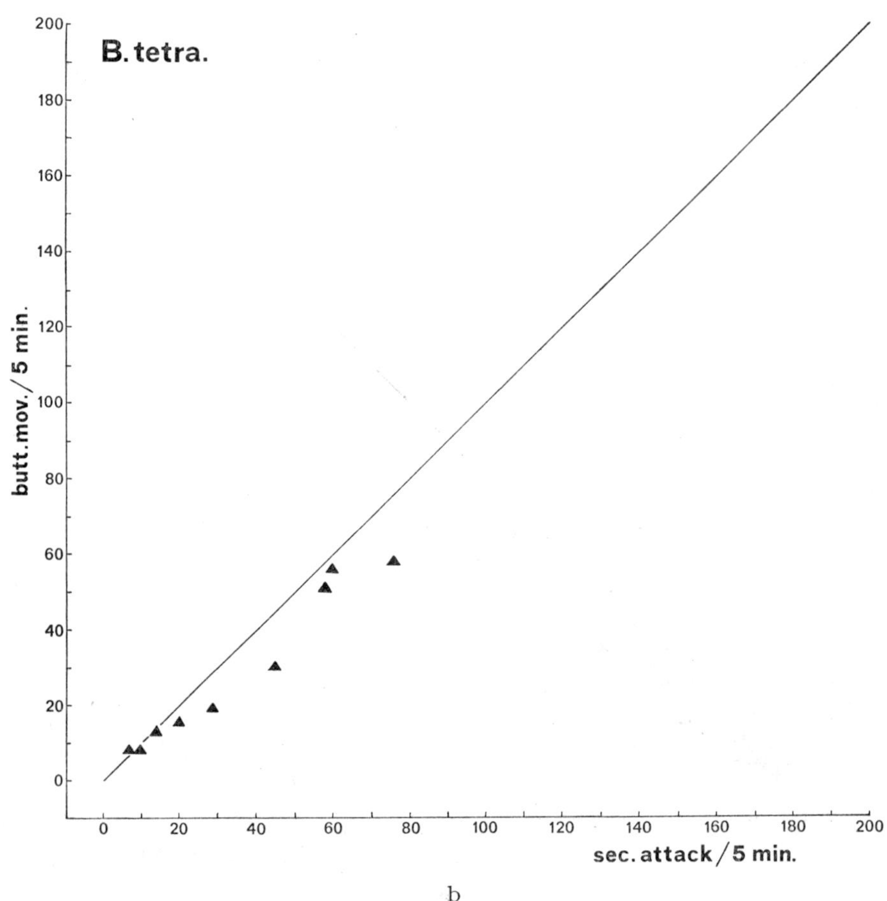

b

Legends fig. 63 see p. 235.

attack) is not obviously dependent on the absolute values of both parameters. (b) there is no consistent difference between the species in the frequency of butting movements. (c) there is no consistent difference between the 3♂♂ and 2♂♂ dominants in the frequency of butting movements. (d) it is by no means clear that — as was found in *conchonius* and *nigrofasciatus* — free inferiors are attacked more than hiding inferiors. (This comparison is possible for the *stoliczkanus* data only). (e) in the 3♂♂ experiments rather few data of *tetrazona* are available to compare with *stoliczkanus*. The

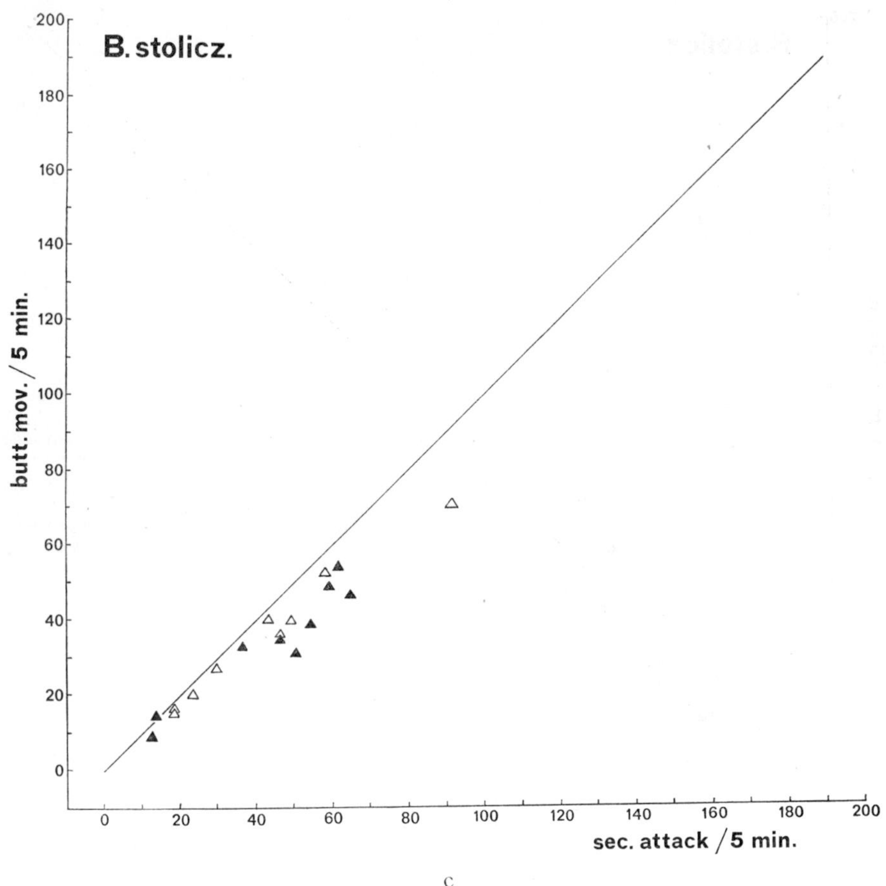

c

Legends fig. 63 see p. 235.

aggressive scores of *tetrazona* present may seem comparatively high if they are compared with only the black triangles in the corresponding *stoliczkanus* graph (a.l. experiments). However, comparably high scores are found in the *stoliczkanus* n.l. experiments (graph 63[c]). At best the aggressive scores of *tetrazona* dominants do not seem to be lower than those of *stoliczkanus* dominants, even if open and half-black triangles are included. This is so even though the inferiors of the former species tend to hide at higher levels in the tank (see tables 10).

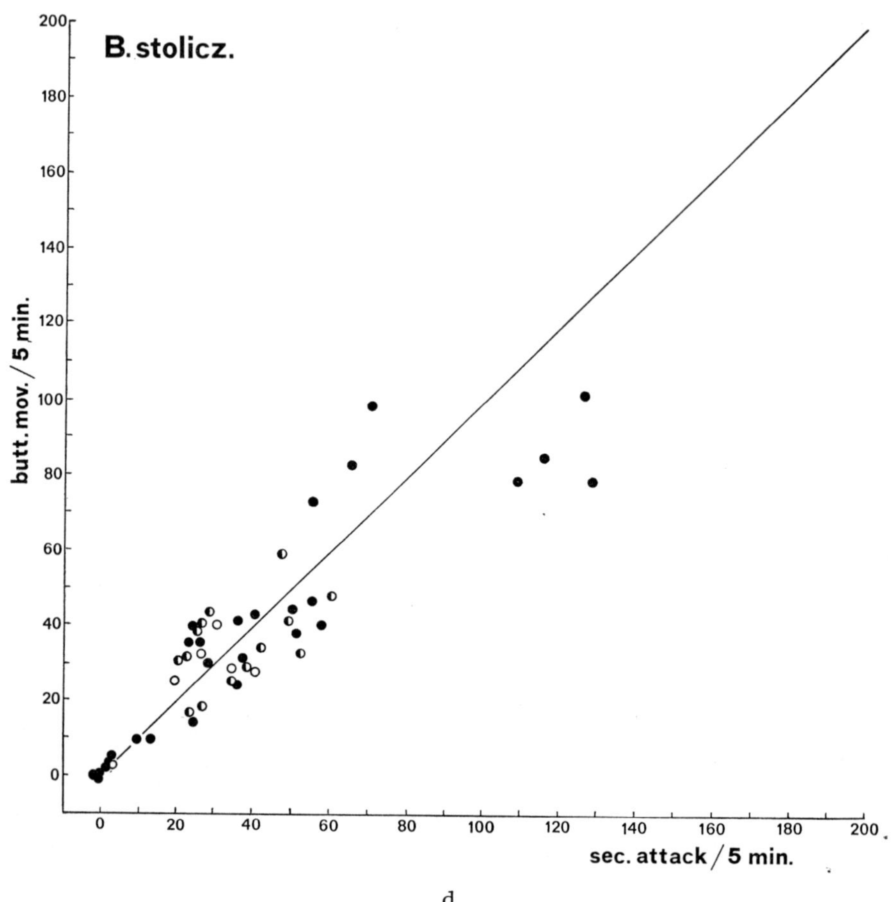

d

Legends fig. 63 see p. 235.

TABLE 12

High and low aggressive scorings of stoliczkanus and tetrazona 2♂♂ dominants (cf. fig. 63[d-g] and text)

	%>60″ attack	%>60 b.m.	%<10″ attack	*n*
stolicz. a.l.	15	15	15	48
tetra. a.l.	23	23	51	35
stolicz. n.l.	22	11	15	27
tetra. n.l.	36	30	24	33
stolicz. +	17	13	15	75
tetra. +	29	26	38	68

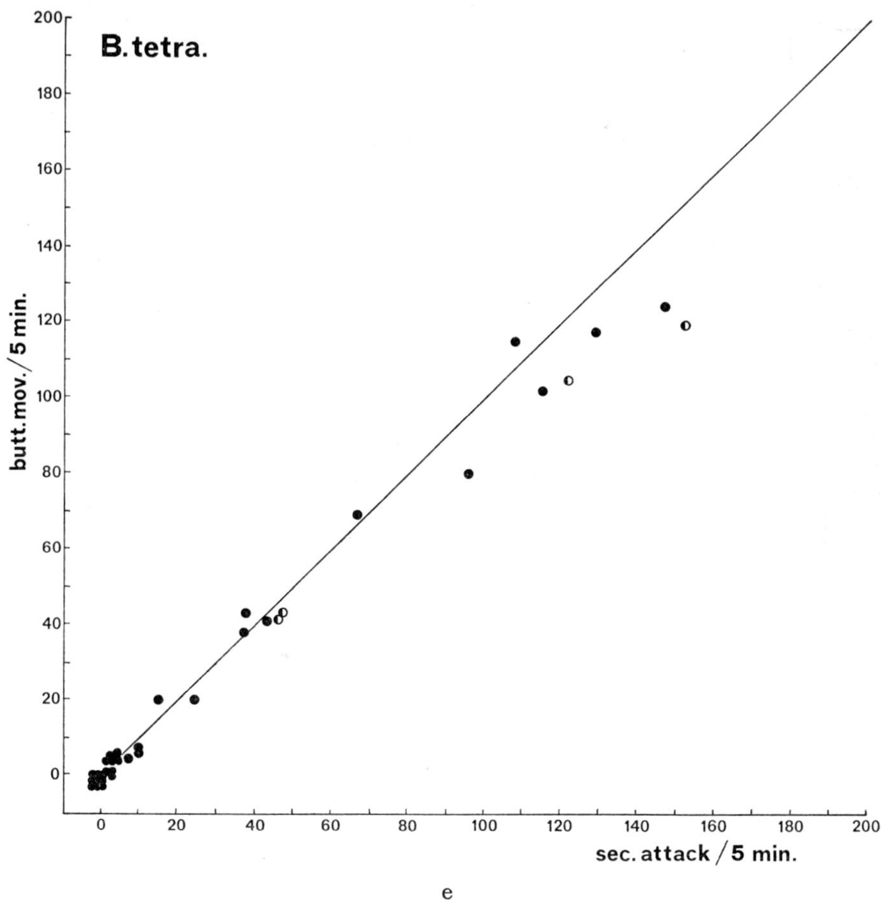

Legends fig. 63 see p. 235.

The results of the 2♂♂ experiments are more clear. Table 12 summarizes some of the characteristic features of the graphs. In comparing the scores of the two species we find that — both in the a.l. and the n.l. experiments — *tetrazona* dominants score above 60″ attack or above 60 butting movements more often than do *stoliczkanus* dominants; the difference is greater for butting scores than for duration of attack. On the other hand, *tetrazona* dominants score very low values (both seconds attack and butting below 10) more often than do *stoliczkanus* dominants. It might be surmised that in the n.l. experiments *stoliczkanus* aggressive scorings (like those of *conchonius*)

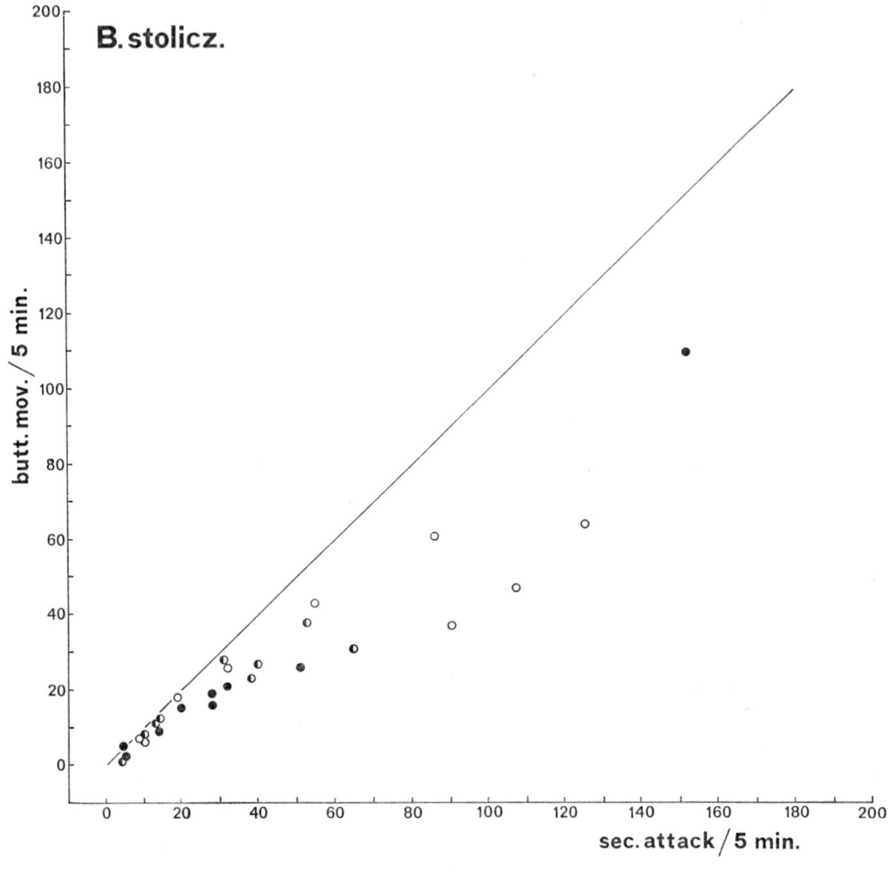

f

Legends fig. 63 see p. 235.

might have been negatively influenced through the absence of early observations. However, the *stoliczkanus* scores of the n.l. and a.l. experiments are very similar to each other. Therefore I assume the two samples to be equivalent and consequently the scores of the a.l. and n.l. experiments have been added up in the table.

It is not surprising that *tetrazona* dominants score very low more often than do *stoliczkanus* dominants because the inferiors of the former species hide both more consistenly and further away from the dominant. In view of the same difference in hiding tendencies, however, the relatively high

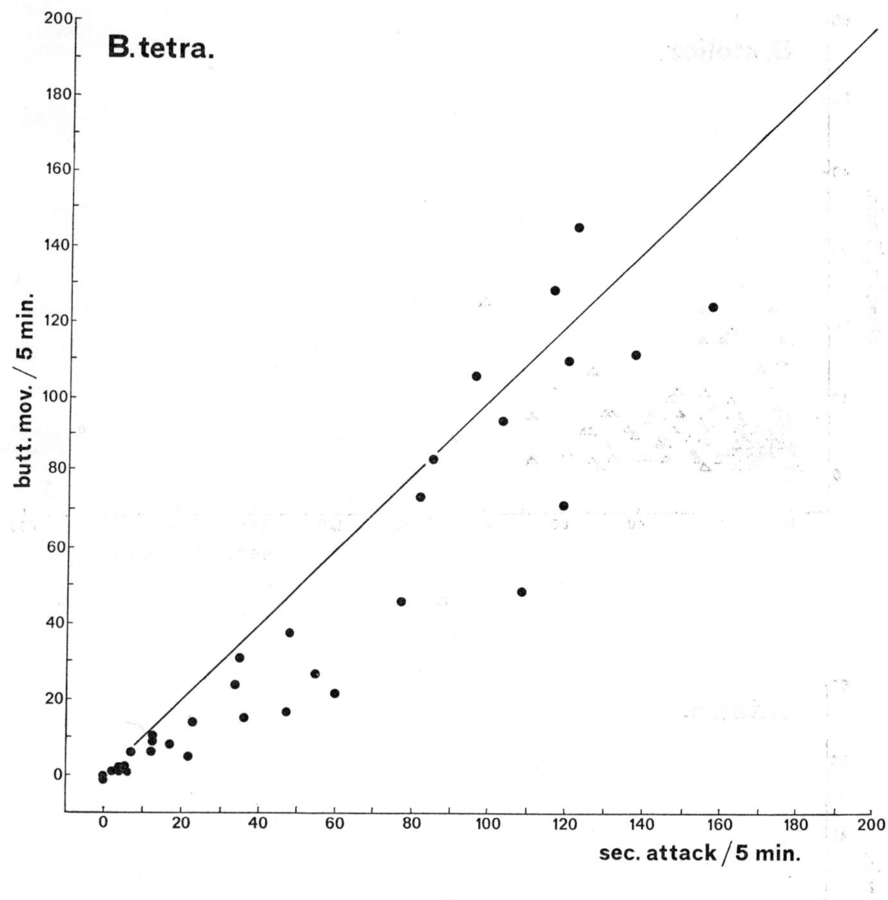

g

Legends fig. 63 see p. 235.

frequency of high aggressive scores in *tetrazona* dominants provides a strong argument in favour of hypothesis (1); for instance all high scores in the *tetrazona* 2♂♂ n.l. experiments (except one at 108″ attack and 49 butting movements) were produced while the inferior hid at 9/10 continuously! It may further be emphasized that the difference in the frequency of high aggressive scorings was found even though the *tetrazona* inferiors hid continuously in almost all periods whereas *stoliczkanus* inferiors were often partly or completely free.

From the observations I got the (unquantified) impression that — though

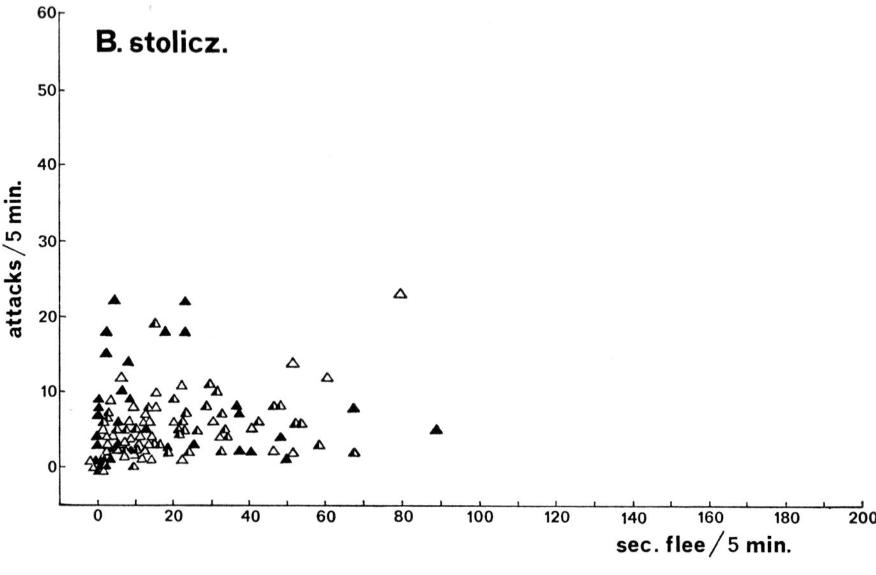

a

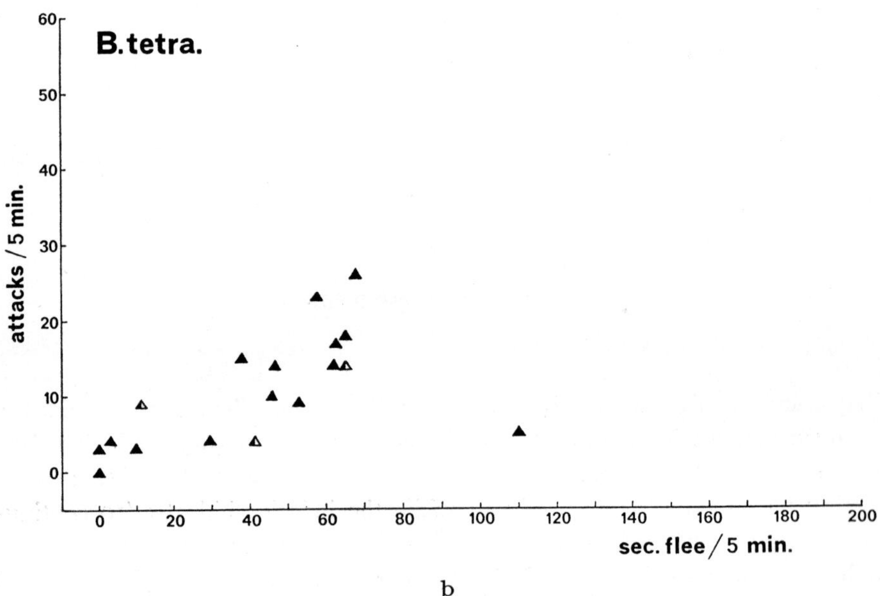

b

Fig. 64[a-d]. *B. stoliczkanus, B. tetrazona*; fleeing and attack scores of inferior males in 3 ♂♂ (triangles) and 2 ♂♂ (circles) groups; n.l. and a.l. series combined; periods with established dominance only.

B. stolicz.
attacks / 5 min.
sec. flee / 5 min.
0
10
20
30
40
50
60
0
20
40
60
80
100
120
140
160
180
200
220

c

B. tetra.
attacks / 5 min.
sec. flee / 5 min.
0
10
20
30
40
50
60
0
20
40
60
80
100
120
140
160
180
200
220

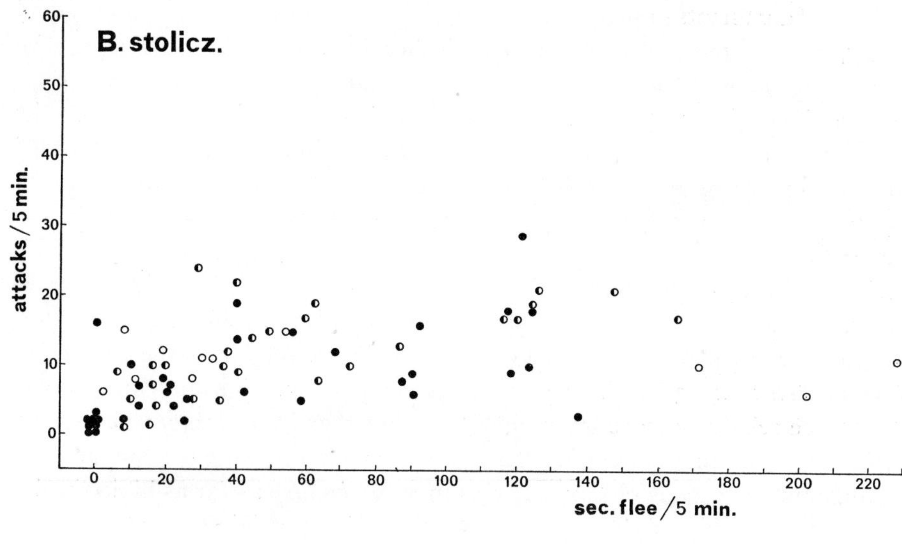

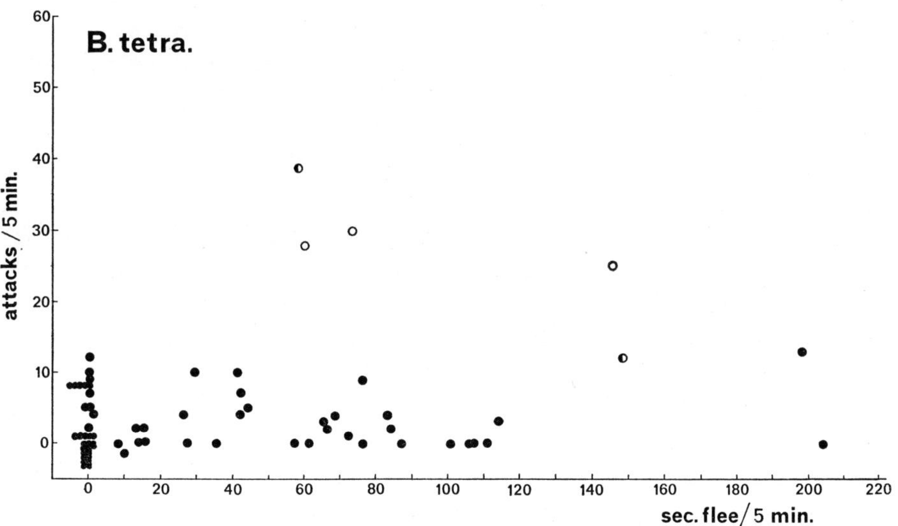

d

also *stoliczkanus* males may at times attack and bite quite vigorously — *tetrazona* dominants much more often show sustained vigorous attack and biting, even if the attacked fish doesn't flee fast. This observation may be taken as another argument in favour of hypothesis (1).

Behaviour of the inferiors. Figg. $64^{a\text{-}d}$ show — for the two species — the number of attacks (full attacks, distance attacks, inhibited attacks all added together) made by each inferior fish on the dominant per 5′ period, plotted against the total number of seconds spent fleeing for the dominant in the same periods. In the 3♂♂ groups the inferiors of both species may exhibit considerable numbers of attacks at the dominant even if they hide continuously. (They differ in this respect from *conchonius* and *nigrofasciatus* inferiors which show little attacking when hiding). As far as the paucity of data on *tetrazona* 3♂♂ inferiors allows comparison, high attacking scores seem to be relatively more common in *tetrazona* than in *stoliczkanus* inferiors. Moreover, the high scores in *tetrazona* go with considerable amounts of fleeing and continuous hiding, whereas in *stoliczkanus* (as far as hiding males are concerned) high aggressive scores seem to be possible only at very low levels of fleeing. Only some of the *completely free stoliczkanus* apparently were able to attain high aggressive scores at the same level of fleeing as did *hiding tetrazona* inferiors.

More data are needed to assess whether (hiding) *tetrazona* inferiors always score few attacks at very low levels of fleeing. The apparent positive correlation between attacking and fleeing in the *tetrazona* graph cannot be due (such as was suggested for the corresponding *nigrofasciatus* graph) to a common correlation of the two measures with hiding. Nor is it probable that it is due to any external synchronization between the males of a group since — judged impressionistically — very little regularity in daily activity rhythms was found in these experiments. It is well possible that the correlation is real and the result of a tendency of dominant and inferior(s) to mutually stimulate each other's aggression. I often saw in the *tetrazona* experiments that an inferior male launched an attack at the dominant immediately after the latter had stopped attacking and the former had returned to its hiding place. Such counter-attacks often elicited renewed attack from the dominant and so on. (The procedure is vaguely reminiscent of 'shuttle' attacks between neigbouring territorial males). This type of interaction might well create a positive correlation between the inferior's attacking and fleeing behaviour. The point couldn't be followed up quantitatively. However this may be, the high aggressive scorings of *tetrazona* inferiors at considerable levels of fleeing suggest a further argument in favour of hypothesis (1).

The possibility cannot be ignored that the arguments drawn from the aggressive behaviour of *tetrazona* dominants on the one hand and inferiors on the other are not mutually independent. It may be envisaged that 1. *terazona* inferiors hide at higher levels because of the aggressive behaviour of the dominant, 2. this brings them farther away from the dominant and 3. this gives them the opportunity, in the periods that they are not attacked, to 'pick up their courage' and attack the dominant. On the other hand such attacks from the inferiors might provoke more aggression in the dominant than when the inferiors would keep still. It is impossible to test these possibilities on the basis of the available data.

Another possible critique of the conclusion drawn from the behaviour of the inferiors is that the aggressive scores of the *tetrazona* inferiors were found relatively shortly after the establishment of dominance whereas those of most *stoliczkanus* inferiors were not. It is unknown what influence the length of the schooling period which preceded the establishment of dominance might have had on the aggressive scoring of the inferiors. In these respects the samples of the 2 species are not quite equivalent. It is impossible to find a satisfactory answer to these questions on the basis of the available data. It is, for instance, unknown at which times dominance was established in the *stoliczkanus* groups. Only it may be mentioned that the aggressive scores of both inferiors in the *tetrazona* 3♂♂b. A experiment on day II were considerably higher in the 14.30 than in the 11.30 observation session (see table 13).

TABLE 13

Attacking scores per 5′ of inferior tetrazona in the 3♂♂ b.A *group*

sessions	5′period	attacks (2 inf. ♂♂)	
II_{11}	1	3	14
	2	9	9
	3	4	4
II_{14}	1	26	14
	2	14	18
	3	5	15

In the 2♂♂ experiments *tetrazona* inferiors in general attacked much less than did *stoliczkanus* inferiors at comparable levels of fleeing. This is so even if only continuously hiding inferiors of the two species are compared. This difference is probably due to the more complete submission of the inferior *tetrazona* (as expressed in the tendency to hide more and at higher levels). However, even in the 2♂♂ situation, some inferior *tetrazona* scored very high numbers of attacks (comparable to the highest scores of *stoliczkanus* 2♂♂ inferiors only). These scores were all from the inferior in the 2♂♂ b.A experiments in the few periods in which it was free (3 periods) or partly free (1 period), shortly after the establishment of dominance. Though small in number, these high scores support the argument for the aggressiveness of *tetrazona* 3♂♂ inferiors.

Figg. $65^{a\text{-}d}$ represent the time spent rolling (ventral *plus* dorsal) and

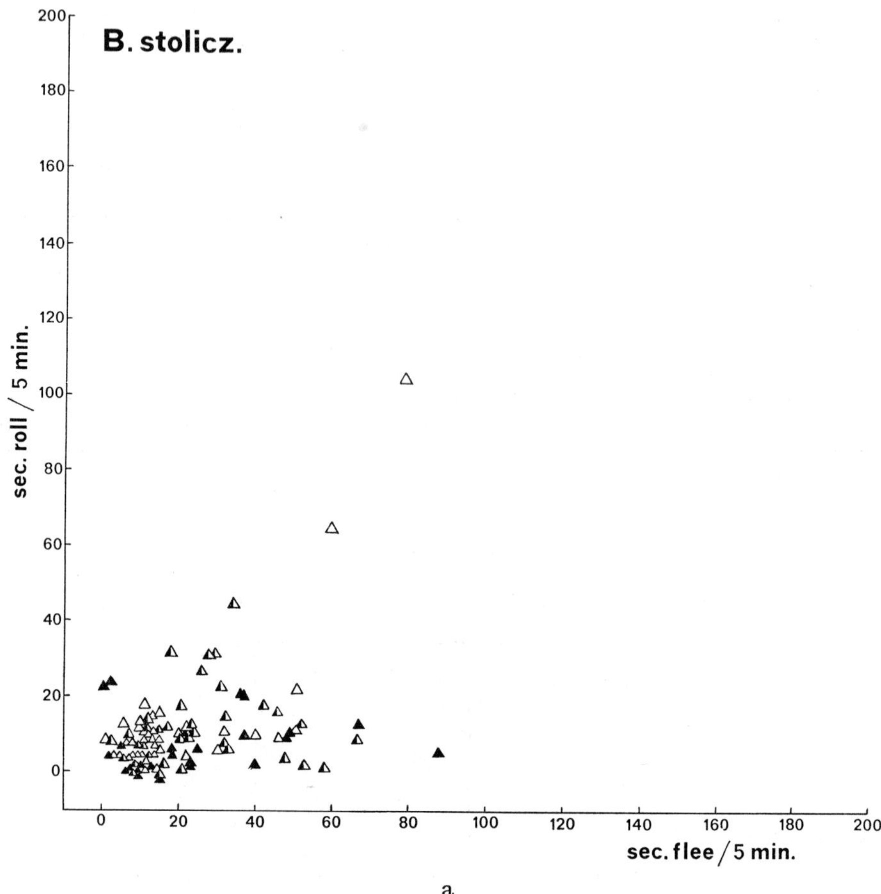

a

Fig. 65a-d. *B. stoliczkanus, B. tetrazona*; fleeing and (total) roll scores of inferior males in 3♂♂ (triangles) and 2♂♂ (circles) groups; n.l. and a.l. series combined; periods with established dominance only.

fleeing for the dominant by the inferiors of the two species. The graphs suggest that at least for *stoliczkanus* there are two rather well separated possibilities: greater amounts of fleeing may go with much rolling (as much as or more than fleeing) or with very little rolling. The same two possibilities may be indicated in the *tetrazona* 2♂♂ graph; only the upper of the two branches is present in the 3♂♂ graph of this species.

Apart from this phenomenon it is clear that in the 3♂♂ groups *tetrazona* inferiors roll much more than do *stoliczkanus* inferiors at comparable levels

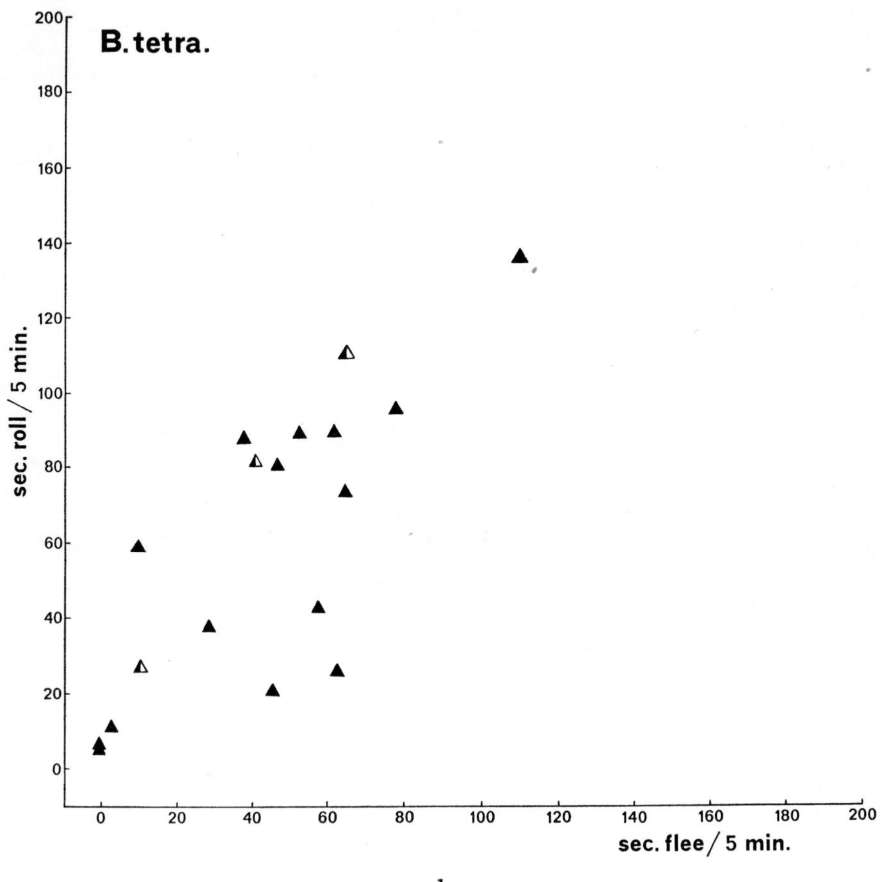

b

Legends fig. 65 see p. 246.

of fleeing. In the 2♂♂ groups there is no obvious difference between the species in the amounts of rolling, especially if only hiding inferiors are compared. This is so in spite of the fact that *tetrazona* 2♂♂ inferiors — as judged from their hiding behaviour — are on the average more completely submitted than are *stoliczkanus* 2♂♂ inferiors. The latter consideration makes that the rolling scores of the 2♂♂ inferiors — as well as those of the 3♂♂ inferiors — plead for hypothesis (1).

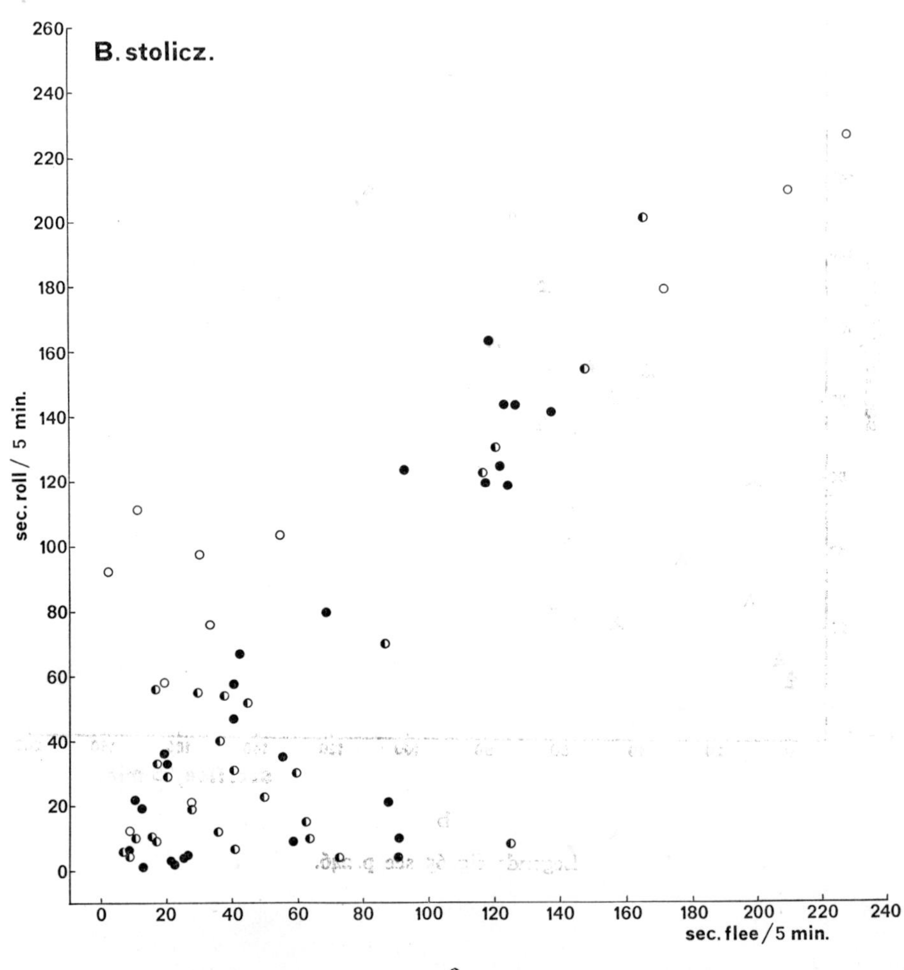

c

Legends fig. 65 see p. 246.

Figg. 66[a-d] represent the amounts of dorsal roll alone shown by the inferiors, plotted against the amounts of fleeing. The differences here are very similar to those found between *nigrofasciatus* and *conchonius*: *tetrazona* inferiors perform much more dorsal rolling than do *stoliczkanus* inferiors, both in 3♂♂ and in 2♂♂ groups. This again pleads for hypothesis (1).

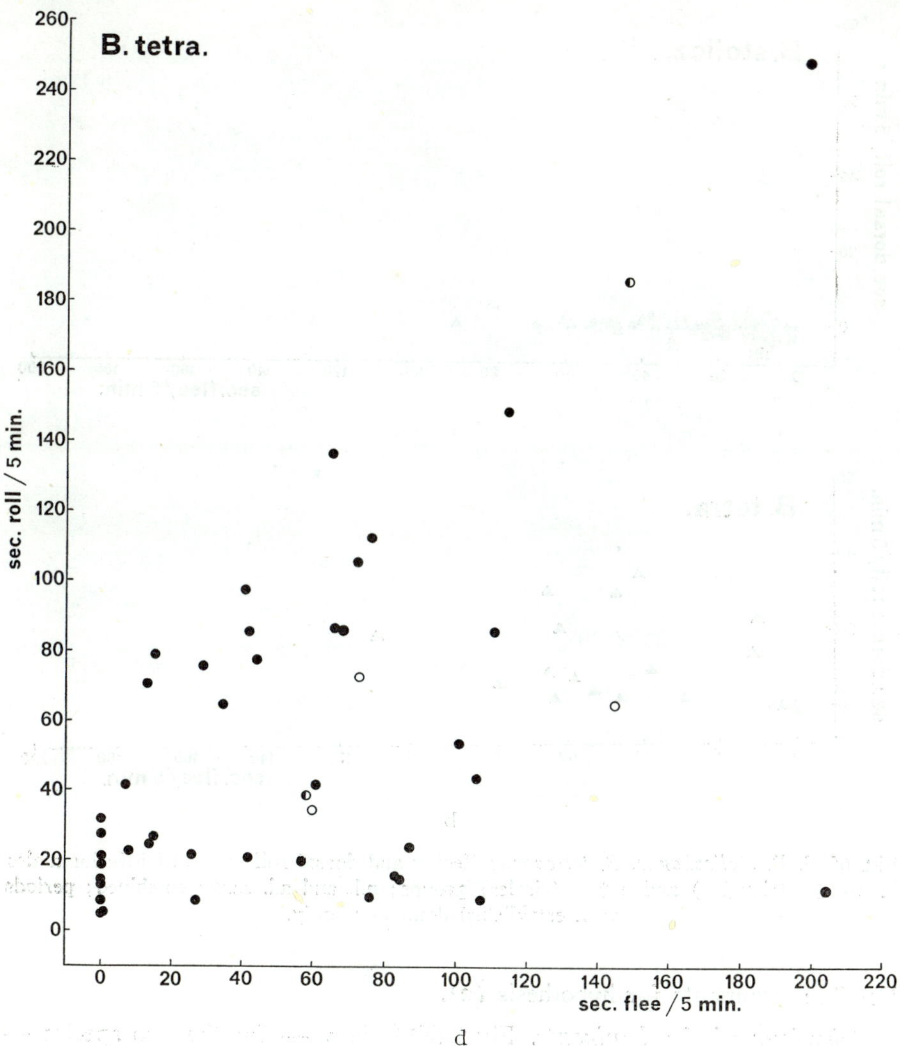

d

Legends fig. 65 see p. 246.

Figg. 67a-d show the frequencies of threat by the inferior fishes towards the dominants per 5′, plotted against the amounts of fleeing for the dominants. The graphs show that threat is rare in both species, both in 3♂♂ and in 2♂♂ inferiors. No appreciable difference between the species may be detected.

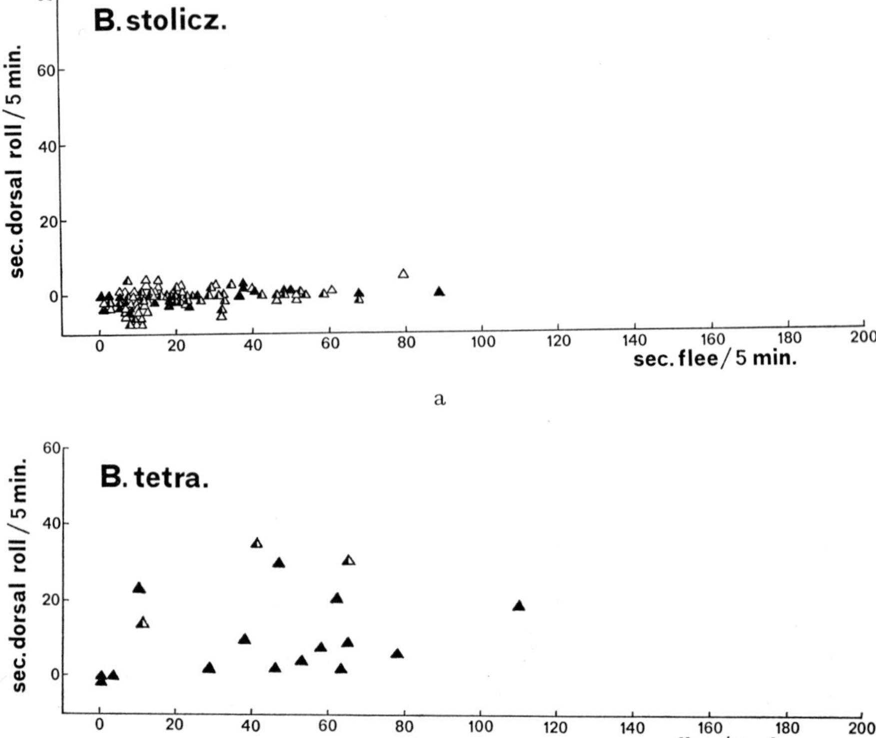

Fig. 66[a-d]. *B. stoliczkanus, B. tetrazona;* fleeing and dorsal roll scores of inferior males in 3 ♂♂ (triangles) and 2 ♂♂ (circles) groups; n.l. and a.l. series combined; periods with established dominance only.

Possible arguments for hypothesis (2).

Behaviour of the dominants. Figg. 68[a-d] show — for the two species — the number of times the dominant flees for the inferior(s), plotted against the time spent in full attack. Neither in the 3♂♂ nor in the 2♂♂ situation is there any obvious difference between the species in the frequencies of fleeing.

Figg. 69[a-d] and 70[a-d] show the frequencies and the total durations of threat by the dominant, plotted against the time spent attacking. Again, no obvious differences may be found between the species. This means that no argument in favour of hypothesis (2) may be found in the quantitative behaviour records of the dominants.

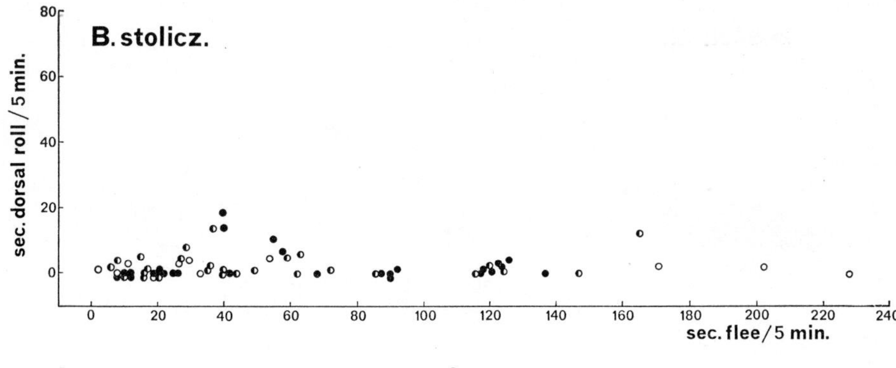

c

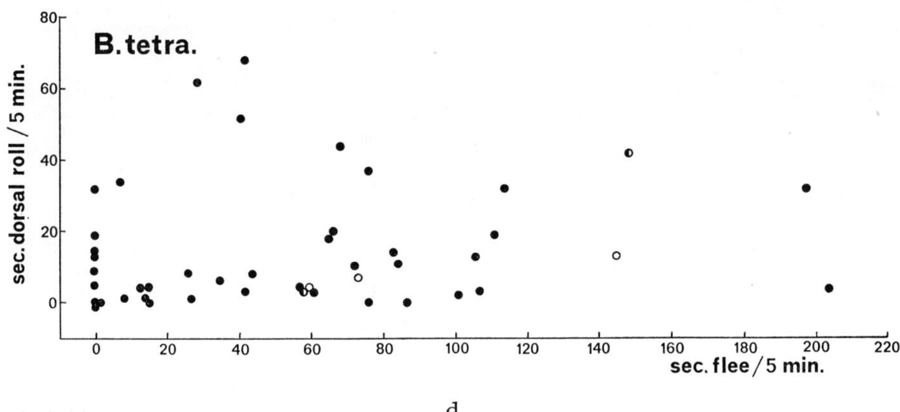

d

Also the fact that, both in the 2♂♂ and the 3♂♂ groups, it takes a *tetrazona* male more time than a *stoliczkanus* male to establish full dominance cannot be accepted as an argument for the hypothesis. This is so for the same reason as was adduced for *nigrofasciatus vs conchonius*: the stronger threat pattern of a *tetrazona* male's conspecifics may well inhibit the development of dominance.

Behaviour of the inferiors. No arguments for hypothesis (2) could be found in the behaviour of the inferiors.

We may summarize this sub-section in the same way as was done with the former: several arguments have been found which plead for hypothesis (1); no arguments could be found to support hypothesis (2). The greater

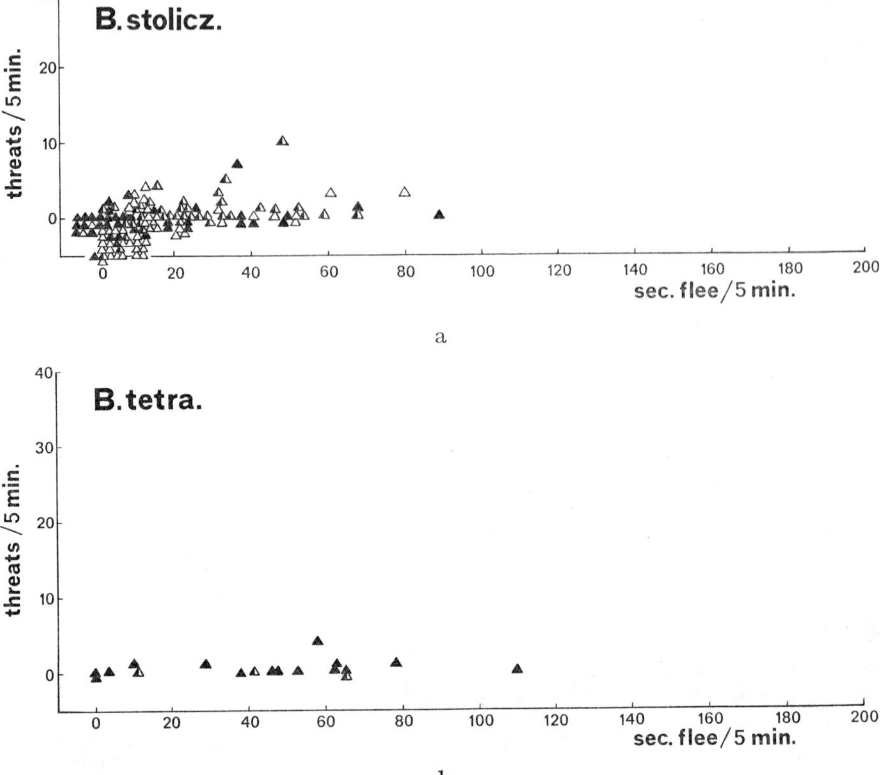

Fig. 67a-d. *B. stoliczkanus, B. tetrazona*; fleeing and threat (frequency) scores of inferior males in 3♂♂ (triangles) and 2♂♂ (circles) groups; n.l. and a.l. series combined; periods with established dominance only.

aggressiveness of *tetrazona* males and the more elaborate threat pattern in the same species, which in some respects counterbalances the effects of the first characteristic, seem to be sufficient to explain all differences found so far between *tetrazona* and *stoliczkanus* males.

Some additional data, analogous to those given for *nigrofasciatus* and *conchonius* in Appendix 2, are presented in Appendix 3.

(c) *conchonius-nigrofasciatus vs stoliczkanus-tetrazona.*

In the sub-sections (a) and (b) it was shown that the differences in behaviour between *conchonius* and *nigrofasciatus* on the one hand and those between *stoliczkanus* and *tetrazona* on the other run very much parallel to each other. This does not mean, however, that the behaviour of *stoliczkanus*

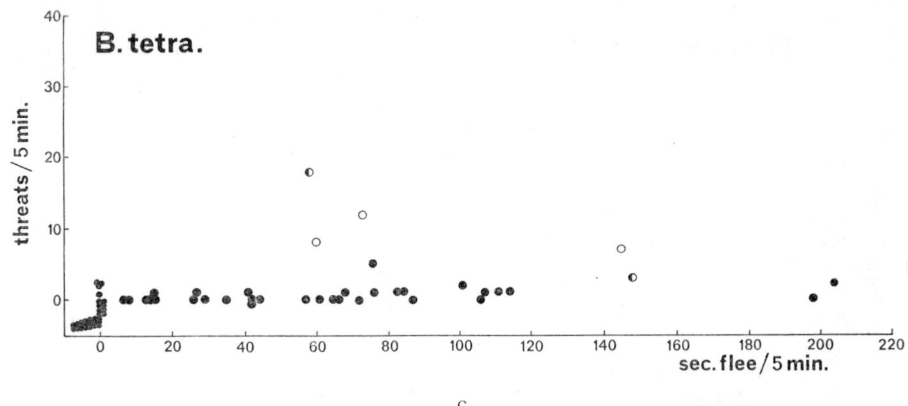

c

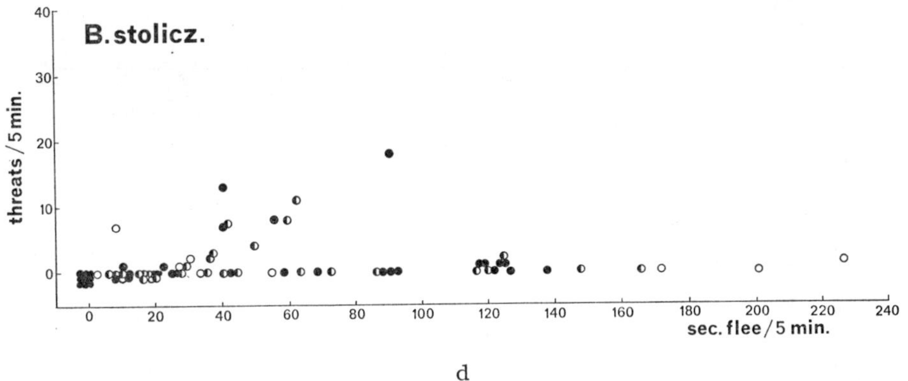

d

is very much like that of *conchonius* or the behaviour of *tetrazona* like that of *nigrofasciatus*. It doesn't mean either that the four species may be ranged on a linear scale. Some differences between the two pairs of species have been mentioned in sub-section (b). One difference — the most important for the present analysis — has not been stated yet. The two pairs of species differ qualitatively in the way in which the inferiors hide. Hiding *conchonius* and *nigrofasciatus* inferiors as a rule hang quite passively, often with the body horizontal or with the head pointing steeply upwards, sometimes facing the walls of the corner in which they hide. In contrast, hiding *stoliczkanus* and *tetrazona* inferiors typically hang in a corner with the head tipped downwards, facing into the open and often following all movements of the

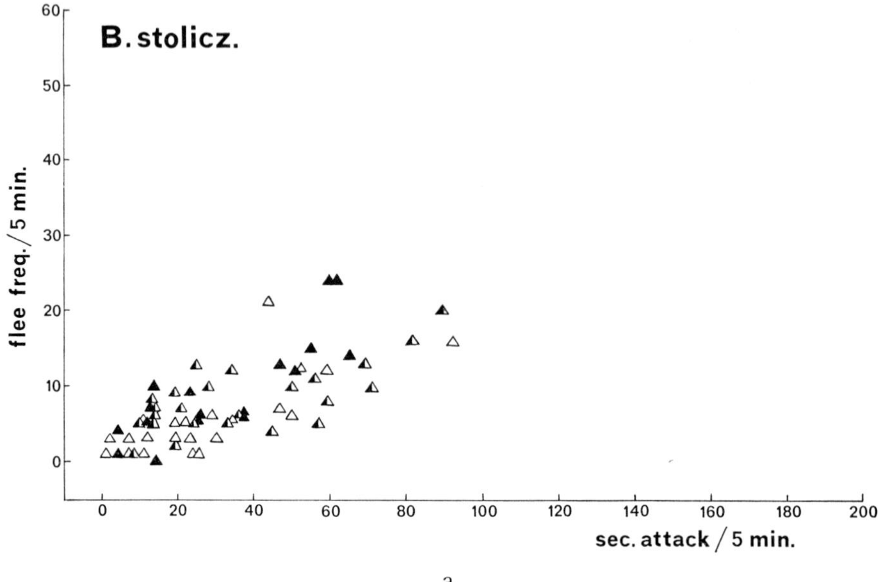

a

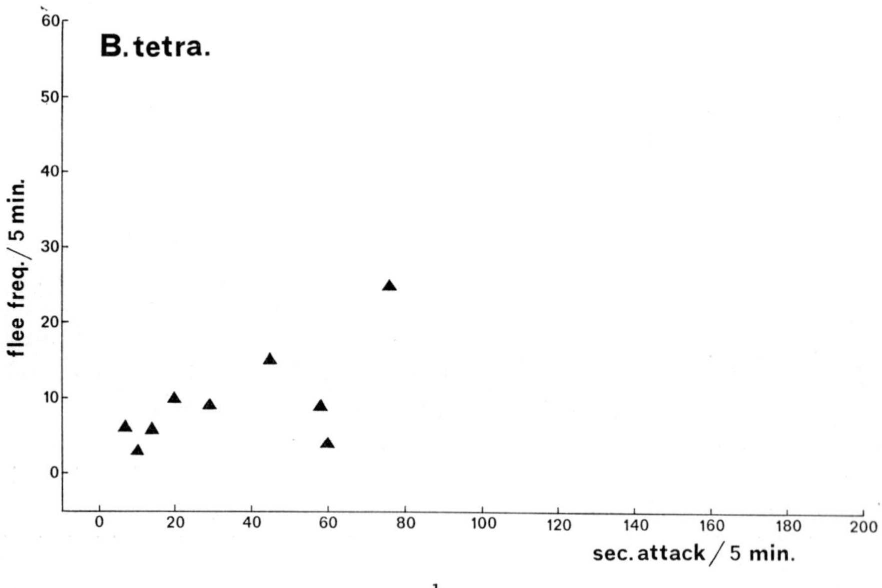

b

Fig. 68a-d. *B. stoliczkanus, B. tetrazona*; attacking and flee (frequency) scores of dominant males in 3♂♂ and 2♂♂ groups; n.l. and a.l. series combined; periods with established dominance only.

B. stolicz.
flee freq./5 min.
sec. attack / 5 min.
0
10
20
30
40
50
60
70
0
20
40
60
80
100
120
140
160
180
200

c

B. tetra.
flee freq./5 min.
sec. attack / 5 min.
0
10
20
30
40
50
60
70
0
20
40
60
80
100
120
140
160
180
200

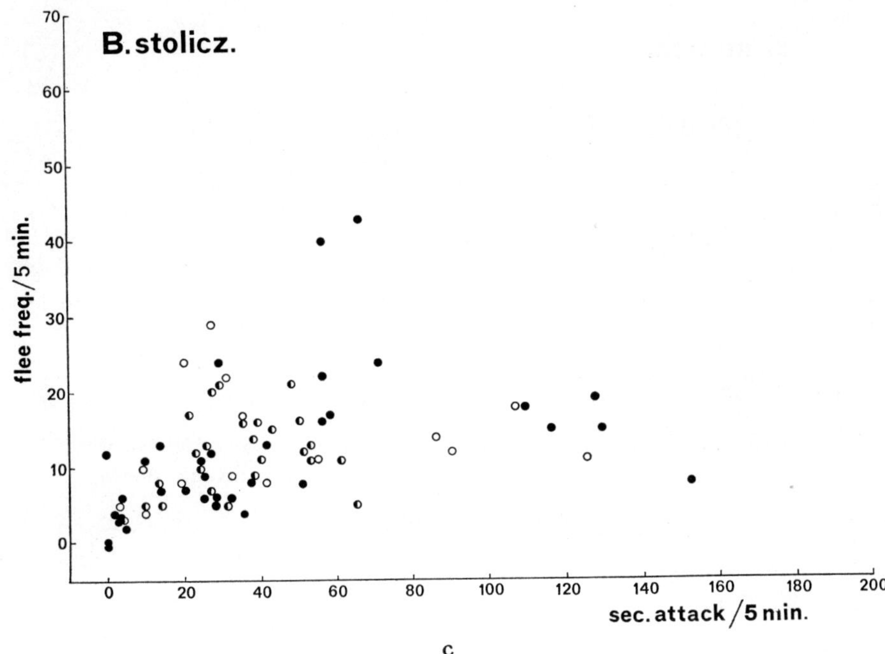

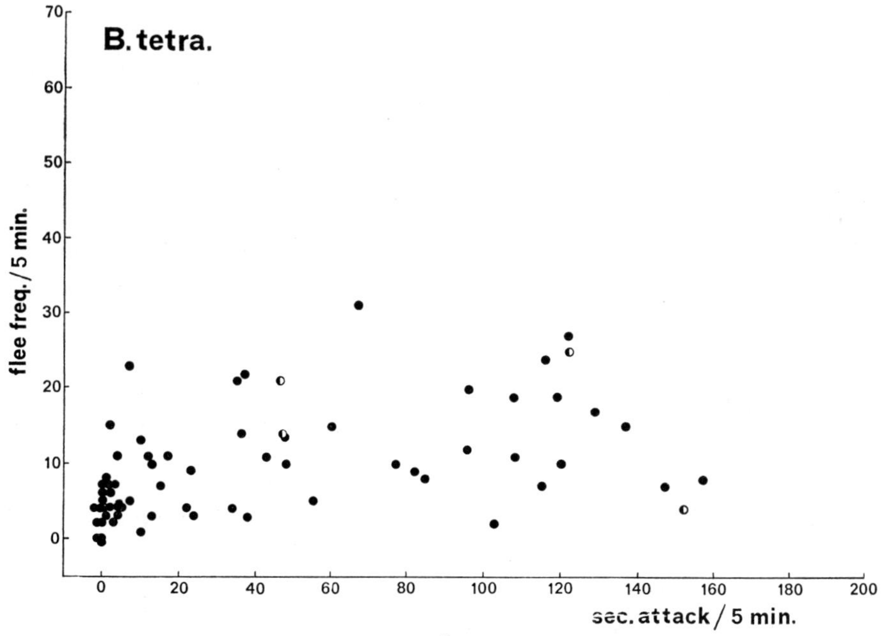

d

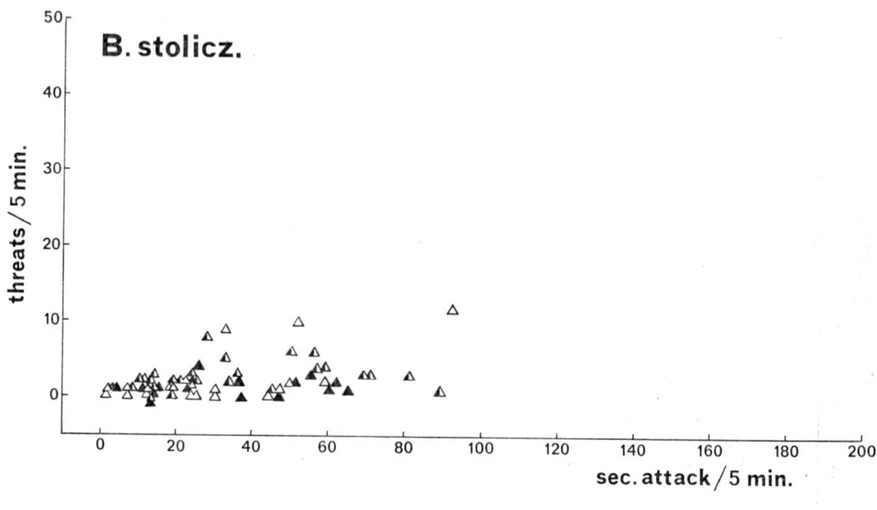

a

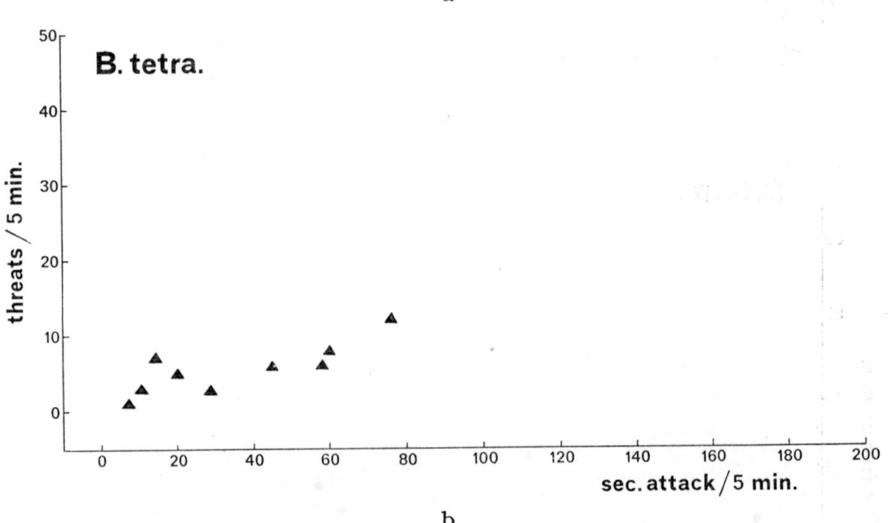

b

Fig. 69a-d. *B. stoliczkanus, B. tetrazona*; attacking and threat (frequency) scores of dominant males in 3 ♂♂ and 2 ♂♂ groups; n.l. and a.l. series combined; periods with established dominance only.

dominant with both eyes, their bodies tense. Only when very seriously harassed by the dominant they may assume a similar passive attitude as do the inferiors of the first two species.

The difference in hiding posture precludes the use of several of the techniques for comparing aggressiveness and fearfulness of different species

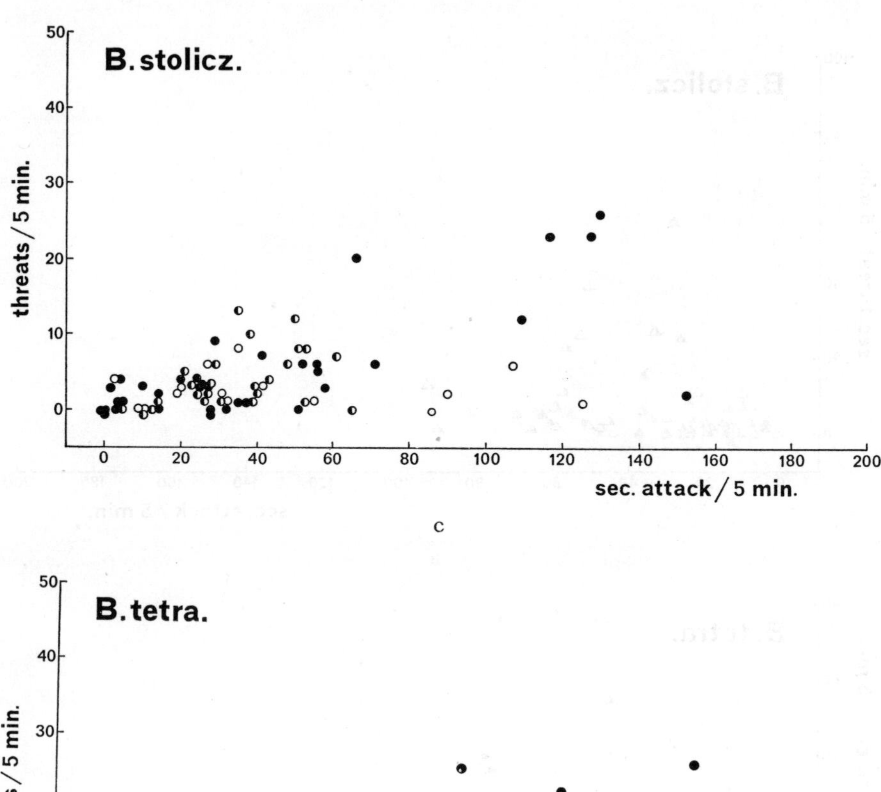

c

d

discussed on p. 199ff. When comparisons across the two pairs of species are to be made, it is senseless to compare aggressive scores of the dominant males even towards hiding inferiors because the stimulus situation is different for different species, even apart from the differences in colour pattern. For the same reason comparisons of fleeing or threat scores of the dominant

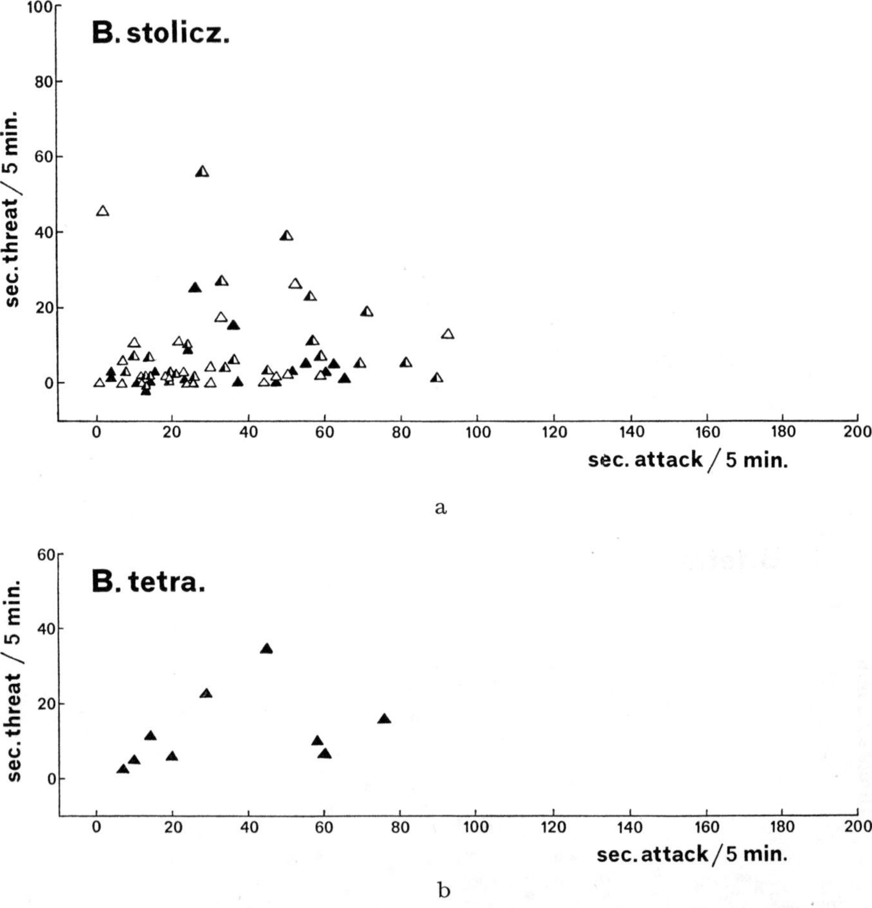

Fig. 70a-d. *B. stoliczkanus, B. tetrazona*; attacking and threat (total duration) scores of dominant males in 3♂♂ and 2♂♂ groups; n.l. and a.l. series combined; periods with established dominance only.

fishes are without sense, at least as far as dominants with hiding inferiors are concerned. Only for periods in which the inferior(s) are completely free the fleeing and threat scores of dominants may perhaps be profitably compared. Likewise, there is probably no reason to discard comparisons between aggressive scores of inferior fishes, either when they are hiding or free. (It must be emphasized, however, that in making these comparisons possible influences from the qualitative differences in colour patterns — such as definition of contrast or green iridescence in faded black markings — are ignored). Consequently, in the following paragraphs we will compare

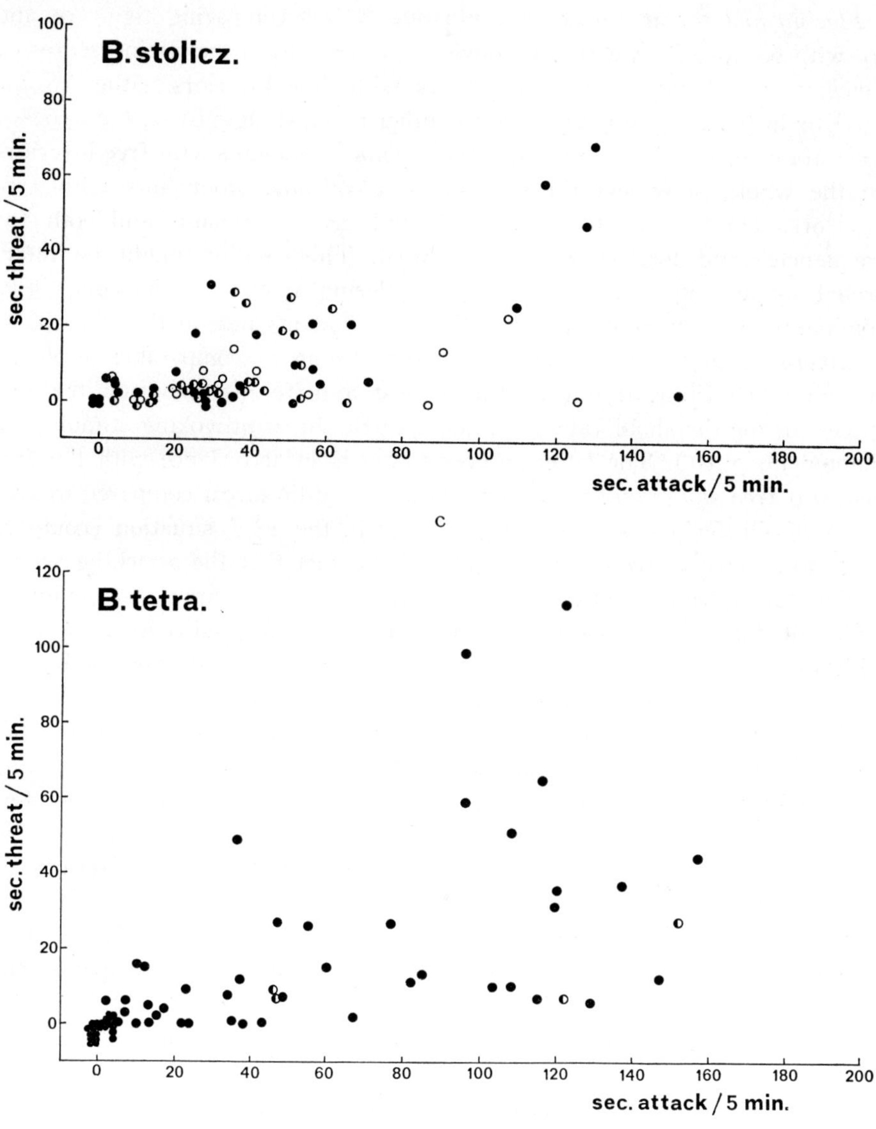

c

d

fleeing and threat scores of *conchonius* and *stoliczkanus* dominants (the only two species in which inferiors are sufficiently often free to allow comparison) and aggressive scores of inferiors of all four species.

Fleeing and threat scores of dominants. When comparing figg. 60[a] and 60[c] with 68[a] and 68[c] we find no obvious differences in the fleeing scores of *conchonius* and *stoliczkanus* dominants with free inferiors, either in the 3♂♂ or in the 2♂♂ situation. On the other hand, if figg. 61[a, c], 69[a, c], 62[a, c], 70[a, c] are compared it is found that *stoliczkanus* dominants with free inferiors on the whole show less threat than do *conchonius* dominants with free inferiors. This is true for both 3♂♂ and 2♂♂ dominants and both for frequencies and total durations of threat. This finding might be interpreted as an expression of a higher fleeing motivation in *conchonius* dominants as compared with *stoliczkanus* dominants but, in the absence of a difference in the frequencies of overt fleeing at comparable levels of attacking behaviour, it might also suggest a difference between the two species in the threshold values for the specific threat-provoking stimuli (see chapter II, p. 81). The latter interpretation is in accordance with the fact that also free *stoliczkanus* inferiors show very little threat compared to free *conchonius* inferiors both in the 3♂♂ and in the 2♂♂ situation (compare figg. 59[a], 67[a], 59[c], 67[c]), this in spite of the fact that the attacking scores of the free inferiors — at comparable levels of fleeing — are not so obviously different for the two species [1]). The interpretation is also in accordance with my (subjective) impression that *conchonius* and *nigrofasciatus* are 'threat specialists' more than are *stoliczkanus* and *tetrazona*.

Aggressive scorings of inferiors. It was suggested in sub-section (a) (p. 210) that inferiors of *conchonius* and *nigrofasciatus* show little attacking when hiding, much less at least than when they are free. Though rather few quantitative data were available to substantiate this suggestion, it is in accordance with the generally passive demeanour of hiding inferiors of these two species. As may be seen from figg. 64[a, c] a similar difference between hiding and free inferiors does not obtain at least in *stoliczkanus*. The comparison cannot be made for *tetrazona* since in this species the inferiors practically always hide but at least the 3♂♂ inferiors of this species did score quite high numbers of attacks even though they hid continuously (fig. 64[b] [2]).

If the 2♂♂ inferiors of the four species are compared (figg. 56[c], 56[d], 64[c], 64[d]) it is found that, at least as long as fleeing scores are low (below some 60″ in *stoliczkanus* and zero in *tetrazona*) hiding inferiors of both *stolicz-*

1) The fact, suggested in the graphs, that hiding or part-time hiding *stoliczkanus* inferiors may occasionally show more threat than do free inferiors of the same species is a different question which cannot be dealt with here.

2) The high attacking scores of some free *tetrazona* 2♂♂ inferiors (fig. 64[d]) are probably typical of a situation in which dominance has only just been established.

kanus and *tetrazona* may score much higher numbers of attacks than do either *conchonius* or *nigrofasciatus* hiding inferiors. A similar comparison cannot be made for the 3♂♂ inferiors of *stoliczkanus* and *conchonius* (because the inferior *conchonius* do not hide) or for the 3♂♂ inferiors of *tetrazona* and *nigrofasciatus* (because of the paucity of periods with little fleeing in the *tetrazona* data).

Somewhat paradoxical results may be obtained if comparisons are made across the two pairs of species. For instance, in comparing the behaviour of hiding inferiors of *stoliczkanus* and *nigrofasciatus* we find that the *stoliczkanus* score more overt attacks both in the 3♂♂ and in the 2♂♂ situation; on the other hand, the dorsal rolling scores are not different in the hiding 3♂♂ inferiors of the two species and are much higher for the *nigrofasciatus* hiding inferiors in the 2♂♂ experiments! Apparently the aggressive tendencies may be expressed in qualitatively different ways in the two species.

An important key to an understanding of the differences between the two pairs of species in the 3♂♂ and 2♂♂ groups may be found in the qualitative differences in hiding behaviour of the inferiors. The tense posture of hiding *stoliczkanus* and *tetrazona* inferiors and their constant eyeing of the dominant accord well with their comparatively high attacking frequencies — which they exhibit especially at low levels of fleeing. If a dominant *stoliczkanus* or *tetrazona* approaches a hiding inferior the latter often opposes him in frontal orientation and indeed may counter-attack. Brief mouth-fighting may even ensue, with the inferior often not threatening at first (see chapter II p. 64). This also accords well with the frequent occurrence of mouth-fighting in duels of *stoliczkanus* and *tetrazona.* As was stated in chapter II, mouth-fighting doesn't occur in duels of *conchonius* and *nigrofasciatus.* All these facts together are in good accordance with my (subjective) impression that *stoliczkanus* and *tetrazona* are 'biting specialists' rather than 'threat specialists' as are *conchonius* and *nigrofasciatus* [1]).

In section II of this chapter it was described that *stoliczkanus* and *tetrazona* males are territorial during reproduction whereas *conchonius* and *nigrofasciatus* are not. In section III it was suggested (p. 184) that hiding inferior males of the former two species in the 2♂♂ 2♀♀ groups may actually defend

---

1) As stated in chapter II, mouth-fighting occurs often but is of short duration in *stoliczkanus* duels; in *tetrazona* duels prolonged mouth-fighting is very common. This difference is another argument in favour of the hypothesis that, of the two species, *tetrazona* is the more aggressive. It is a strong argument since the mirroring effect (see p. 200) is obviously absent in duels.

a minute territory. This suggestion would also fit well the hiding behaviour of inferior *stoliczkanus* and *tetrazona* in the 3♂♂ and 2♂♂ situations. This would be in accordance with the generally alert and aggressive demeanour of hiding males of the two species. The incidental occurrence of mouth-fighting and 'shuttle'-attacks between the dominant and the inferior would also fit in with this view. It was suggested in section II (p. 144) that duels including mouth-fighting occur in 6♂♂ 6♀♀ groups of *tetrazona* mostly during the phase that territories are being established or when territory borders are shifted. The inferior males of the 3♂♂ and 2♂♂ groups, though standing their ground on many occasions, may also often be chased from their 'territory' by the dominant fish (witness the high attacking scores of some dominants). The whole situation may be interpreted as an extreme case of two neighbouring territorial males which are not equally strong (see section II p. 139).

The connection between the kind of hiding behaviour and territorial behaviour is confirmed by the behaviour of *conchonius* and *nigrofasciatus* inferiors during hiding. Their passive demeanour does not suggest any territorial tendencies. It is not clear why *conchonius* and *nigrofasciatus* inferiors never hide in the 2♂♂ 2♀♀ groups (section III, p. 184) whereas they may do so in the 3♂♂ and 2♂♂ groups. One possibility is that the larger number of individuals doesn't lead to hiding in the inferior male; another possibility is that the presence of females and/or the occurrence of courtship by the males prevents hiding. No experiments have been done with 4♂♂ or 2♂♂ 1♀ groups. It may however be reminded that in *nigrofasciatus* it takes a male more time to establish full dominance in a 3♂♂ group than in a 2♂♂ 2♀♀ group (p. 199). A similar effect was not found in *tetrazona* (p. 232). This may indicate a difference between the two species in the effect of the presence of females on at least the initiation of reproductive behaviour.

It is obvious that the 'territorial' demeanour of hiding *stoliczkanus* and *tetrazona* inferiors involves a state of considerable and continuous conflict. The conflict may be expected to be most acute in *tetrazona* inferiors. *Tetrazona* inferiors may be severely harassed by a dominant male and may indeed die from injury or from the infections following these. However, they may also die within a day or two after dominance has been established (in a 2♂♂ group at least) without any considerable physical damage having been done. Once they have started to hide they often do not take any more food do without food for several days, it is probably the continuous state of stress itself which causes death in the latter case. Neither of the two causes but, as *tetrazona* and other species can under more normal circumstances of death has ever been testified in any of the other 3 species (see also section III, *cumingi,* p. 270).

The situation in the 3♂♂ and 2♂♂ groups apparently is stressful for the dominant *tetrazona* as well. Though often quite active, they may sometimes go and stand in one place somewhere near to the bottom of the tank. Their only activity then consists of quite stereotyped, almost neurotic, intention movements of approach and retreat, first directed at the corner in which an inferior is hiding, later more or less independent from the direction in which the inferior is to be found. It is probably not accidental that in the 3♂♂ and 2♂♂ experiments it happened twice that one of the *tetrazona* males jumped out of the tank. In the one case it was known to be the dominant male; in the other it is unknown whether dominance had perhaps already been established late in the evening. Jumping out did not occur in any of the other species.

Tetrazona males may die from shock when moved from one tank, where they have been living together with conspecifics for a long time (say 6 months), to another. They sink to the bottom, fall over and are dead within half a minute. Typically, as they die and thereafter, the dorsal halves of their bodies are tinged quite dark, the ventral halves completely light.

In section II (p. 148) I have described how, in a more natural situation, acute conflicts may arise between schooling and reproductive behaviour. The aggressive tension in a group which is about to switch from schooling to reproduction, the tenseness of the fishes when mobbing a predator and their hectic activity during reproduction, together with the facts described above, point to the conclusion that *tetrazona,* more than any of the other species, also under as nearly natural conditions as could be reached in the laboratory, is an almost constantly high-strung animal which — if the reader will excuse my speaking so anthropomorphically for a moment — as none of the other species 'plumbs the abysses of conflict and stress or rides the peaks of ecstatic excitement'.

(d) *B. cumingi.*

Situational data.

The situations obtaining during the observations are shown in table 14. The table is arranged in the same way as tables 8 and 9; the same symbols are used. The 2♂♂ pl.A experiment was broken off after the observation sessions of day I. The inferior fish was taken out of the tank at 16.45 of that day because by then it had already been very seriously damaged by the dominant. The inferior fish of the 2♂♂ b.A experiment was taken out of the tank at 17.00 on day II for the same reason.

TABLE 14

B.cum.

sessions ↓ / groups →	3♂♂pl.A	3♂♂pl.B	3♂♂b.A	3♂♂b.B
I9	◬◬◬ ◬◬◬ ◬◬◬	△△△	◬◬◬ ◬◬◬ ◬◬◬	◬◬◬
I11	◬◬◬ ◬◬◬ ◬◬◬	◬◬◬	△△△ △△△ △△△	◬◬◬
I14	◬◬◬ ◬◬◬ ◬◬◬	◬◬◬	△△△ △△△ △△△	◬◬◬
II9	◬◬△ ◬◬△ ◬△△	◬◬◬	◬◬◬ ◬◬◬ ◬◬◬	△△△
II11	◭◭◭ ◭◭◭ ◭◭◭	△△△	◬◬◬ ◬◬◬ ◬◬◬	△△△
II14	▲▲▲ ▲▲▲ ▲▲▲	△△△	◬◬◬ ◬◬◬ ◬◬◬	△△△

a

B.cum.

sessions ↓ / groups →	2♂♂pl.A	2♂♂pl.B	2♂♂b.A	2♂♂b.B	2♂♂b.n.l.
I9	● ● ● ● ● ●	⊙ ⊙	⊙ ⊙ ⊙ ⊙ ⊙ ⊙	⊙ ⊙	⊙ ● ⊙ ● ● ●
I11	● ● ● ● ● ●	○ ○	● ● ● ● ● ●	⊙ ⊙	● ● ● ● ● ●
I14	● ● ● ● ● ●	⊙ ⊙	● ● ● ● ● ●	⊙ ⊙	● ● ● ● ● ●
II9		● ●	⊙ ● ● ● ● ●	⊙ ⊙	● ● ● ● ● ●
II11		● ●	● ● ● ● ● ●	⊙ ⊙	● ● ● ◐ ● ◐
II14		● ●	● ● ● ● ● ●	⊙ ⊙	◐ ◐ ◐ ◐ ◐ ◐

b

No transient changes in dominance, such as were found in one of the *conchonius* experiments, took place during any of the observation sessions. Two cases of permanent dominance reversal obtained, one in the 3♂♂ pl.B and another in the 3♂♂ b.A experiment.

Full dominance had not yet been established at the beginning of the first observation session in 3 out of the 4 3♂♂ groups and in 4 out of the 5 2♂♂ groups. Dominance was not established before the second day in 2 3♂♂ groups and in only 1 2♂♂ group. In one of the 2♂♂ groups both fishes kept schooling and fluttering also throughout day II. In the 3♂♂ groups established dominance was found in 58 out of the 144 5′ periods (40%); in the 2♂♂ groups in 89 out of the 114 periods (78%). This indicates that the establishment of dominance was less strongly inhibited in the 2♂♂ than in the 3♂♂ groups. (Similar results are obtained if observation sessions are chosen as units instead of 5′ periods). The delays in the establishment of dominance in the 3♂♂ groups are of the same order as those found in *nigrofasciatus* and *tetrazona* groups. The delays in the 2♂♂ groups are equalled by the *tetrazona* 2♂♂ groups only.

An interesting feature not found in the other four species is that in several cases the fishes returned to the schooling phase after dominance had already been established. This happened in 2 out of the 4 3♂♂ groups, the same groups in which dominance reversal took place. In the one case (3♂♂ b.A) one male first became dominant on day I. Shortly after the last observation session of that day he was beaten by another in a duel. The latter male was very weakly dominant for the rest of the day and the whole of day II. He asserted full dominance only in the course of the third day. In the other case (3♂♂ pl.B) one male had already established full dominance prior to the first observation session. In the course of the first day, however, his dominant position became less and less clear. This situation was continued into day II until, between the first and the second observation session of that day, another male became fully dominant. It is not known whether a duel accompanied this reversal of dominance. The two cases suggest the existence of some causal relationship between the return to the schooling phase and dominance reversal. A third case of return to schooling phase obtained in the 2♂♂ pl.B experiment. After one male had become dominant prior to the second observation session of day I, both males again mainly schooled and fluttered during the third session. During this session the original dominant still showed some aggressive behaviour (12″ attack, 10 butting movements and 2″ threat) whereas the other male scored nothing but 5 very short inhibited attacks (amounting to about 2″ together). No dominance reversal took place in this group.

Hiding was relatively rare in the 3♂♂ inferiors. Both inferiors were free during 40 out of the 58 5′ periods with established dominance (69%); they hid continuously and for part of the time in 9 periods each (16%). On the other hand, hiding was very common in the 2♂♂ inferiors. Continuous hiding obtained during 79 out of the 89 5′ periods with established dominance (89%); the inferior was free in only 2 periods (2%). Thus the difference between the hiding tendencies of 3♂♂ and 2♂♂ inferiors is larger than in any of the other four species.

The heights at which inferiors hid are shown in tables 15[a, b]. For each observation period both the minimal and maximal heights are given. It follows from the table that *cumingi* inferiors, when hiding, often hide near to the bottom, more often than any of the other four species. In fact, the hiding inferiors often rested on the bottom for much of the time. On the other hand, if hiding is not near to the bottom it is mostly near to the surface. Intermediate positions are rare. Hiding inferiors often did not

TABLE 15

Lowest and highest values for height of hiding inferior(s)

(explanation see text)

a. *cumingi* 3 ♂♂ groups

group 3 ♂♂ pl. A only; 2 sessions day II: both inferiors hiding at 0.

b. *cumingi* 2 ♂♂ groups

groups:	2 ♂♂ pl. A	2 ♂♂ pl. B	2 ♂♂ b. A	2 ♂♂ b. B	2 ♂♂ b. n.I.
sessions					
I9	0—9/10	—	—	—	0—9/10
I11	0—9/10	—	0	—	0
I14	9/10	—	0—9/10	—	0
II9		0	0	—	0
II11		0	0	—	0
II14		0	9/10	—	0

stand in a corner but lay on the bottom or hung against the surface at rather arbitrary places.

In all cases the demeanour of the hiding inferiors was rather passive; the body axis was kept about horizontal both at the 0 and the 9/10 position. They did not face the dominant fish as did *stoliczkanus* and *tetrazona* hiding inferiors.

We may summarize the data presented so far as follows. (1) it takes a *cumingi* male about as much time as it does a *tetrazona* or (in a 3 ♂♂ group) a *nigrofasciatus* male to establish full dominance (2) once dominance has been established, *cumingi* 2 ♂♂ inferiors hide as much as do *tetrazona* or *nigrofasciatus* 2 ♂♂ inferiors; *cumingi* 3 ♂♂ inferiors hide even less than do *stoliczkanus* 3 ♂♂ inferiors (3) hiding is much more common in 2 ♂♂ than in 3 ♂♂ inferiors (4) it takes a male more time to establish dominance in a 3 ♂♂ than in a 2 ♂♂ group (5) hiding by the inferiors is near to the bottom more often than in any of the other four species (6) hiding inferiors are rather passive, more like *conchonius* and *nigrofasciatus* than like *stoliczkanus* and *tetrazona* inferiors.

Conclusion (1) is further supported by many unstandardized observations of small all-male groups in similar tanks. No such data are available to either support or contradict the other conclusions.

Conclusion (2) is in accordance with what has been reported on the behaviour of inferior males in 2 ♂♂ 2 ♀♀ groups only as far as the 2 ♂♂ inferiors are concerned (*cf* section III p. 184). Rather surprisingly, 3 ♂♂ inferiors seem to hide less than do either 2 ♂♂ inferiors or 2 ♂♂ 2 ♀♀ inferior males. This suggests that for the hiding tendencies of the inferior males it is mainly the number of males that counts. If in *cumingi*, as was suggested for *nigrofasciatus,* (p. 262) the presence of females stimulates the reproductive behaviour also of the inferior male, such an effect is completely overridden by the extreme aggressiveness of the dominant male.

Conclusion (1) contrasts with my experiences with small ambisexual groups (2♂♂ 2♀♀ mostly) in similar tanks. In such groups dominance is usually found to be established early in the morning after introduction. Intensive reproductive behaviour is then shown, except if a female is the dominant fish — which happens relatively often in this species. Complete reproductive behaviour was also observed on the first morning after introduction in the 6♂♂ 6♀♀ group described on p. 154 (section II).

We may conclude that (7) like in *conchonius* and *nigrofasciatus,* establishment of dominance and the development of reproductive behaviour is quicker in ambisexual than in all-male groups (8) like in *stoliczkanus* and *tetrazona* and unlike in *conchonius* and *nigrofasciatus,* inferior males in ambisexual groups often hide and do not partake in courtship. This is so in spite of the fact that 3♂♂ inferiors are often completely free.

Conclusions (3) and (4) are qualitatively analogous to those stated for all other four species. It may be clear that the combination of the other six conclusions sets *cumingi* apart from both the *conchonius-nigrofasciatus* and the *stoliczkanus-tetrazona* types.

Behaviour records.

In order to interprete and supplement the situational data we will now turn to the quantative behavioural data of the individual fishes. The same data will be presented for this species as were given for the other four. The data will be compared with those of the other four species as far as possible.

Aggressive scores of dominants. Figg. 71[a, b] represent — for the 3♂♂ and 2♂♂ experiments separately — the time spent in full attack and the number of butting movements by the dominants in all 5′ periods. For the 3♂♂ experiments the activities of the dominant directed at both inferiors have again been added together. The data from the 2♂♂ b. n.l. experiment have been discarded because in this experiment full attack and distance attack had not been properly distinguished. The data from all 5′ periods without established dominance have been left out.

The following trends are apparent from the graphs: (a) for both the 3♂♂ and 2♂♂ dominants there is a distinct positive correlation between the two measures; the frequency of the butting movements (per seconds attack) is practically independent from the absolute amounts of both. (b) there is no obvious difference between the 3♂♂ and the 2♂♂ dominants in the frequency of butting movements. (c) there is no obvious difference between

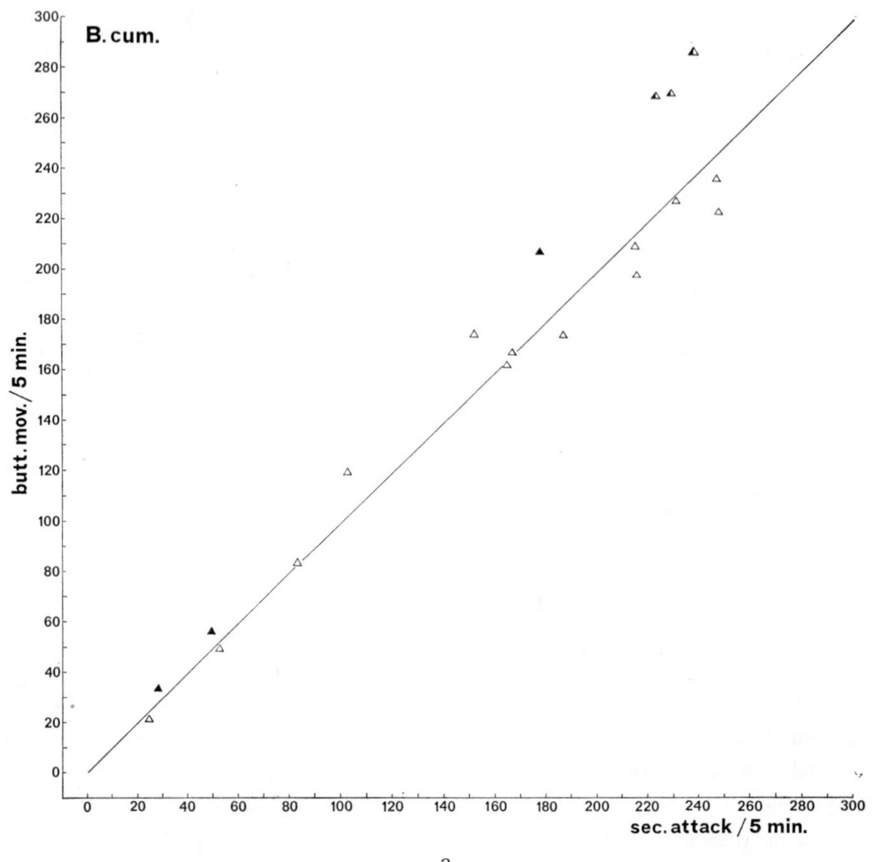

Fig. 71a-b. *B. cumingi*; aggressive scorings of dominant males in 3♂♂ (triangles) and 2♂♂ (circles) groups; 45° diagonal for reference; a.l. series only; periods with established dominance only; specially marked points graph b: inferior hiding at 9/10 continuously.

this and the other four species in the frequency of butting movements. (d) It is difficult to assess whether hiding inferiors are attacked less than are free inferiors but the presence of such a difference does not seem very likely: some of the highest scores of the 3♂♂ dominants are made on continuously or part-time hiding inferiors; although the 2♂♂ inferiors nearly always hid continuously and the 3♂♂ inferiors were mostly free, the aggressive scores of the 2♂♂ dominants are not lower than those of the 3♂♂ dominants. (e) Both aggressive scores are higher than those of any of the other four species; the highest attacking scores approach the maximum score of 300″/5′. This is true for both the 3♂♂ and 2♂♂

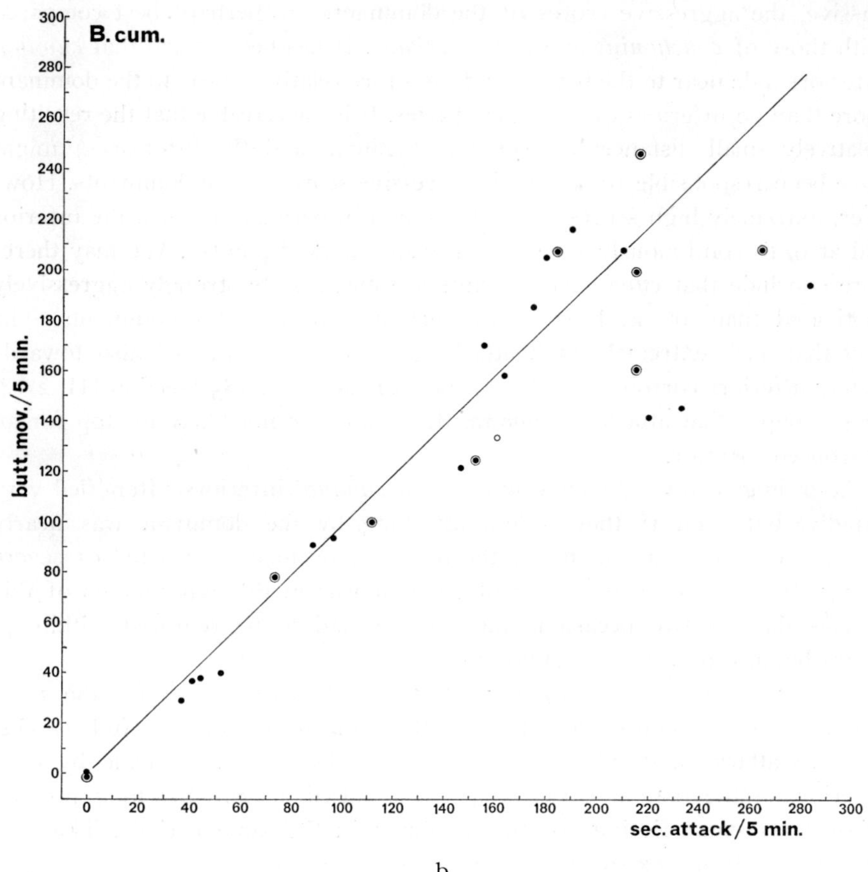

b

dominants and for dominants with hiding inferiors as well as for dominants with free inferiors.

Point (e) requires some critical evaluation. Because of the mirroring effect, the high aggressive scores towards free inferiors may indicate a comparatively high aggressive motivation in the dominant or a high fleeing motivation in the inferiors or both. The high aggressive scores attained also at hiding inferiors plead for the first interpretation. However, as we have seen in the fore-going subsections, also the place and the manner of hiding must be taken into account. Since hiding *cumingi* inferiors are rather passive, the aggressive scores of the dominants are perhaps best compared with those of *conchonius* and *nigrofasciatus*. It has been stated that *cumingi* inferiors hide near to the bottom and therefore relatively near to the dominant more than do inferiors of the other species. It is conceivable that the resulting relatively small distance between the dominant and the inferior(s) might have been responsible for the high aggressive scores of the dominants. However, extremely high scores were also found in periods in which the inferior hid at 9/10 continuously (see 71[b], specially marked points). We may therefore conclude that *cumingi* dominants are much more strongly aggressively motivated than are at least *conchonius* or *nigrofasciatus* dominants. The fact that such extremely high attacking scores were attained also towards hiding inferiors corroborates the statement made on p. 185 (section III, 2♂♂ 2♀♀ groups) that attacking *cumingi* dominants are not likely to stop for an ensconced inferior.

According to my subjective impression *cumingi* inferiors often fled very rapidly, but even if they didn't, attacking by the dominant was nearly always very vigorous; many of the butting movements consisted of severe biting. It is no doubt not accidental that it was in the experiments of this species that on two occasions the inferior had to be removed within 48 hours because of serious woundings.

Aggressive scorings of inferiors. Figg. 72[a, b] represent — for the 3♂♂ and 2♂♂ experiments separately — the number of attacks (full attacks, distance attacks and inhibited attacks all added together) made by each inferior on the dominant per 5′ period, plotted against the total number of seconds spent fleeing for the dominant in the same periods. The data of the 2♂♂ b. n.l. experiment have here been included.

In comparing these data with those of the other four species we will again ignore the possible influences of the qualitative differences in the colour patterns of the species (*cf.* p. 258). The data of 72[a] suggest that in the 3♂♂ experiments hiding inferiors exhibit less attacking than do free inferiors. Neither in the 3♂♂ nor in the 2♂♂ groups hiding inferiors showed any considerable amount of attacking at low levels of fleeing. In these two respects the *cumingi* inferiors resemble the *conchonius* and *nigrofasciatus* and differ from the *stoliczkanus* and *tetrazona* inferiors. The data are in accordance with the generally passive demeanour of the hiding inferiors. The free *cumingi* 3♂♂ inferiors display considerable numbers of attacks on the dominant — more than do *conchonius* inferiors (at the higher levels of fleeing at least) but not as many as do free *nigrofasciatus* or hiding *tetrazona* 3♂♂ inferiors. The *cumingi* 2♂♂ inferiors show little attacking on the dominant. Like in the *nigrofasciatus* and the *tetrazona* 2♂♂ inferiors

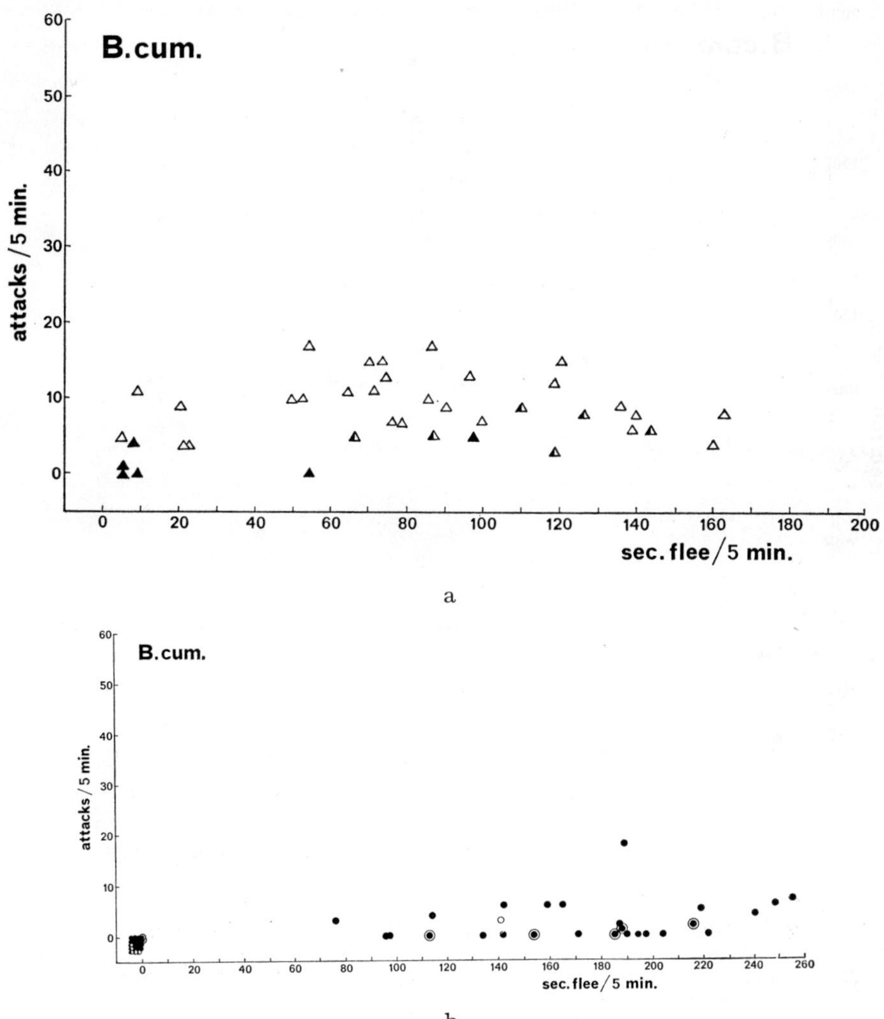

Fig. 72a-b. *B. cumingi*; fleeing and attack scores of inferior males in 3♂♂ (triangles) and 2♂♂ (circles) groups; n.l. and a.l. series combined; periods with established dominance only.

this is probably due to mere complete submission by the dominant as compared with the 3♂♂ inferiors. This interpretation is supported by the fact that 2♂♂ inferiors both flee and hide much more than do 3♂♂ inferiors. An interesting feature, not found in the *nigrofasciatus* or *tetrazona* 2♂♂ inferiors, is the apparent positive correlation between fleeing and attacking in the *cumingi* 2♂♂ inferiors. This correlation is not easy

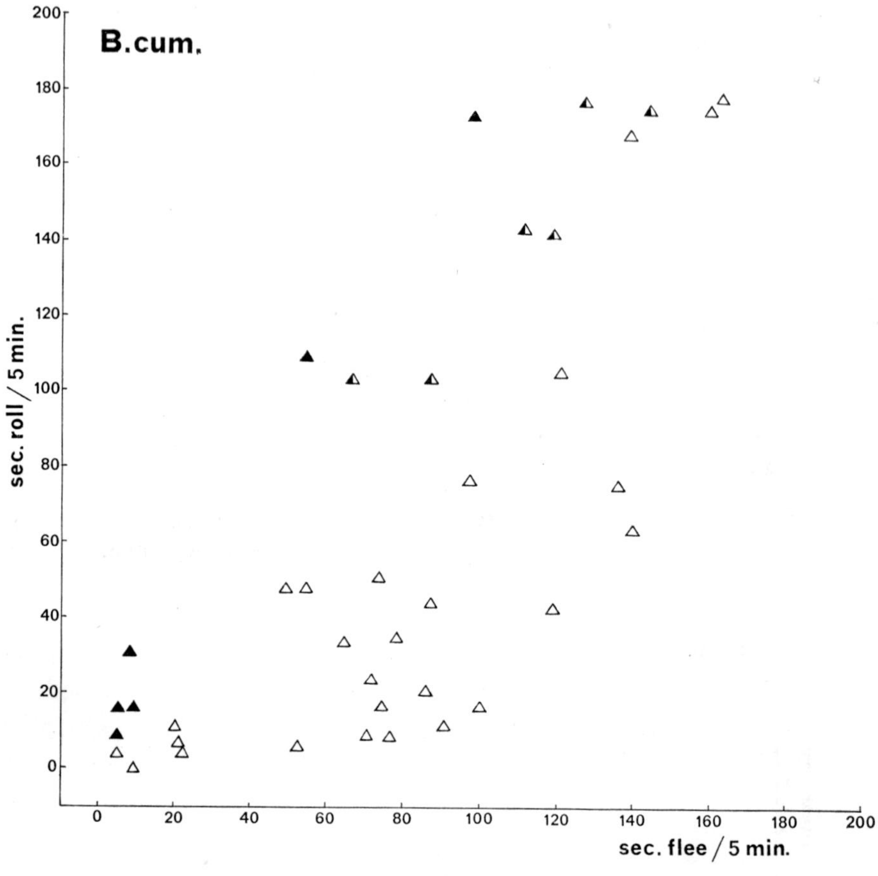

a

Fig. 73[a-b]. *B. cumingi*; fleeing and (total) roll scores of inferior males in 3♂♂ (triangles) and 2♂♂ (circles) groups; n.l. and a.l. series combined; periods with establish dominance only; specially marked points graph b: inferior hiding at 9/10 continuously.

to interprete but it may be due to a tendency of inferiors to launch short counter-attacks just after the dominant has stopped attacking. This might be interpreted as vestigial 'shuttle'-attacking.

Figg. 73[a, b] and 74[a-b] represent the total amounts of rolling (ventral *plus* dorsal) and the amounts of dorsal rolling alone displayed by the inferior(s) towards the dominant per 5′, both plotted against the total amounts of fleeing for the dominant. In the 3♂♂ groups most of the inferiors which are free display relatively little rolling (less than fleeing in most cases) and almost no dorsal roll. On the other hand, the inferiors which hide

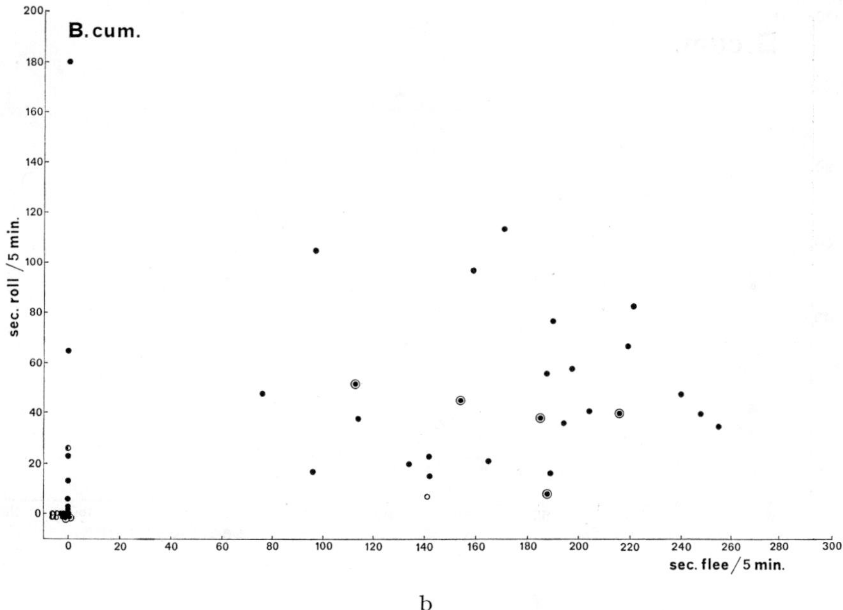

b

(continuously or part-time) show considerable amounts of rolling (more than fleeing in all cases) and a considerable portion of this is dorsal roll. Thus we have the surprising situation that, whereas the free inferiors attack more than do the hiding ones, they roll less at comparable levels of fleeing. A similar phenomenon was so far found in the *nigrofasciatus* 3♂♂ inferiors at low levels of fleeing only. However, in *nigrofasciatus* the difference is mainly due to the hiding inferiors displaying considerable amounts of rolling but almost no overt attacks. This may be readily interpreted as the result of suppression of overt attack but not of ambivalent movements in the completely submitted hiding inferiors. In the *cumingi* inferiors, on the other hand, the difference is also due of the free inferiors displaying considerable numbers of attacks but relatively little rolling. It seems that, at least in free *cumingi* inferiors, the frequent occurrence of both fleeing and attacking (or the factors underlying these) does not facilitate the occurrence of rolling to the same degree as it does in the other four species.

The 2♂♂ inferiors are similar to those of the more aggressive species *nigrofasciatus* and *tetrazona* in that they roll less than do the 3♂♂ inferiors

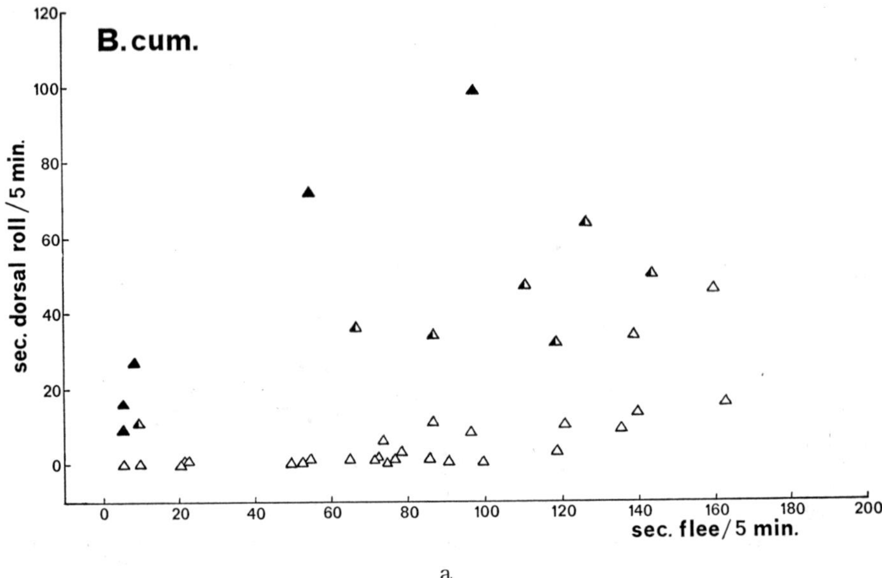

a

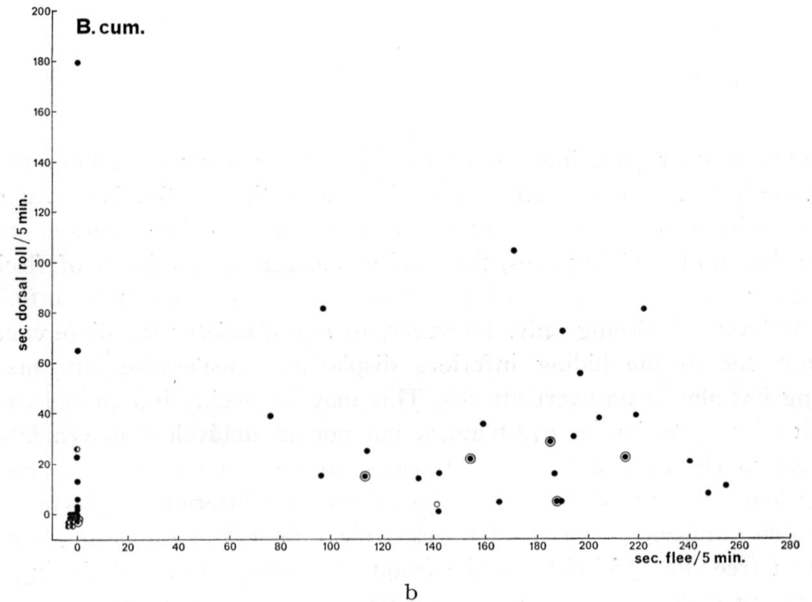

b

Fig. 74[a-b]. *B. cumingi*; fleeing and dorsal roll scores of inferior males in 3♂♂ (triangles) and 2♂♂ (circles) groups; n.l. and a.l. series combined; periods with established dominance only; specially marked points graph b: inferior hiding at 9/10 continuously.

and that much of this rolling is dorsal roll. It is interesting to note that they do not roll much less — at comparable levels of fleeing — than do the free 3♂♂ inferiors of the same species; the amounts of dorsal roll are even much higher than in the free 3♂♂ inferiors. In this they differ from *nigrofasciatus* inferiors. (Comparison with *tetrazona* inferiors is not possible since these are never free). Since the rolling and attacking scores of the *cumingi* 2♂♂ inferiors do not differ essentially from those of the *nigrofasciatus* or *tetrazona* 2♂♂ inferiors, the impression is strengthened that it is the free 3♂♂ inferiors which are peculiar in showing relatively little rolling.

The large amounts of dorsal roll scored by hiding *cumingi* inferiors are in accordance with the observation that inferiors hiding pressed against the bottom often show very strong dorsal roll as a first reaction to approach by the dominant; this may occur even if the dominant passes at a considerable distance (>10 cm). The total rolling and dorsal rolling scores of those inferiors which hid continuously near to the surface are specially marked in the graphs. It is evident that their total rolling scores are not very high and their dorsal rolling scores are comparatively low. When approached by the dominant, inferiors hiding at 9/10 often do not roll at all but only flee fast as a first reaction.

Figg. 75[a, b] represent the number of threats by the inferior(s) towards the dominant, plotted against the total time spent fleeing for the dominant. It is evident that threat scores are practically nil in almost all periods. The scores are as low as those of *nigrofasciatus* inferiors and perhaps even lower than those of *stoliczkanus* and *tetrazona* inferiors.

The data on the behaviour of the *cumingi* inferiors do not allow a simple quantitative comparison with any of the other four species. The particularly strong tendency to hide near to the bottom suggests the presence of considerable aggressiveness in the inferiors, both because they are relatively near to the dominant and because they prefer the lowest levels of the tank [1]). This feature is particularly eloquent in view of the extreme aggressiveness (both quantitatively and qualitatively) of the dominant. The same may be said on the high incidence of dorsal roll both relative to the amounts of fleeing and of ventral roll. (It may be emphasized that the preference for dorsal roll and the tendency to hide low may not be mutually independent). On the other hand, the passive behaviour of the hiding *cumingi* inferiors does not suggest any territorial tendencies. This precludes any simple quantitative comparison of their attacking scores with those of *stoliczkanus* and *tetrazona*. Comparison with *conchonius* and *nigrofasciatus*

1) For an interpretation of depth preference see section II, p. 112ff.

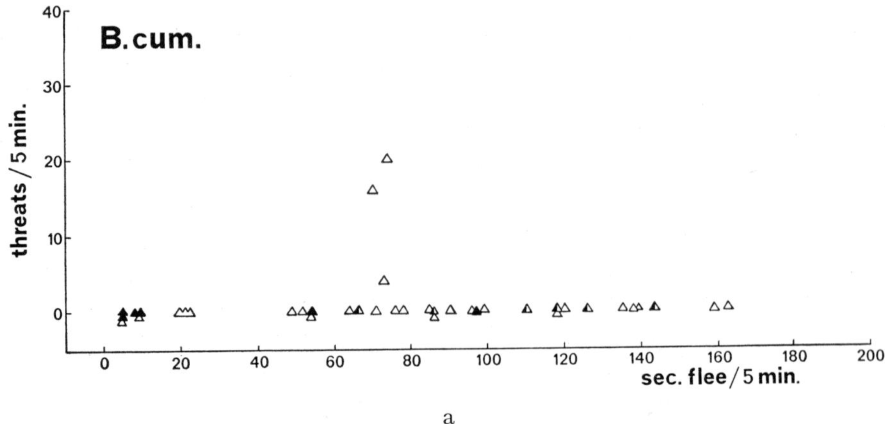

a

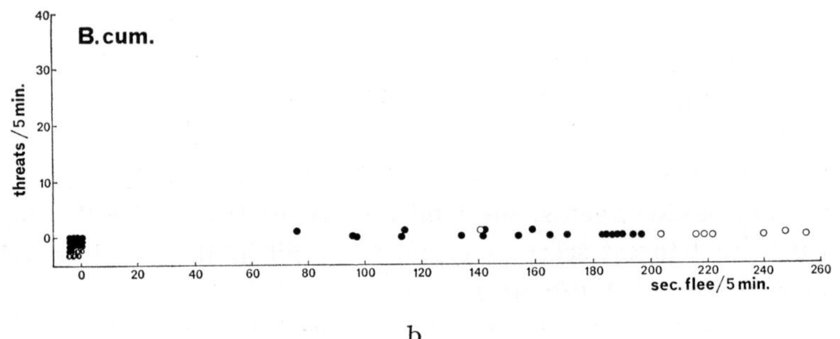

b

Fig. 75a-b. *B. cumingi*; fleeing and threat (frequency) scores of inferior males in 3♂♂ (triangles) and 2♂♂ (circles) groups; n.l. and a.l. series combined; periods with established dominance only.

may be more fruitful. As far as the aggressive scores of the dominants and the behaviour of the hiding inferiors are concerned, *cumingi* seems to present itself as an even more aggressive version of *nigrofasciatus*. However, the behaviour of the free 3♂♂ inferiors does not fit this interpretation. Both their total rolling scores and their dorsal rolling scores are lower — at comparable levels of fleeing — than those of the free *nigrofasciatus* 3♂♂ inferiors. Also their attacking scores, though considerable, are lower than the corresponding *nigrofasciatus* scores.

It might be supposed that the aggressive scorings of the free *cumingi* 3♂♂ inferiors are comparatively low because the attacks of the *cumingi* dominants are more intensive than those of the *nigrofasciatus* dominants, but this wouldn't explain why their rolling scores are lower than those of the hiding inferiors of the same species. Moreover, the question arises why the free *cumingi* inferiors, if they are attacked so fiercely, do not hide.

The fact that they do remain free, in spite of the fierce attacking of the dominant, may itself be interpreted as an expression of aggressiveness [1]). We must, therefore, conclude that the aggressive tendencies of the inferior *cumingi* and *nigrofasciatus* are expressed in qualitatively different ways. *Nigrofasciatus* 3♂♂ inferiors may express their aggressiveness by remaining free and scoring much rolling, dorsal rolling and attacking or, if they hide, by scoring much rolling only. *Cumingi* 3♂♂ inferiors, on the other hand, express their aggressiveness in the first place by remaining free and scoring considerable numbers of attacks; they score little rolling in this case. If they hide they express their aggressiveness by scoring considerable amounts of rolling and dorsal rolling but they score little attacking.

It was stated in section II (p. 158) that *cumingi* males are territorial throughout reproductive periods in near-natural circumstances. In section III (p. 184) it was stated that no traces of territorial behaviour could be detected in the inferior males of the 2♂♂ 2♀♀ groups. Apparently the same holds for the 3♂♂ and 2♂♂ inferiors. For the 2♂♂ 2♀♀ inferior males the absence of territorial tendencies was ascribed to the unrestrained aggression by the dominant (p. 185)). It is implied that the aggressive bombardment by the dominant did not leave to the inferior male any possibilities but passive hiding. The results of the 3♂♂ experiments suggest that this interpretation is incomplete. If the 3♂♂ inferiors manage so often to remain free in spite of the attacks of the dominant, it is not logical to assume that their territorial tendencies are suppressed as a result of complete subordination. In this respect their demeanour differs markedly from that of *tetrazona* inferiors which almost always hide and yet very often retain some territorial tendencies.

It is impossible to asses, on the basis of the present evidence, exactly which factors are responsible for the absence of territorial behaviour in *cumingi* inferiors in small groups. However, the following remarks may be relevant. First, it was stated in section II (p. 159) that the territorial behaviour of reproductive *cumingi* males differs considerably from that of *tetrazona* and *stoliczkanus* males in that one-sided or mutual penetration by the males into each other's territories was much more common in the first species. Second, *cumingi* is obviously not a 'biting specialist' such as are *stoliczkanus* and *tetrazona*; frontal postures (including frontal threat) which

---

1) For an interpretation of preference for seeking open spaces instead of vegetation see section II, p. 112ff.

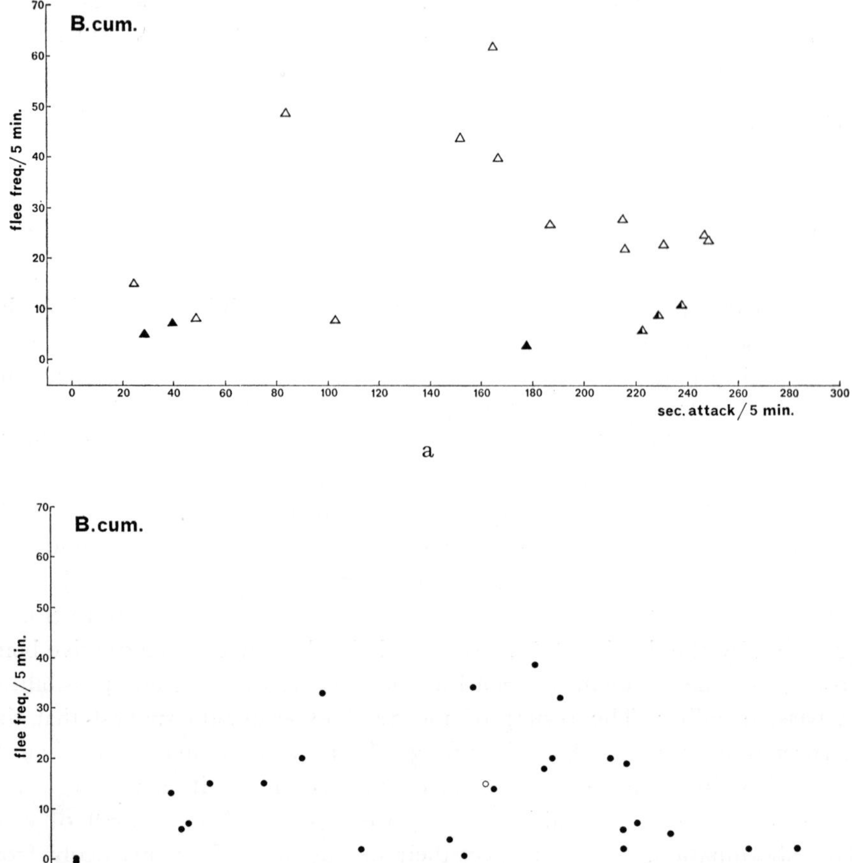

Fig. 76a-b. *B. cumingi*; attacking and flee (frequency) scores of dominant males in 3♂♂ and 2♂♂ groups; a.l. series only; periods with established dominance only.

are common in the latter two species, are rare in *cumingi*. Also mouth-fighting is virtually absent in this species (chapter II, section I, p. 64; chapter III, section I, p. 106). Thus *cumingi* lacks a category of behaviour which plays an important part in the 'territorial' defence of inferior *stoliczkanus* and *tetrazona*. Third, it may be reminded that in the 2♂♂-2♀♀ groups the *cumingi* dominants — like *stoliczkanus* and *tetrazona* and unlike *conchonius* and *nigrofasciatus* dominants — attained very low mating scores only (section III, p. 186). This in spite of the fact that they did

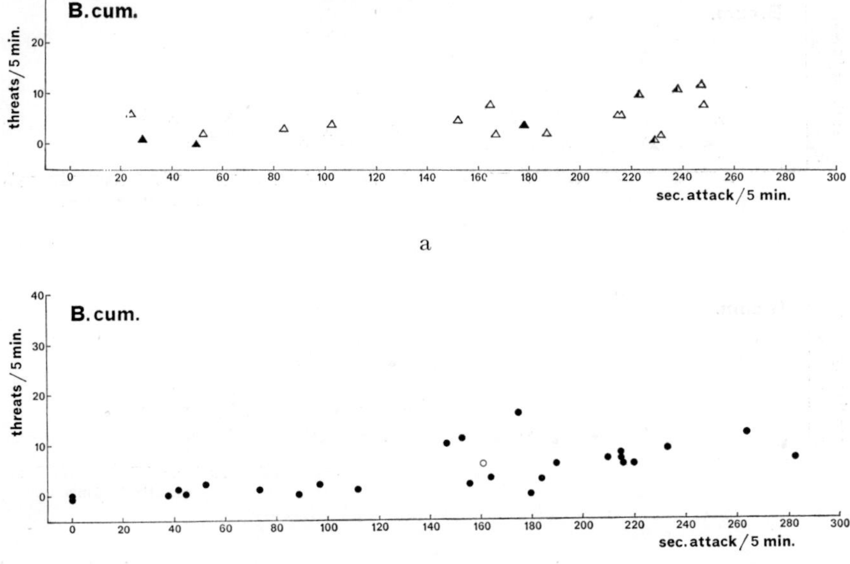

Fig. 77a-b. *B. cumingi*; attacking and threat (frequency) scores of dominant males in 3♂♂ and 2♂♂ groups; a.l. series only; periods with established dominance only.

attain comparatively high mating scores and proved comparatively immune to severe inter-male competition in the 6♂♂ 6♀♀ situation (section II, p. 156). This may suggest that it is either the crowded condition or the small size of the tanks *per se* which limits the full deployment of reproductive behaviour and therefore of territorial behaviour in both the dominant and the inferior *cumingi* males in the small groups.

Flight scorings of dominants. Figg. 76a, b, 77a, b and 78a, b represent for all 5′ periods — for the 3♂♂ and 2♂♂ experiments separately — the number of times the dominant fled for the inferior(s), the number of times he threatened the inferior(s) and the total duration of threat towards the inferior(s), all three plotted against the total duration of full attack on the inferior(s). For the 3♂♂ experiments the activities of the dominant directed at both inferiors have again been added together. The data from the 2♂♂ b. n.l. experiment have again been discarded. Only 5′ periods with established dominance have been included.

Because of the qualitative differences in hiding behaviour, flight scorings of the dominants with hiding inferiors cannot be directly compared with the analogous scorings of *stoliczkanus* or *tetrazona* dominants. Comparison with the flight scorings of *conchonius* and *nigrofasciatus* dominants

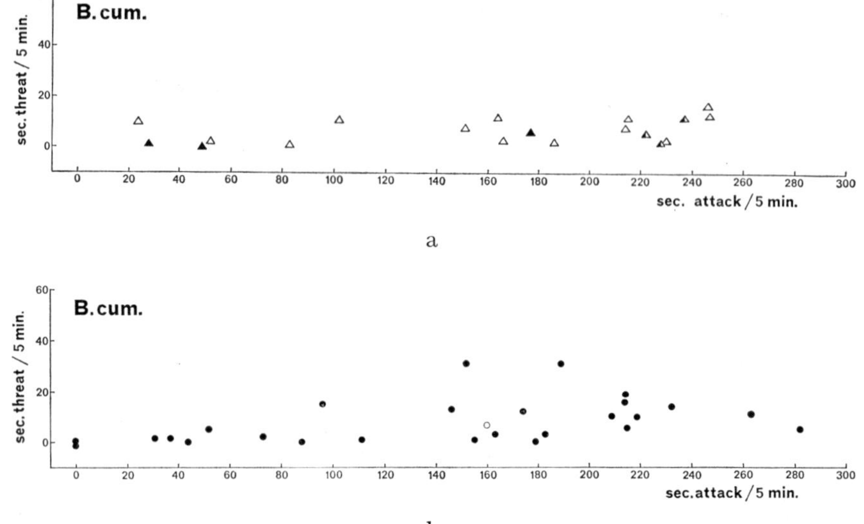

Fig. 78a-b. *B. cumingi*; attacking and threat (total duration) scores of dominant males in 3♂♂ and 2♂♂ groups; a.l. series only; periods with established dominance only.

may be made; possible consequences of the difference in the height of hiding will be discussed below. Flight scorings of dominants with free inferiors, on the other hand, may be compared with the analogous scorings of all four other species. In making the comparisons we will again ignore the possible influences of the qualitative differences in colour patterns between the species.

In the *cumingi* data situations with free inferiors are practically restricted to the 3♂♂ groups. In comparing the fleeing scores of the dominants from these situations with the analogous fleeing scores of the dominants of the other four species we find that — as far as the attacking scores allow comparison — the *cumingi* dominants on several occasions scored much higher values than did *conchonius, nigrofasciatus* or *stoliczkanus* dominants. (Comparison with *tetrazona* is not possible because the 3♂♂ inferiors of this species always hid). The fleeing scores of the *cumingi* dominants do however show an enormous variation; several scores — again as far as the attacking scores allow comparison — fell well within the ranges of the other three species or even were comparatively low.

The few 3♂♂ *cumingi* dominants with hiding inferiors scored very low values for fleeing. No sufficient data are available to allow comparison with *conchonius* or *nigrofasciatus*. The fleeing scores of the *cumingi* 2♂♂

dominants with hiding inferiors again show an enormous variation. On several occasions the scores were much higher — at comparable levels of attacking — than those of either *conchonius* or *nigrofasciatus* 2♂♂ dominants with hiding inferiors.

In contrast to the fleeing scores, the threat scores of the *cumingi* dominants are very low, both in frequency and in total duration per 5′. The 3♂♂ dominants with free inferiors scored considerably less threat — at comparable levels of attack — than did *conchonius* and *stoliczkanus* 3♂♂ dominants with free inferiors, especially as far as the total duration of threat is concerned. Only the *nigrofasciatus* 3♂♂ dominant with free inferiors scored as low as did the *cumingi* dominants. The 2♂♂ *cumingi* dominants with hiding inferiors showed less threat, both in frequency and in total duration, than did the 2♂♂ dominants with hiding inferiors of any of the other four species.

The most striking feature in the *cumingi* dominant's flight scorings is the comparative scarcity of threat in spite of the fact that attacking and fleeing scores are often much higher than in the other species. Thus it seems that, in the *cumingi* dominants, the frequent occurrence of both attacking and fleeing (or the factors underlying these) does not facilitate the occurrence of threat to the same degree as it does in the other four species. The relative paucity of threat may be explained by assuming that the sensitivity for the specific threat-provoking stimili is particularly low in this species (*cf.* chapter II, section II, p. 81). Alternatively, however, it may be a result of a special kind of interaction between the attacking and fleeing systems, typical for this species, which leaves little room for the occurrence of ambivalent states. The latter interpretation seems more attractive because it could also explain the relative paucity of ambivalent behaviour in other categories of behaviour. For instance, it was found (p. 273) that the free 3♂♂ inferior *cumingi* displayed comparatively little rolling even though their fleeing and attacking scores were relatively high. Likewise, rapid oscillation between fierce attacking and swift leading without much intervening ambivalence was found to be typical for *cumingi* courtship (section II, p. 160). Nothing in the behaviour of the other four species compares with the sudden qualitative change in behaviour exhibited by a *cumingi* male which from one moment to the next switches from fierce chasing or close courting posture to wide and swift leading, hardly managing to approach the female again between successive leading movements (see section I, fig. 43, p. 106). The relative rarity and the comparatively low degree of formalization of the duels in this species also fits in with this view (see section I, p. 106; section II, p. 160). It is well possible that it is the paucity or the short duration of

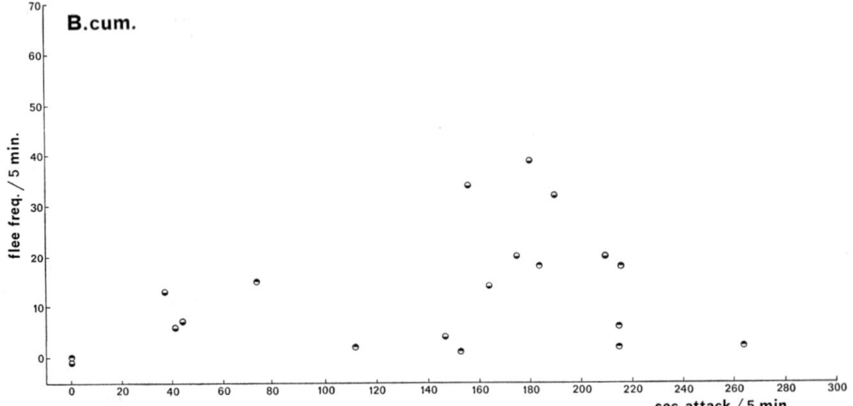

Fig. 79. *B. cumingi*; attacking and flee (frequency) scores of dominant males in 2♂♂ groups (*cf.* fig. 76b); circles: upper half black: inferior hiding at 9/10 continuously; lower half black: inferior hiding at 0 continuously.

ambivalent states that is responsible also for the paucity of threat in the dominants of the 3♂♂ and 2♂♂ groups of this species, because it does not leave the specific threat-provoking stimuli sufficient time to exert their effect.

The high frequencies of overt fleeing displayed by the dominant males at least in some periods may indicate a comparatively high fleeing motivation in this species. However, a few alternative interpretations must be considered. First, as far as the dominants with hiding inferiors are concerned, it may be reminded that *cumingi* inferiors hid near to the bottom and therefore relatively near to the dominant much more often than did *conchonius* or *nigrofasciatus* inferiors. The relative proximity of the hiding inferiors might well have provoked relatively high fleeing frequencies in the dominants. Fig. 79 represents the fleeing scores of the 2♂♂ dominants which were faced by inferiors hiding at 0 and 9/10 positions respectively. The graph may suggest that fleeing scores of the dominant are higher if the inferior hides at the 0 position at least in the higher categories attacking.

However, this interpretation cannot account for the high fleeing scores of the 3♂♂ dominants with free inferiors (see however, below). One possible interpretation of the high fleeing scores of dominants both with free and with hiding inferiors is the assumption that the fishes in the 3♂♂ and 2♂♂ groups remembered their experiences with the highly aggressive conspecifics prior to the experiments and as a result were more liable to flee for them even though they were dominant in the present situation. All fishes used in the experiments must have had ample opportunity to collect such experience in the stock tanks from which they were taken for the experiments. On the basis of this assumption, however, one might also have expected the *nigrofasciatus* and *tetrazona* dominants to flee more frequently than the *conchonius* and *stoliczkanus* dominants — which they didn't. Nevertheless the possibility cannot be excluded for *cumingi*.

Another possible explanation is that the *cumingi* dominants do sometimes flee more than the other species because they threaten less. If, as was suggested in chapter II (p. 85), a fish is more or less bound to the lateral orientation during threat, it is conceivable that the performance of threat may prevent overt flight (or attack). This would imply that the absence of threat may facilitate the occurrence of overt fleeing in a dominant fish.

Finally, it seems hardly conceivable that the species-specific colour patterns are responsible for the high fleeing scores of the *cumingi* dominants because *cumingi* has less black markings than *nigrofasciatus*. It may be emphasized, however, that the effects of the red colours of *conchonius* and *nigrofasciatus* (which are lacking in *cumingi*) are as yet unknown (see also chapter IV, p. 292). It is difficult, without further experiments, to decide between the several interpretations suggested here.

A final possibility, analogous to the one suggested for the hiding 2♂♂ inferiors, is that the dominants flee so often because the free inferiors — at comparable amounts of aggressive behaviour by the dominant — stay on the average lower and/or nearer to the dominant than do the free inferiors of the other species. The data do not allow testing of this possibility.

It may be emphasized that even if this explanation would be true, *cumingi* still are pretty unique among the other species in that the dominants respond with flight to an aggressive activity (staying near or low) by the inferior rather than countering it with more attack.

It may be noted that the above suggestion actually implies a special case of what has been called the secondary mirror effect (see p. 201).

Similar difficulties are met in the attempts to interprete the comparatively long delays in the establishment of dominance. It was suggested in subsections (a) and (b) (p. 228, 251) that the long delays found in *nigrofasciatus* (as compared to *conchonius*) and *tetrazona* (as compared to *stoliczkanus*) were at least partly due to the more elaborate black marking patterns of these two species. In the case of *cumingi* the black marking patterns may hardly be held responsible for the long delays found in this species: *cumingi* has less black marking patterns than *nigrofasciatus* and yet shows longer delays than does *nigrofasciatus*, at least in the 2♂♂ groups. (It must again be emphasized, though, that the effects of the red colours of *conchonius* and *nigrofasciatus* males are unknown). Two alternative explanations of the long delays may be suggested. First, it is possible that *cumingi* males do have a stronger fleeing motivation than have *nigrofasciatus* males. Second, the reminiscene of the aggressiveness of conspecific males prior to the experiments may well have disencouraged males to become dominant. Again, on the basis of the present evidence, it is impossible to decide between these explanations. It may be emphasized that the factors causing the long delays need not be the same as those causing the high fleeing scores in the dominants.

If the reminiscence of conspecific aggressiveness plays a role in inhibiting the establishment of dominance in *cumingi* small all-male groups, a similar factor may have played a role in causing the long delays found in *nigrofasciatus* and *tetrazona*. In addition, in *tetrazona* the strong schooling tendency typical of that species may have further prolonged the delays. A similar factor cannot be held responsible for the comparatively long delays found in *cumingi* because this species is not a particularly strong schooler.

Conclusions.

As a general conclusion to the last two sections of this chapter we may state that the situations found in the small groups reflect rather well the qualitative differences in behaviour found to exist between the species in the 6♂♂-6♀♀ groups. A few notable exceptions are the absence of obvious territorial tendencies in the *cumingi* inferior males and the differences between ambisexual and all-male groups in the speed of initiation of full reproductive behaviour in *cumingi* and *nigrofasciatus*. The quantitative data of the behaviour of the individual fishes in the small groups corroborate several of the semi-quantitative observations described in section II.

To conclude this chapter I want to emphasize that it apparently is very difficult, if not impossible, to compare degrees of aggressiveness of different species on a simple linear scale — even within a group of closely related species. Within each species or within some pairs of species relatively simple correlations may be found between the different agonistic behaviour patterns, suggesting the existence of a rather unitary aggressive motivation in each of the species. For instance, within the pair *conchonius-nigrofasciatus* or the pair *stoliczkanus-tetrazona* the different aggressive tendencies to attack, bite, seek open spaces and seek low levels seem to be rather strictly correlated with each other (see also Kortmulder, in prep.). Nevertheless, many of these correlations seem to break down when species not belonging to the same pair are compared. Different species may be specialized in biting, threat, staying in open spaces or avoidance of ambivalent states.

If we assume that the different properties of the aggressive system and this control must be of considerable complexity, involving qualitative as are under genetic control, it is evident from the results of this study that this control must be of considerable complexity, involving, qualitative as well as quantitative effects. The data discussed in chapter I suggest a similar complexity of genetic control for the colour patterns of the 5 species. Natural selection must have worked differentially on different parts of these genetic systems to give rise to the present species. Yet it seems that within each species there must have been selection for integration of the aggressive sub-systems into one more or less unitary system.

At a higher level of integration, it seems that — at least in some of the species — certain behavioural characteristics are not restricted to one category of behaviour only (such as for instance fighting or courting) but more or less completely pervade the whole organization of the behaviour of the individual. Examples of this may be found in the acuity of all conflicts in *tetrazona* (p. 262, 263) or the comparative lack of ambivalence in *cumingi* (p. 281). It is this kind of phenomena that induced me to give each of the species an additional characteristic name in english (see chapter III, section I).

Finally, it may be surmised that, at the highest integrative level of the individual, there is at least *functional* integration of different behaviour patterns and of morphological and behavioural characters into one adaptive whole. An analysis of such functional interrelations between colour patterns and behaviour in each of the 7 species will be the main theme of the next chapter.

CHAPTER IV. SYNTHESIS

'The pattern of the elements'

SECTION I. THEORETICAL CONSIDERATIONS

Perhaps the most important conclusion from chapters I (sections III and VIII) and III of this study is the close concordance between the morphological (mainly referring to colour patterns) and ethological classifications of the 7 species. Table 16 summarizes most of the characteristics which were used for these classifications.

For the following discussion we will assume that the number of species chosen for this study is large enough to exclude the possibility that the concordance between the two classifications is fortuitous. Further we will assume that the species used were not selected in order to create the concordance. At least I am sure that *I* did not make such a selection. The only principle of my selection was to obtain a number of species which differed as much as possible from each other in colour patterns and yet were closely related to one another. (For phylogenetic relationships between the species see chapter I, sections VII and VIII). However, there is no doubt that my choice was to some extent limited by the availability of species on the market. This availability, of course, depends on such factors as the popularity of species with fish-fanciers and the possibilities of breeding them in captivity. It is impossible to say whether or in what way this limitation may have influenced the final results of the study. The only way out of this problem would be to find other species of the same sub-group and, popular or not, have them imported or to observe their behaviour in the natural habitats. Finally, we will assume that the number of characters upon which the classifications are based is large enough and that the characters chosen are sufficiently relevant to make both classifications realistic ones.

Having made these assumptions, several hypotheses may be erected to account for the concordance found.

1. It is possible that the differences between the species in behaviour are merely a direct result of the differences in the colour patterns of the conspecifics. Several applications of this hypothesis will be discussed — and rejected — in the next section.
2. The concordance between the morphological and behavioural classifications may be genetically determined through a close linkage between those genes controlling the development of the colour patterns and those

Review of morphological and behavioural characters of the 7 species

morphol. characters	*conchonius*	*nigrofasciatus*	*stoliczkanus*	*tetrazona*	*cumingi*	*phutunio*	*gelius*
sex dichromatism	+++	+++	+	+	(—)	(—)	(—)
iridescence on back	+++	++♂ +♀	—	—	—	—	—
♂ fin colours	++	++	+	+	(—)	(—)	—
♂ fin colour accent	D; A; V; (P)	D; A; V; (P; C)	D; ((V))	D; (V)	(D)	(D); ((V))	—
silver background	+	+	++	+++	(—)	+	—
markings contrast	+	+	+++	+++	+	+	+
green below markings	(—)	—	++	++	+	—?	—
number, size markings	(+)	+++	+	+++	++	++	++
gold front tail mark	+	—	+	—	+ juv.	+	—
gold front pect. mark	—	—	—	—	—	+	—
lateral compression	+	++	++	+++	++	++	++
transparency	—	—	—	(—)	(+)	+	++
behaviour characters							
reproduction day-day	each day	each day	irregular	irregular	2/3 days	consec. days	consec. days
diurnal rhythm	fixed	fixed	irregular	irregular	± irregular	—↗	—↗
reproduction in bouts	++	++	++	++	++	—↗	—↗
onset of bouts	smooth	smooth	explosive	explosive	smooth	? ↗	? ↗
social structure reprod.	group	group	territory (?)	territory	territory	territory!	'solitary'?
sensitivity competition	little	little	strong	strong	little	? ↗	little?
♂ attacks courted ♀	—	—	++	++	++	++	++
♀ 'emancipated'	—	(—)	+	++	+	?	?
pattern movement reprod.	smooth	smooth	hesitating	jerky	± smooth, unrestrained	jerky, unrestrained	unrestrained, hovering
conflict school-reprod.	smooth	smooth	acute	acute	smooth	?	?
accent on threat	++	++	+	+	(+)	(—)	—
accent on bite	—	(—)	+	++	+, attack	+, attack	+, attack
accent on roll	—?	(—)?	+?	++?	(+)?	(—)?	—?
head-down threat	++	—	++	++	++, +head-up	?	?
transition	smooth		abrupt	abrupt	smooth		
mouth-fight	—	—	(+)	++	(—)	—?	—?
formalization duel	++	++	++	++	+	(—)	—
aggressiveness	moderate, balanced	strong, balanced	moderate, unbal.	strong, unbal.	very strong, unrestrained	unrestrained	most unrestrained
dorsal *vs* ventral roll	—	+	—	+	+	?	?
delay dominance 2♂♂ 2♀♀/2—3♂♂	—/—	—/+	—/—	+/++	—/++		
dominance 2♂♂ 2♀♀/2—3♂♂	(+)/+	(+)/++	+/+	++/++	+++/+++		
defeat inf. 2♂♂ 2♀♀/2—3♂♂	—/+	—/++	(+)/(+)	(++)/(++)	+++/++, +		

controlling the development of behaviour, or through pleiotropic effects of genes acting on the development of both colour patterns and behaviour. It is difficult to prove or disprove this hypothesis without for instance doing differential selection experiments on colour pattern characteristics and on behavioural characters. However, even if such a hypothesis were proven true, the question of why the characters became thus linked in the phylogeny of the species would remain a relevant one.

3. It might be that the specific environment of each species acts in a specific way on the ontogeny of both colour patterns and behaviour. This would mean that the concordance between the morphological and behavioural classifications is nothing more than a reflection of a classification of environments. However, the fact that all species are easily bred under the same artificial circumstances, without loosing either their species-specific colour patterns or their species-specific behaviour pleads strongly against this hypothesis.
4. Another possible hypothesis is that one of the categories of characters is strongly genetically determined (in the sense that they remain constant in a wide range of different environments) and that the other category develops, through intraspecific interactions during ontogeny, as a result of the first. For instance, it is conceivable that (a) the colour patterns of each species are genetically determined in the above defined sense and that the species-specific behaviour develops as a result of interactions with conspecifics bearing these patterns. Or, conversely (b) it would seem possible that the social behaviour of each species is more strongly genetically determined and that the interactions between conspecifics — different according to the species — would, through hormonal processes, lend each species its specific colour patterns.

An appropriate way of testing these hypotheses would be to disengage social interaction from the growth of the colour patterns by means of Kaspar-Hauser experiments. To my knowledge no such experiments have been conducted in this group. However, there is at least one observation which pleads against hypothesis (a). Albino *B. tetrazona* (lacking the black and green pigments but retaining the red) have been bred and have been available on the market for several years. In spite of their lack of pigmentation these fishes exhibit (at least qualitatively) normal species-specific behaviour. This observation is relevant to hypothesis 4[a] only if such albinos have grown up among other albinos only. Albinos have not been bred in our laboratory but, considering the average techniques used by fish breeders, it seems very unlikely that the albinos have been mixed with normal *tetrazona* during any stage of their ontogeny.

No direct arguments can at present be offered against the converse hypothesis (b) (see, however, hypothesis 6 and p. 294).

5. It is possible that both the behavioural and the morphological classification are a simple reflection of phylogenetic relationship between the species. It is conceivable that more closely related species have had less time to diverge both in morphological and in behavioural characters.
6. Yet another possibility is that the two classifications do not necessarily reflect phylogenetic relationships but that morphological and behavioural characters (both under some degree of genetic control) are functionally interdependent in such a way that a change in for instance the colour patterns of a species will automatically bring about a change in the selective pressures acting upon behaviour and *vice versa.*

   This hypothesis is not necessarily incompatible with all of the other hypotheses, nor will it be easy to test against some of them. As the functional hypothesis has been taken as a working hypothesis of the present study it is well to look somewhat more closely at its content, at the way it should be tested and at its relationships to the other hypotheses.

To facilitate the discussion I will first outline the theoretical background and define some concepts that will be used in this chapter. In doing so I will largely follow the views of WILLIAMS (1966) and NICHOLSON (1960).

I will say that a certain character (either morphological or ethological) has *adaptive value* if it — under natural conditions — in some way or another contributes to the survival and eventually to the reproductive success of the individual carrying that character. Reproductive success is measured in numbers of fertile offspring. By way of short-hand I will call a character *adaptive* if it has adaptive value in the above defined sense. If I say that a character is more adaptive than an alternative character I mean that the former character has a higher adaptive value than the latter. *The way in which* an adaptive character contributes to the individual's reproductive success I will call the *function* of that character.

If an adaptive character is under any degree of genetic control, it is open to natural selection ('genic selection' — WILLIAMS, *l.c.*, p. 97). If this is the case, the probability of occurrence of the character in the following generation will be determined by its adaptive value and by the adaptive values of alternative characters present in other individuals in the same population. As a result of natural selection the characters with the higher adaptive values will survive in succeeding generations at the cost of the alternative characters, provided that the average numbers of individuals in the population remain constant in successive generations (NICHOLSON, *l.c.*, p. 477-480).

I will call a character an *adaptation* if there is reason to believe that it has evolved in its present form through natural selection as a result of its function. According to WILLIAMS (*l.c.*, p. 3-19) the criterion for calling a certain character an adaptation is that it looks as if 'designed' for its function.

It is obvious that in this theoretical frame-work the number of resultant fertile offspring is the only important measure of adaptive value. Individual survival is important only insofar as it is a condition for successful reproduction (WILLIAMS, *l.c.*, p. 26). It is equally obvious that natural selection is nothing more than the tautology that individuals with the more adaptive characters leave relatively more offspring than do the others (WILLIAMS, *l.c.*, p. 25, 109; NICHOLSON, *l.c.*, p. 479).

In the above definitions I have deliberately emphasized the importance of the survival and reproductive success of the *individual* as opposed to the survival of a whole *population* or *species*. This is in accordance with the views of WILLIAMS, who has argued that any interpretation of a character as an adaptation to group survival ('biotic adaptation') should be rejected as long as the character may be understood as an adaptation to individual survival ('organic adaptation' — WILLIAMS, *l.c.*, p. 4, 5, 19, 97, 108). I am aware that my definition of function (and — by implication — that of survival value) differs from that used by WILLIAMS in that he attributes function to characters which are adaptations only (*l.c.*, p. 8-9) whereas according to my definition function may be attributed to any character as long as it does contribute to reproductive success and as long as it is in some way passed on to progeny. I think it is rather unfortunate to restrict the concept of function to adaptations only. The problem is mainly a practical one. The restriction of the concept of function to characters with obvious 'designs' is all right as far as organs such as eyes, lungs or lateral line systems are concerned, which have, so to speak, their designs 'written all over them'. However, there are many structures which are demonstrably or probably or conceivably adaptive and yet do not invite the predicate of 'design'. For instance, as to the barbs, I would perhaps find it possible to detect 'design' in the total array of colour patterns and colour changes of an individual of a species but I would find it very hard to detect 'design' in a single colour pattern such as for instance a pectoral marking, irrespective of how well the adaptive value of such a marking might be demonstrated.

In the above definitions of survival value and of function I have consciously avoided the exigency that the characters concerned are under some degree of genetic control. This exigency is implicit, however, in the discussion of natural selection. The degree to which characters contribute to an individual's reproductive success is relevant to the mechanism of natural selection only insofar as the character is genetically determined and thus passed on to progeny. And it is only with evolution through natural selection that we are concerned here. However, theoretically there is at least one other way in which an adaptive character may play an important role in evolution, namely if it is passed on to progeny by tradition and is therefore open to what one may call 'cultural selection'. Therefore, I can see no logical reason why the concepts of adaptive value and function should be reservated for genetically determined characters only. Cultural selection will probably play and have played some part in the evolution of higher mammals. It is probably largely irrelevant to the present discussion.

The essence of the functional hypothesis is that, given a certain configuration of colour patterns in a population of a species and dependent upon that configuration, some modes of behaviour are more adaptive than others and *vice versa*. Likewise, behavioural as well as morphological characters will be functionally interdependent among each other.

One implication of the hypothesis is that, given a certain configuration of colour patterns in the population and all other things (such as population density, physical structure of the environment, predators) being equal, one (or several) modes of behaviour will be optimally adaptive and thus be promoted by natural selection. Another implication is that if, by whatever cause, a change occurs in the colour patterns in the population (and again all other things being equal) there will be a shift in the relative adaptive values of alternative modes of behaviour and that, as a result, different modes of behaviour will be favoured by selection. Both implications may also be read the other way round: if a change should take place in the mode of behaviour, selective pressures on colour patterns will also change. A third implication is that, if two species differ in colour patterns, again all other things being equal, natural selection will have caused the two species to have different modes of behaviour and *vice versa,* the differences being functionally understandable in relation to each other.

The *crux* of the last implication is that 'all other things' are rarely equal in two different species. However, it seems sensible to assume that population density, structure of the environment, colour pattern configurations and modes of behaviour all are functionally interdependent (probably the structure of the evironment will be the least dependent of the four). Therefore we may still expect to find functionally understandable relationships between colour patterns and behaviour. It is equally clear, however, that in interpreting these relationships the other factors will have to be taken into account as well. I will return to this point later in the discussion.

Let me first examine how the functional hypothesis should be tested. In principle this might be done in two ways which, ideally, should supplement each other. One method is to measure reproductive success of individuals with differing morphological and ethological characters. This would involve a laborious breeding programme under as natural as possible conditions. Such a programme has not been undertaken so far in our laboratory. The discussion of reproductive successes of different types of *nigrofasciatus* males under various circumstances (chapter III, p. 133ff.) might be taken as an example of an attempt in this direction. The other way to test the hypothesis is to investigate the effects of certain characters (for instance the effects of colour patterns on the behaviour of conspecifics or predators) preferably under standardized laboratory conditions and subsequently to reconstruct whether and in what way these effects contribute to survival and reproduction under natural circumstances. The second procedure was followed by DUNHAM *et al.* (1968) in a study of the functions of lateral threat and ventral roll in *B. stoliczkanus*. The same procedure is being followed in the present study. In addition to both procedures it should be assessed whether the characters concerned are under any degree of genetic control. Evidence on the genetic control of both colour and behaviour patterns in the barbs is very incomplete as yet. However, it seems very probable — and for this study it will be taken for granted — that those species-specific colour and behaviour patterns which will be discussed in this chapter are largely genetically transmitted. As was already discussed in chapter III (p. 123) there is no evidence on genetic determination of variations of either colour patterns or behaviour within any one species of barbs.

Direct evidence concerning the effects of colour patterns in the barbs is also incomplete as yet. In general, colour patterns may have many kinds of effects and functions. The fact that in the barbs colour changes occur mainly during social interaction and that some colours display their greatest brilliancy during such interaction suggests that the colour patterns play an important role in social communication. Model experiments are being conducted in the laboratory to investigate the effects of different colour patterns on the behaviour of the barbs. These experiments are not yet in a stage to be extensively reported upon here. This will be done in a future paper. However, some relevant points may be mentioned. It is clear from the experiments done so far that (a) a barb (*nigrofasciatus* ♂) responds very differently to models with varying configurations of white, grey and black bars. (b) Models with dark cross-bars as well as generally darker models seem to be less readily attacked than models without bars and generally lighter models.

The behaviour of the males involved in the model experiments was clearly of the self-actualizing type (see chapter II, p. 96). The models were stationary or were moved around in a narrow circle by means of a small machine, *i.e.* the models did not 'flee' or 'hide'. This means that the results of the model experiments are primarily relevant to

situations in which a male is met by a rival of roughly equal strength and thus to the situations in large groups described in chapter III, section II, and in natural situations.
The results are less relevant to the situations of dominant males in small groups and even less to inferior males. (See further chapter III, p. 203).

Since detailed results of the model experiments are not yet available, in the present study mainly the second step of the functional analysis (reconstruction of functions under natural circumstances) is open to investigation. Therefore, in the following section, I will try to find as many plausible functional connections between colour patterns and behaviour as possible.

The model experiments so far were set up to investigate the effects of colour patterns on reproductive behaviour. However, it is quite possible that the colours of the barbs have other, non-reproductive, functions as well. These will be discussed in sub-section IIB. Some remarks on possible extra-specific functions (in relation to predators) will be made in sections IIB and III.

Section III will mainly deal with the following question: If two species differ from each other in colour patterns and in modes of behaviour in a functionally understandable way, what has induced the one species to 'choose' for one combination of colour patterns and behaviour and the other species for another combination? As was argued above, the colour patterns and the behaviour of a species are not the only factors determining the functional interrelation between the two. Factors such as population density and the structure of the environment (including predators) probably play an important role as well. As the structure of the extra-specific environment seems to be the least dependent of all functionally interdependent variables, it seems most likely that it is the differences in the extra-specific environments of the species which are ultimately responsible for the differences in colour pattern-behaviour combinations. TINBERGEN (1958), citing from the work of CULLEN (1957) on the Kittiwake, and BAERENDS & BLOKZIJL (1963, p. 526), in a comparative study of two *Tilapia* species, also suggest that the ultimate causes of a species' "choice" of adaptations has to be sought in the extra-specific environment. As a special case, the differences between closely related species may be adaptations to prevent hybridization (NICHOLSON, *l.c.*, p. 517-518; WILLIAMS *l.c.*, p. 145-146). However, since none of the 7 species seem to be sympatric with each other (even *phutunio* and *gelius*, though perhaps sympatric with each other and with *conchonius* in a broad geographical sense, probably inhabit different waters) this is not likely to be the explanation in this case.

Still another question that will be briefly dwelt upon in this chapter (section IV) is what have been the substrates on which selection could work to produce the present species.

Finally, the relationships of the functional hypothesis to the other hypothese.

No single functional correlation between a colour pattern and a behaviour pattern found within one species, however well documented it may be, will suffice to prove the functional hypothesis or to disprove any of the others. This is so because both single colour patterns and single behaviour elements vary between the species in a quantitative rather than in a completely qualitative way, which makes it difficult to describe either of them in terms of 'design'. However, the more different kinds of functional correlations between colour patterns and behaviour that are found and the greater the number of species for which these correlations are valid, the more plausible the functional hypothesis will become relative to some of the others.

Finding a large number of functional correlations may disprove hypothesis 2 as a single explanation but it would not prove that the essentials of hypothesis 2 are false. Rather, the functional hypothesis might supplement hypothesis 2 by explaining *why* linkage of the characters has evolved. Likewise, confirmation of the functional hypothesis would disprove hypothesis 4^b as a single explanation, but it could also supplement it by explaining why *functionally adaptive* colour patterns result from the hormonal processes alluded to above.

Even if numerous functional correlations are found, it will be difficult to distinguish the consequences of the functional hypothesis from those of hypothesis 4^a. Learning how to react most adaptively to the genetically determined species-specific colour patterns might conceivably mimick the effects of the functional hypothesis. (Of course, for reactions to species-specific colour patterns to become adaptive through learning processes some genetic programming of behaviour is necessary but such programming is of a much more general level than that implied in the functional hypothesis).

Finally, proof of the functional hypothesis would not mean that similarities between species do not reflect phylogenetic relationships (hypothesis 5). It would only mean that phylogenetic relationship is no sufficient explanation for the differences and similarities found between the species and, conversely, that phylogenetic relationships may not be simply deduced from the degrees of similarity.

It may be concluded that some of the other hypotheses are no true alternatives to the functional hypothesis: proof of the functional hypothesis will not automatically disprove all others. Testing of several of the other hypotheses is possible only by the methods indicated on p. 288.

Finally, it is possible that none of the hypotheses discussed is universally true. Different relationships between colours and behaviour may be explained by different hypotheses.

I hope that by this discussion I have sufficiently discredited the functional hypothesis with the reader to allow him to judge the following sections without prejudice.

SECTION II. FUNCTIONAL RELATIONSHIPS BETWEEN COLOUR PATTERNS AND BEHAVIOUR

A. Colour patterns and reproductive behaviour.

1. *B. conchonius vs nigrofasciatus; stoliczkanus vs tetrazona.*

Conchonius and *nigrofasciatus* are very similar to each other both in colour patterns and in behaviour, and so are *stoliczkanus* and *tetrazona.* The two species of each pair mainly differ from each other in that one has more black patterns and is more aggressive than the other. The model experiments indicate that black patterns lower the probability of attack by conspecific males. As has been argued in chapter III (p. 204) this finding eliminates the possibility in this case that the differences between the species in behaviour are a direct result of the differences in colour patterns. Therefore I hypothesize that in these two pairs of species the amount of black patterns and the degree of aggressiveness are functionally interdependent. It is a plausible assumption that the males of a more aggressive species need stronger threat colours to prevent attack from and to compete successfully with conspecific males than do males of a less aggressive species. On the other hand, it may be disadvantageous for a male of a relatively unaggressive species to carry too strong threat colours because threat colours might interfere with successful courting behaviour (see chapter III, p. 129ff.). Further, too strong threat colours relative to aggressiveness might interfere with schooling behaviour or might break up the spawning aggregations which seem to be essential for successful reproduction in *conchonius* and *nigrofasciatus* males.

2. *B. conchonius - nigrofasciatus vs stoliczkanus - tetrazona.*

In contrast to the species belonging to the same pair, the two pairs of species differ from each other in a considerable number of behaviour and colour characteristics. One might surmise that some of the behavioural differences are a direct result of differences in colour patterns. For instance, the low probability of attacking the courted female in *conchonius* and *nigrofasciatus* as compared with *stoliczkanus* and *tetrazona* males and the comparatively low degree of female 'emancipation' in the former two species might be ascribed to the extremely strong sexual dichromatism found in the former two species as compared with the other two. At least for the behaviour of the males, however, several arguments may be raised against this supposition: (1) A group of fully adult *conchonius* × *stoliczkanus* hybrid males, which had not yet developed any appreciable nuptial colours, courted and fought each other in a *conchonius*-like manner. The fishes typically showed prolonged courting sequences, interrupted only by the keeping-off of other individuals; the tendency to attack the courted fish was apparently very low. Thus these fishes showed the same discrimination between courted fish and rival as do *conchonius* and *nigrofasciatus* males, even though the courted fish and the rivals did not differ in colour patterns. This suggests that the 'unaggressive' courtship of *conchonius* males is not primarily dependent on sexual dichromatism (though of course sexual dichromatism probably helps in directing courtship towards females and not towards males). (2) If it is supposed that *stoliczkanus* and *tetrazona* males often attack courted females because the colour patterns of the sexes are similar, one might expect them also to court males more than do *conchonius* and *nigrofasciatus* males. As has been shown in chapter III, male *stoliczkanus* and *tetrazona* in large groups hardly ever court other males. In 2♂♂ 2♀♀ groups *stoliczkanus* dominant males practically never courted the other male; *tetrazona* dominants did do so occasionally in some observation series but, whereas the females were often attacked nearly as much or even more than the other male, the latter was always courted very little even if one or both females were courtel for considerable amounts of time (see fig. 48[c], p. 172). (3) A final argument is a comparative one. Sexual dichromatism in *conchonius* and *nigrofasciatus* is mainly caused by the males getting brilliant red and black colours rather than by the females loosing colour. On the other hand, in *stoliczkanus* and *tetrazona* both males and females change relatively little in colour during reproductive periods. This would suggest that the comparative lack of sexual dichromatism in the latter two species is due to a lack of typically male colours rather than to the females looking like males. *Ceteris paribus* this fact would again lead one to expect that the males of the latter two species would court males rather than attack females. It may be argued that it is the presence of red and black colours rather than the degree of deviation from the 'neutral' a-reproductive pattern which makes a typically male pattern. Also in this case, however, the same argument would hold at least for *stoliczkanus* because in this species the females, like in *conchonius* and *nigrofasciatus*, carry little if any red colours. On the basis of these arguments I conclude that the behavioural difference in question cannot be simply a result of the difference in colour patterns.

Several other possible dependencies of behavioural differences on colour differences could be brought up, but these won't be discussed here.

An interpretation of the various behavioural and colour differences between the two pairs of species in terms of the functional hypothesis may be made as follows.

The *behavioural characteristics* may be grouped, in a functionally meaningful way, around one central feature: the formation of spawning aggregations in *conchonius* and *nigrofasciatus* and individual territoriality in *stoliczkanus* and *tetrazona*. (1) The strict *daily rhythm* of reproductive activity in *conchonius* and *nigrofasciatus* may be interpreted as an adaptation to spawning in aggregations on special spawning grounds. Such rhythmicity would seem to be unnecessary for territorial species. (2) The relative *immunity to competition* — *i.e.* the ability of a male to mate frequently in spite of severe competition — found in *conchonius* and *nigrofasciatus* is no doubt adaptive in a species spawning in aggregations. A similar immunity was not found in *stoliczkanus* and *tetrazona*. (3) In a territorial species, in which a male can hardly mate without being able to defend a territory, it would seem adaptive for a male if its *first reaction* to a conspecific is to attack it, even at the risk of attacking females now and then. On the other hand, it would seem that in an aggregating species a male's primary concern should be to court, even at the risk of courting males, and to fight only when it is seriously hindered in its courting activities. This may be the functional explanation why *stoliczkanus* and *tetrazona* males often attack females as well as males even when courting tendencies are strong, whereas *conchonius* and *nigrofasciatus* do not; why *nigrofasciatus* males start reproductive behaviour much sooner in ambisexual than in unisexual groups (p. 199); why *conchonius* and *nigrofasciatus* inferiors in 2♂♂ 2♀♀ groups do not hide but often court successfully (p. 199); and why homosexual males were found in *nigrofasciatus* and possibly also in *conchonius* (p. 171) and not in the other two species even though sexual dichromatism is strongest in the first two species. The low tendencies of normal *conchonius* and *nigrofasciatus* to court other males will be further discussed below. (4) The *threat specialism* found in *conchonius* and *nigrofasciatus* males (p. 206) would seem to be adaptive for males courting in dense aggregations where it is important for a male to keep-off rivals and at the same time not to loose contact with the courted female. On the other hand, the *biting specialism* of *stoliczkanus* and *tetrazona* (p. 261) may, in combination with a high propensity for attack, be more adaptive in territorial species in which males have to defend their grounds not only against neighbours but also against varying numbers of schooling unattached individuals. Biting specialism may also contribute to the frequent occurrence of strong appeasement rolling followed by renewed attack found in some *tetrazona* males (chapter III, section II, p. 141; and chapter II, section II

p. 93). Strong rolling apparently was essential for ♂3 in playing its role of successful underdog (p. 140-142). The relation between a great tenacity in defending even a minute territory and biting specialism has been demonstrated in section IV^c of chapter III. (5) The territorial status of the males and their high propensity for attack make it important for *stoliczkanus* and *tetrazona* females to actively go and visit a male in his territory and to make herself known as a willing female, if not by colours then by behaviour. This may have led to the comparatively *'emancipated'* behaviour of the females of these two species. Such behaviour may be just impedimental to successful courtship in aggregating species.

We can now proceed to a functional interpretation of the differences in the *colour patterns* of the two pairs of species. In doing so I will again take the group- or territorial structure of spawning behaviour as a focal point. (6) It may be tempting to interprete the strong *sexual dichromatism* found in *conchonius* and *nigrofasciatus* as an adaptation at the population or species level. In an aggregating species it is no doubt of advantage for a male to be able to distinguish quickly between males and females lest time should be lost in courting males. A similar exigency, though valid also for territorial species, would seem much less stringent there because there is more time for a male to recognize the sex of an intruder and for a willing female to make herself known as such. This lesser stringency would leave room for any selection pressures in the direction of similarity in colouration between males and females (see below, sub-section B). An adaptation, at the species level, to quick sex recognition is suggested by NOBLE (1938, p. 137) for guppies. The interpretation, however, implies that males of an aggregating species are coloured different from females in order to help other males in distinguishing between the sexes. Clearly, this would be an example of group adaptation. According to the principles discussed in section I of this chapter, such an interpretation should be rejected if an interpretation in terms of individual adaptation is possible. This can be done along the following lines.

As was argued above, in the aggregating species it would be adaptive for a male if its first reaction to a conspecific is to court it. Several arguments were given that this is indeed a typical feature of *conchonius* and *nigrofasciatus* males, especially as compared with *stoliczkanus* and *tetrazona* males. From the observations of *nigrofasciatus* groups containing one or more homosexual males it was clear that normal males were very seriously hampered in their own courting activities as long as they were courted by a homosexual male. Homosexual males are very tenacious in sticking to a courted male (or female), much more so than was ever seen in attacking males.

Even mutually threatening males only rarely get 'trapped' in a duel (chapter II, section II, p. 85). As a rule all kinds of agonistic interactions between males are very short-lasting compared to courting sequences of homosexual (and normal) males. From this it follows that it is advantageous for a male to prevent being courted even at the risk of being attacked. It is not known whether homosexual males occur also in nature but the high courting propensity of normal *conchonius* and *nigrofasciatus* males would no doubt easily lead to males being courted if they did not look different from females. Therefore I hypothesize that a function of the brilliant colours of *conchonius* and *nigrofasciatus* males is to prevent that they are courted by other males. The fact that at least the black colours apparently act as threat colours suggests that courting by other males is prevented through the elicitation of flight or counter-threat which in its turn suppresses courting. Even if some of the male's colours would elicit attack from its rivals it would be better off than when it is courted. In the aggregating species defeat costs less time than being courted and time means opportunity to court females — which means reproductive success.

It would follow from these considerations that it is the very existence of the bright male colours in *conchonius* and *nigrofasciatus* which prevents that normal males exhibit their high courting propensities also towards other males (see above, p. 297). It also follows that because of the existence of sexual dichromatism normal males can afford even better to have courting as a primary reaction because the risk of courting a male is virtually reduced to zero.

For the females of aggregating species it would clearly be advantageous, also at the individual level, to look as different from males as possible. In the territorial species the absence in the males of courting as a first reaction would make it unnecessary for males to look very different from females. For the females of the latter species it may still be adaptive to look different from males but, like was argued above, this exigency is less strict for females of territorial species than for those of aggregating species. Other reasons may be present for females not to look too different from males (see below, sub B; see also p. 300).

Another function of the male colours may be suggested. When a male is confronted with another male or with a model, its initial responses mainly include keeping away, fleeing and threat. After a few minutes, at least if only one rival is present, these responses are gradually replaced by attacking behaviour. The initial fear response is particulary strong if the two males do not know each other. If the exposure is maintained for hours or days, also the aggressive responses wane gradually. It may well be that the intimidating effect of the male colour patterns in *conchonius* and *nigro-*

fasciatus males is enhanced by the fact that during reproductive periods their general appearance is so very different from what they look like outside reproduction. Such an effect is likely to be present especially in aggregating species in which males have little opportunity to learn to know each other individually. On the other hand, in territorial species the surprise effect is likely to be much less strong as far as competition between neighbouring males is concerned. The tendency of males to return to the same territory site on consecutive days gives them ample opportunity to learn to know their neighbours (see chapter III, p. 140). Therefore it is conceivable that *stoliczkanus* and *tetrazona* males, by looking essentially the same all the time, 'speculate' on the waning of the aggressive responses of their neighbours rather than on the surprise effect.

It may follow from these considerations that for the females — apart from exigencies to look different from males, see above — it is adaptive to look essentially the same both during and outside reproductive periods. This is in fact what the females of all species do (see also below, sub B).

(7) The comparatively *silvery back-ground colour* combined with the *sharp definition of contrast* of the black markings make *stoliczkanus* and *tetrazona* conspicuous from a distance. This would seem to be adaptive in territorial species in which a male's colours may serve to advertise its presence as a territory owner (*cf.* TINBERGEN, 1951, p. 153-154; LORENZ, 1963, chapters I, II). Accordingly, the black body markings of *stoliczkanus* and *tetrazona* males tend to fade much less strongly, even during intensive reproductive behaviour, than do those of *conchonius* and *nigrofasciatus* [1]). It may be surmised that the presence of *green iridophores* in the body markings serves to maintain the contrasts even if the black colour fades. Also the bright red and black *dorsal fins* of *stoliczkanus* and *tetrazona* may well serve as a flag, conspicuous as they are from a distance. The colours of the anal and ventral fins are much less conspicuous. It is not quite clear why the *colour accent* should be on the dorsal fin, apart from the fact that it is the largest fin available and the uppermost one and therefore likely to be most strongly illuminated under natural light conditions. It is conceivable that the presence of similar red-black colour contrasts in the ventral and anal fins [2]) would spoil the flag-like appearance of the whole fish and would make it tend towards somatolysis. It is also possible that the red ventral fins of

---

1) In the latter species the body markings may be dark but then the wole of the body is dark.

2) Blanck and red colours are present in the anal fin of *tetrazona* too but these are not conspicuous at all. The black triangle at the base of the fin looks like an extension of the first tail bar.

tetrazona serve other than agonistic purposes. In *B. tetrazona partipentazona* the ventral fins are yellow.

In *conchonius* and *nigrofasciatus* the definition of contrast between the black body markings and the back-ground is much less than in the former two species. Also the other colours of the fins and body tend to shade into each other rather than forming sharp lines of contrast. There is no colour accent on any single fin; the dorsal, anal and ventral fins are equally coloured black. This probably makes the fins particularly effective in increasing the apparent size of the fish during threat. This feature is no doubt related to the threat specialism found in *conchonius* and *nigrofasciatus*.

Because of the absence of sharp contrasts, the colours of *conchonius* and *nigrofasciatus* males, though brilliant, are not very conspicuous from a distance. Flag-like colours such as found in *stoliczkanus* and *tetrazona* are probably not adaptive in aggregating species. They may even be disadvantageous. Courting behaviour, especially the later phases just preceding mating, attracts a lot of attention — and competition — from other males (see *e.g.* chapter III, section II: 'sneaking'). In this situation flag-like colours might attract even more rivals.

(8) The above discussion suggests a functional interpretation of one more behavioural feature. The flag-like character of *stoliczkanus* and *tetrazona* male colours is no doubt enhanced by the rather jerky movements typical of reproductive males of these species. Also the comparatively angular shape of the fins and the body may contribute to this effect. On the other hand, in *conchonius* and *nigrofasciatus* males conspicuousness is further reduced through the smoothness of movements during reproductive behaviour, may be in combination with the comparatively rounded shape of body and fins. Significantly, in *tetrazona* males the pattern of movement becomes relatively smooth only during intensive courtship (see chapter III, section I, p. 105) when attracting attention from distant rivals may become a disadvantage.

It may be concluded that a considerable number of plausible functional relations between colour patterns and reproductive behaviour characters may be found in the present four species. It must be emphasized that some of these functional relations might be not mutually independent. The various morphological and behavioural features might share common causal factors. This would reduce the number of identifiable functional relations. For instance, the comparatively jerky movements of *stoliczkanus* and *tetrazona* males may be causally connected to the comparatively agonistic structure of courting behaviour of these species. It has been stated that the movements of *tetrazona* males become more smooth during prolonged uninterrupted courtship. In *nigrofasciatus* the pattern of movement during courting is less smooth in males which show relatively high agonistic tendencies (see chapter III, section II, types of males). Likewise, some of the postulated functional relations might be invalidated if it could be shown that certain behavioural characters were a direct and simple result of the colour patterns of the species. However, these questions cannot be further discussed here.

3. *B. cumingi, phutunio, gelius.*

The specific combination of colour patterns and behaviour found in *cumingi* would seem to upset many of the functional rules suggested by the discussion of the first 4 species. In accordance with the territorial status of reproductive males, male *cumingi* show much aggressive behaviour towards females even during bouts of intensive courtship; the accent of male agonistic behaviour is on straight attack and biting rather than on threat and the females exhibit comparatively 'emancipated' behaviour. However, whereas in *stoliczkanus* and *tetrazona* the movements of the males are rather jerky, those of *cumingi* males are much more smooth, especially during courtship. In contrast to *stoliczkanus* and *tetrazona* males, *cumingi* males are comparatively immune to severe competition. *Cumingi* males display none of the flag-like colour patterns such as are typical of *stoliczkanus* and *tetrazona*: the contrasts are poorly defined; the background colour is greyish; the fin colour accent is on the dorsal fin but even this fin is not brightly coloured. Though there is a layer of green iridocytes in the tail spot, this does not make the spot any more conspicuous when the black colour fades. Although a simple quantitative comparison of the aggressive behaviour of *cumingi* and the other species is not possible (see chapter III) it may be stated that *cumingi* have surprisingly little black patterns for its unrestrained aggressiveness.

Much of what has been said for *cumingi* is valid for *phutunio* as well. Though the reproductive males of the latter species are more strictly territorial than those of any of the other species, they lack most of the same characteristics typical for *stoliczkanus* and *tetrazona* which were found to be absent in *cumingi* males. The general pattern of movement, though more jerky than in *cumingi*, does not display the erratic characteristics typical for *stoliczkanus* and *tetrazona*. Rather, the jerkiness results from the extreme swiftness of forward movement interrupted by sudden brakings. This, in combination with the comparatively dull colours, does not make a male particularly conspicuous from a distance.

Gelius males apparently are not territorial during courtship. Therefore their lack of conspicuous colours does not contradict the suggestion that territorial males have flag-like colours. It may seem surprising, however, that they so often attack females even when courting tendencies are high. On the other hand, the fact that they do not seem to form spawning aggregations either may put less premium on having courting as a primary response such as was found in *conchonius* and *nigrofasciatus* males. Like in the territorial species, more time may be available — as compared with species courting in dense aggregations — for a male to distinguish willing females from other conspecifics and for willing females to make themselves known as such. Like in the territorial species the tendency of males to attack a conspecific as a first reaction rather than to court it would make it unnecessary for a male to look very different from females.

So far the specific combination of colour patterns and behaviour of *gelius* does not seem to contradict the rules suggested for the first four species. However, like in *cumingi* and *phutunio,* reproductive male *gelius* would seem to carry very little black patterns for the extremely fierce and unrestrained aggressiveness of their conspecific rivals. (For colour patterns during reproduction see chapter III, p. 168).

Cumingi, phutunio and *gelius* share many of those characteristics in which they differ from both the *conchonius-nigrofasciatus* and the *stoliczkanus-tetrazona* pair (see table 16). Therefore, in seeking an explanation for the specific colour pattern-behaviour combinations of the three species I will first try to find a common explanation for all three. Subsequently, I will discuss each of the species separately in order to see whether additional explanations for each of the species may be found.

The absence of flag-like colours, the absence of sexual dichromatism (in the sense that the males are brightly coloured and the females not) and the relative paucity of black markings all three contribute to one characteristic feature of the males: inconspicuousness. Also the transparency of the body, especially in *gelius,* would seem to contribute to this feature. On the other hand, the unrestrained nature of agonistic behaviour, the relative paucity of threat and rolling and the low degree of informalization or the absence of duels all centre around a low propensity for threat and other 'ambivalent' postures. The same unrestrained quality is found in the courting behaviour of *cumingi* and *phutunio.* A suggestion as to the functional connection between the two groups of characters may be found in the observation that in *cumingi, phutunio* and *gelius* (in that order) attacking males are progressively less impressed by threat postures of conspecifics. As has been discussed in chapter III, section IV (p. 203; see also DUNHAM *et al.,* 1968) threat colours seem to loose much of their threatening effect if the attacker is very aggressive. It is possible even that threat colours, like the threat posture, mainly stimulate attacking behaviour in a very aggressive opponent. Even if such a dual effect is not a property of threat *colours*, in the absence of an intimidating effect of its colours a male would seem to be better off without any eye-catching colours at all (see above: absence of flag-like colours in aggregating species). I therefore hypothesize that the inconspicuousness of the males of the three species is adaptive because they are not able to keep their highly aggressive conspecifics from attacking even by means of the strongest threat colours; and because under these conditions any conspicuous colours would only attract competitive attention of rivals.

The absence of threat colours might bear some relation to the relative paucity of threat posturing. However, it is difficult to say whether the latter trait is merely another aspect of the unrestrained nature of agonistic behaviour in general, or a specific adaptation parallel to the first one. In both cases it may be asked, if threat is ineffective or even provokes attack, why threat posturing does occur at all in the three species. It may be answered that threat need not be ineffective in all situations. (For instance, *gelius* males were often seen to be quite successful in keeping-off rivals by means of their hovering keep-off manoeuvres combined with threat). A detailed analysis of the situations leading to threat in the three species will be needed to answer this question satisfactorily. Appeasement rolling in response to attack is comparatively rare in *cumingi* (see chapter III, section IV, p. 273) and probably also in the other two species. It is more or less replaced by pure flight. Again, the relative rarity of rolling may be a mere result of the generally unrestrained nature of agonistic behaviour, or a specific adaptation to the unrestrained aggressiveness of the conspecific which is not likely to stop attacking even in response to rolling (*cf.* DUNHAM *et al.*, 1968).

I will now proceed to a discussion of each of the three species separately. In interpreting the colour pattern arrangement of *cumingi* it must be recalled that (a) *cumingi* is the only one of the 7 species in which males may mate frequently both in a territorial and in a non-territorial situation (see chapter III, section II, p. 159) (b) territoriality in *cumingi* males differs from that of *stoliczkanus* and *tetrazona* in that centres of territories are often closer together (at least 6♂♂ could hold territories in the 130 cm tank) and in that mutual penetration is much more intense. Both conditions lead to situations more or less similar to those of the aggregating species. This may mean that, like in the aggregating species, inconspicuousness from a distance is more adaptive than flag-like colours. The same points would explain the smoothness of especially courting movements and the comparative immunity of the males to severe competition. At the same time, the high propensity of the males for attacking rather than courting would make sexual dichromatism unnecessary. These considerations, combined with those presented for all three species, would seem to provide a fair explanation of the colour and behaviour characteristics of the species.

Phutunio is the most strictly territorial of the 7 species. Penetration is comparatively rare. Therefore the specific considerations made for *cumingi* cannot be applied to this species. However, in *phutunio* the inconspicuousness of the males may have another additional function. The practice of collecting the eggs in a nest-like structure and thus concentrating them in one place would make the eggs comparatively vulnerable to conspecific or allospecific predation, especially because the male spends a considerable part of his time manoeuvring around the 'nest' (see chapter III, section II, p. 165). It is well possible that under these conditions a reproductive male cannot 'afford' to bear conspicuous colours. Even a comparatively defensible fish like a male three-spined stickleback looses a good deal of its conspicuous colours as it passes from sexual to parental behaviour (SEVENSTER, 1961, p. 12).

The almost complete absence in *phutunio* of specifically male colours together with the extremely swift and unrestrained attacking of any conspecific intruder by the male would seem to place on the female the burden of making herself known by her behaviour more than in any of the other species. This requirement seems to be met by the specialized slightly head-up display shown by females of this species as they enter a male's territory. Another adaptation in this species is the specific inhibition in the males to eating their own eggs. This adaptation seems to be unique among the 7 species studied; it is indispensible in a species in which the reproductive males are constantly confronted with a concentration of their 'own' eggs and in which the males do not alternate bouts of reproductive behaviour with non-reproductive periods in which feeding can take place.

Since *gelius* are neither territorial nor aggregating during reproduction and because of the high attacking propensity of the males, no specific explanations are needed for the absence both of brilliant male colours and of flag-like colours. On the other hand, the utterly unrestrained nature of their aggressive behaviour may have evoked a number of specific behavioural adaptations. The hovering manoeuvring technique displayed during keeping-off may well have evolved as a device to prevent straight attacking at the rival, which almost certainly would result in the male loosing contact with the courted female. Likewise, the tendency of willing females to stay put as long as the male is engaged in keeping-off seems to be a specific adaptation preventing the loss of contact between male and female. A similar adaptation may be seen in the hovering leading movements which keep the male close to the female. It might be envisaged that if in *gelius*, as is the case in *cumingi* and *phutunio*, leading movements were as swift and unrestrained as attacking and fleeing behaviour, the male would again easily loose contact with the female. The hovering displayed by *phutunio* might serve similar functions as in *gelius*. However, in the former species the territorial attachment of the males would also seem to facilitate the maintenance of contact between male and female.

B. Other than reproductive functions of the colour patterns.

Much less can at present be said on possible non-reproductive functions of the colour patterns. The fact that at least the black body markings are the same in males and females of all 7 species suggests that these patterns play a role in schooling behaviour or perhaps in a more general mechanism of species recognition. This interpretation is supported by the fact that the markings are already present in juveniles long before sexual maturity is attained. Juveniles above a certain age spend much of their time in schooling. In all species males and females school together outside reproductive periods. This suggests that in all species there may be selective advantage for males and females to look the same at least outside reproductive periods. This advantage may conflict with the exigency (discussed above) for males and females to look different at least during reproduction. In *conchonius* and *nigrofasciatus* this conflict is resolved by the males displaying brilliant colours only during reproductive periods; in *tetrazona* by reproductive males and females deviating slightly to either side of the average non-reproductive colour pattern. In the former two species, because of the brilliant colours of the males, there is no need for females to change colour during reproduction; in *tetrazona* the females apparently do have to change their colour in the opposite direction as do the males in order to attain marginal differentiation.

Similarity of the colour patterns of both sexes would be particularly important in *tetrazona* because of their strong schooling and 'mobbing' behaviour (see chapter III, p. 147) (LORENZ, 1963, p. 44-45) has suggested aposematic function for *tetrazona* colour patterns in connection with the mobbing habit. It is well possible that this is the functional reason why *tetrazona* is the only species in which also the females bear appreciable red colours.

In both the *conchonius-nigrofasciatus* and the *stoliczkanus-tetrazona* pair not only the males but also the females of the more aggressive species have more black markings. Apart from the fact that also female *nigrofasciatus* and *tetrazona* may be somewhat more aggressive than female *conchonius* and *stoliczkanus* respectively, the non-reproductive functions of the black markings seem to provide the most probable explanation of this fact.

As to the inconspicuousness of *cumingi, phutunio* and *gelius* males (and females), various non-reproductive functions — in addition to those discussed above — might be suggested. For instance it is conceivable that these species cannot afford any conspicuousness because of the presence of particularly efficient predators. However, there is no way of testing such suggestions except by studying the natural ecology of these species.

C. **Discussion.**

I must leave to the reader the conclusion whether or not the considerations presented above are sufficiently suggestive of the functional hypothesis.

It must be emphasized that, if the above interpretations are correct, the functional connections between colour patterns and behaviour do not obey simple rules. NOBLE (1938, p. 147; 154) has suggested that in Cyprinids and other families males typically are conspicuously coloured in territorial species (the females are not) whereas both sexes have neutral colours in most school breeders. The conclusions reached in this section are not only at variance with this general rule but they are also more complicated. First, brilliancy of colouration has to be distinguished into shading and flag-like patterning. Second, this study suggests that flag-like colours (going with little sexual dichromatism) are typical for territorial species, *provided* that there are no other reasons to avoid conspicuousness. Third, brilliant but shading male colours are typical of aggregation breeders. No doubt, study of more species will reveal exceptions also to the last rule. Many Cyprinid species of the temperate zones (*Leuciscus, Cyprinus, Abramis, Blicca etc.*). seem to breed in aggregations and yet have little or no sexual dichromatism. Only detailed studies may reveal the functional reasons for this.

SECTION III. COLOUR PATTERN—BEHAVIOUR COMBINATIONS AND ENVIRONMENT

It has been argued in section I of this chapter that a species' environment is probably ultimately responsible for its specific 'choice' of a colour pattern—behaviour combination. In the absence of detailed information on the habitats of the various species (see chapter I, section II) assessing functional correlations between the colour pattern—behaviour combinations on the one hand

and the environment on the other is a matter of sheer speculation. The structure of the physical environment may well have been decisive for the choice between aggregation or territorial types of reproductive behaviour. For instance the availability of extended shallow waters near to banks, separated from deeper feeding grounds, may have facilitated aggregation; clear open waters of moderate depth, structured by clumps of vegetation and rocks may have evoked the evolution of territorial behaviour. The latter conditions *plus* the presence of predators may have led to the development of mobbing behaviour in *tetrazona*. This might have induced contrasting colour patterns which in their turn may have facilitated the development of territorial behaviour.

In *cumingi* territorial behaviour does not seem to be essential to male mating success. It is conceivable that territoriality in this species is in the first place a device for a fish to keep oriented in fast flowing waters. Also the unrestrained nature of *cumingi* reproductive behaviour may well be interpreted as an adaptation to living in fast flowing and turbulent water. Ambivalent posturing would seem to be inefficient under those conditions. This may, in this species, have been the decisive factor in the development towards neoteny. The development of unrestrainedness in *phutunio* and *gelius* must have had other reasons. Both species probably live in stagnant water; their hovering way of locomotion would seem to bear testimony of this. Unrestrained attacking and courting behaviour in *phutunio* may well serve the efficiency of territorial defense and the prevention of interference with courtship by rival males. The development of territoriality may again have been dictated by the physical environment. It has opened the way towards nesting behaviour which in its turn must have accentuated the necessity of efficient territorial defense. Prolonged duelling could easily expose the eggs to conspecific or aliospecific predation. Finally, it is possible that demographic circumstances in any or all of the latter three species have primarily induced the shift towards neoteny.

SECTION IV. SUBSTRATES OF SELECTION

A few concluding remarks may be allowed on the substrates on which selection could work to produce and maintain the characters of the present species. The data presented in chapter I (section V, p. 26) suggest that the differences between the species in the number and size of black body markings (and several other black pigmentations) are of a quantitative rather than of a qualitative nature. The analyses of chapter III suggest that within each of the *conchonius-nigrofasciatus* and the *stoliczkanus-tetrazona* pairs aggressiveness is largely a quantitative trait whereas differences between other species are partly of a qualitative nature (see also chapter III, section IV, p. 284). If the close resemblances between the two species of each pair also reflect close phylogenetic relationships and if the common ancestors of each pair did not differ in any more qualitative ways from the present species, one may assume that evolution could proceed from the ancestral to the present forms largely by selection on a few quantitative characters. On the other hand, the transition from an aggregating to a territorial social structure or *vice versa* or the development of either of these types from ancestral forms must have involved selection on many more characters, many of which are more or less qualitative ones [1]).

The shift towards neoteny which I suppose to have taken place to different degrees in the evolution of *cumingi, phutunio* and *gelius* from a presumably *nigrofasciatus*-like ancestor, may be interpreted as primarily due to selection on one quantitative character. However, the analyses of section II of this chapter suggest that selection for a number of more or less qualitative characters has been necessary as well to make the members of each of the species an adapted whole.

The latter 3 species do by no means seem to be less specialized animals than are the other four species. This suggests that development towards neoteny need not always lead to escape from specialization (*cf.* HARDY, 1958).

Though the behaviour of the 7 species was found to be very similar at the behaviour element level, it is difficult at the present stage to assess whether the similarities reflect true behavioural homology. In most of the interspecific comparisons made in this study behavioural homology was taken for granted. However, more information is available on this matter in the colour patterns, especially the black body markings. It was argued in chapter I that here homology is only partial in some cases. Some of the markings were not homologous at all, even within the group of 7 species. The most diverse structural origins were found in the pectoral marks of these and some other species. In *stoliczkanus, cumingi* and *nigrofasciatus* it is formed out of the normal scale pigmentation; in *tetrazona* and *arulius* it is first formed out of scattered melanophores in the deeper layers of the skin; in *oligolepis* and *semifasciolatus* (and probably in other species as well, see chapter I, p. 47) it begins its development as a small round cluster of melanophores in the deeper layers of the skin, centred around the lateral

1) This distinction between quantitative and qualitative traits may seem to be at variance with what has been said on p. 294. It must be emphasized, however, that deciding what traits are quantitative or qualitative is often a matter of taste.

line; finally, in *gelius* it grows out of peritoneal pigmentation peeping through a hole in the ventral membrane. This diversity of structures may suggest two conclusions. (a) The presence of a pectoral marking would seem to be functionally important in many species. Natural selection displays its opportunism in shaping such a marking out of any material available. (b) The structural diversity, in even closely related species, of roughly topologically similar markings may serve as a warning against a too quick homologizing also of behaviour patterns.

APPENDIX I

(*ad* p. 162; *B. cumingi*, 6♂♂ 6♀♀ group).

A remarkable type of behaviour was seen on some days in the evening (8/5, 16/5, 23/5, 27/5). The fishes move through the tank, mostly in the surface layers. Most of the time they stay in a group, sometimes divided into sub-groups. In swimming around they often repeatedly cross the bubble stream coming from an airstone and let themselves be carried upwards by it for a small distance. They are constantly on the move and there is no trace of territories. Frequent surface nipping (probably without actually taking food) may occur. While in the school the males show a rather chaotic mixture of (often long) attacks on both males and females and courting. There may be some inter-male keeping-off. Courting only consists of a rather aggressively tinged type of courting posture. During courting both male and female keep moving on quickly. Probably no matings occur. Both attacks and courting have the same vigorous and unrestrained character they have in normal reproductive periods, but yet the whole pattern of movement is much more smooth and continuous probably because abrupt brakings are absent or at least much less frequent. The movements of the fishes are rather erratic, but yet they seem difficult to disturb from the outside. Males may carry nuptial colouration, but often their black body markings as well as the whole bodies are very light.

Repeated swimming through an air stream was also observed once from a single male *nigrofasciatus*. In doing so it displayed the same type of continuous movement as was described above for *cumingi*. It displayed the same imperturbability to external disturbance and it also became quite light on the whole body; even the black body markings were almost completely faded. This kind of "play" has been noticed in fishes by other authors (see EIBL-EIBESFELDT, 1967: p. 243). I agree with EIBL-EIBESFELDT's doubts as to whether this type of behaviour may be put into one category with playing behaviour of higher mammals. However, another question is whether it is right to put together all kind of higher mammal playing into one category. It would certainly seem more worthwhile to look for behaviour patterns in higher vertebrates which could be comparable to air-stream playing in fishes, rather than enquiring whether the fish behaviour fits into a set of rules set up for mammal playing. Though not an expert in mammal behaviour I do have the impression that some kinds of "voluptuous solo-playing" in mammals may be comparable to air-stream playing in fishes [1]). The chaotic nature of the behaviour displayed by *cumingi* may be related (fore-shadowing?) to the "free disposal of behaviour elements" which is characteristic of higher mammal playing.

Little can be said about the causation of the kind of behaviour displayed by *cumingi* in the 6♂♂ 6♀♀ group. In three out of the four cases in which it was observed it followed an inactive day [2]). However, these were days on which late observations were preferably made. In all four cases it was followed by an active day but, because most days were active days, again little may be concluded from this. Further, I cannot

---

1) In analyzing the phenomenon of air-stream playing a possibly more promising line of research might lead to the physiological aspects of states of "trance" or "high", rather than directly relating it to other kinds of playing.

2) On the fourth day (27/5) it was reckoned as a second reproductive period (see p. 157.

be sure that the same kind of behaviour didn't occur on many more days. In one of the four cases (16/5) it was observed during the last minutes only before the lights went out. Such late observations were only made on this day and there is no guarantee at all that it was unique. Finally, the fact that the kind of behaviour was never seen earlier than 19.00 *might* suggest that its occurrence was due to the unnaturally long days (14 hrs.) extant during the observations.

APPENDIX 2

(*ad* p. 228; *B. conchonius, B. nigrofasciatus*)

Two more activities of the dominant males of the 2♂♂ and 3♂♂ groups are worth mentioning: distance attack and courting posture. (Unlike in ambisexual groups, 'normal' males of all 5 species do court males in small all-male groups; this courting mainly consists of courting posture; mating attempts are virtually absent; leading *s.l.* may be rather frequent but is excluded from the present analyses).

No consistent differences between the 2 species could be found in the amounts of distance attack displayed by the dominant males.

For the sake of this analysis the distance attack and courting posture scores may be added together. Courting posture as displayed in these observations may be considered as another kind of low intensity attack. (The performer often follows the courted male from a distance, like in distance attack; if it is on close quarters mild butting movements are frequent). Alternatively, courting posture may be considered as an expression of the sexual tendency which may only express itself if the aggressive tendency is relatively weak (as measured by the intensity of attack). (See further Kortmulder, in prep.).

Figg. 80a-d show the total number of seconds spent per 5 minutes in distance attack *plus* courting posture, plotted against the (full) attack scores. As may be seen in the graphs *conchonius* males, both 3♂♂ and 2♂♂ dominants, attain much higher scores for distance attack *plus* courting posture than do *nigrofasciatus* dominants.

The point is not easy to interpret. As was described on p. 74, distance attack results from the attacker slowing down rather than from the fleeing fish speeding up. However, from this it may not be concluded that the attacking *motivation* is lower during distance attack than during full attack; as was elaborated on p. 200, the chased fish's speed of fleeing may influence the strength of the aggressive tendency in the chasing fish. From direct observation I got the impression that distance attack and courting posture (in the 3♂♂ and 2♂♂ groups) are associated with slower progress of both fishes as compared to full attack. Distance attack and courting posture in this situation may then result from an interplay of chaser and chased: slower fleeing of the chased fish lowers the aggressive tendency in the chaser and thus leads to distance attack and courting; distance attack and courting posture of the chaser elicit slower fleeing of the chased fish than does full attack — and so on; it is impossible to say which of the two influences is more important.

Extrapolating this interpretation to inter-specific comparison, one might suspect that *conchonius* inferiors on the average flee more slowly than do *nigrofasciatus* inferiors and/or that *conchonius* dominants attack with less speed than do *nigrofasciatus* dominants. I cannot say that this is true — without precise measurement of speeds — but it seems a reasonable hypothesis. It is clear that under this interpretation the data do not decide between the hypotheses (1) and (2) as proposed on p. 199.

Finally, it is obvious that, for the purpose of the analyses of chapter III section IV, distance attack and courting posture scores cannot be simply added to the full attack scores because it is unknown what precisely the role of the sexual tendency is. It is quite conceivable that, once courting has been elicited through a lowering of the aggressive tendency in the attacking fish, the sexual tendency is largely responsible for its continuation.

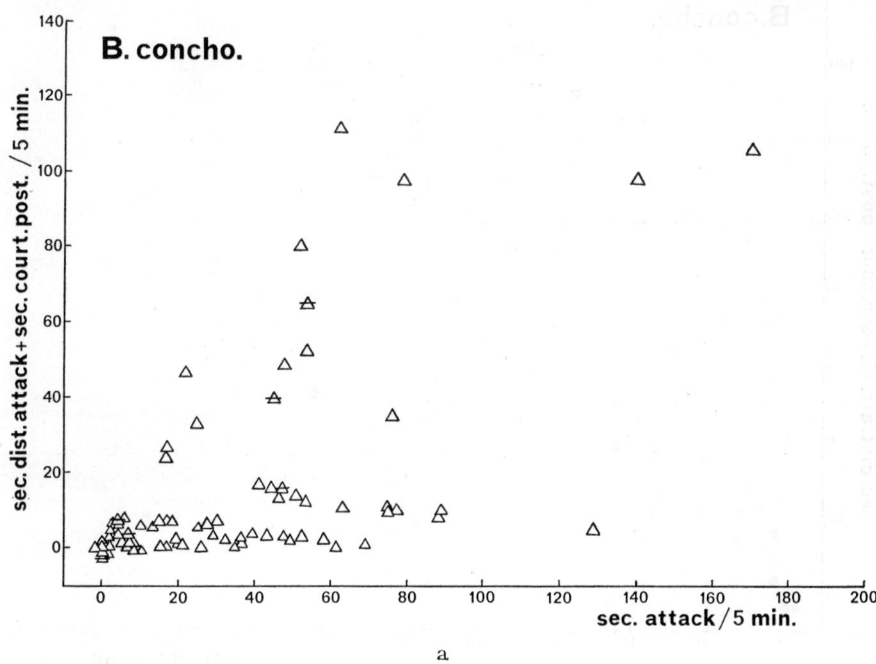

a

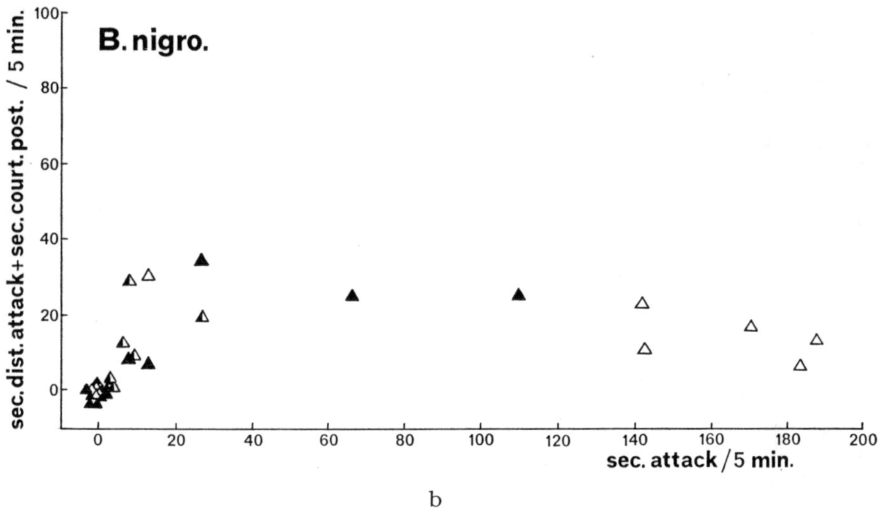

b

Fig. 80a-d. *B. conchonius, B. nigrofasciatus*; attacking and distance attack + courting posture scores of dominant males in 3♂♂ (triangles) and 2♂♂ (circles) groups; a.l. and n.l. series combined; periods with established dominance only.

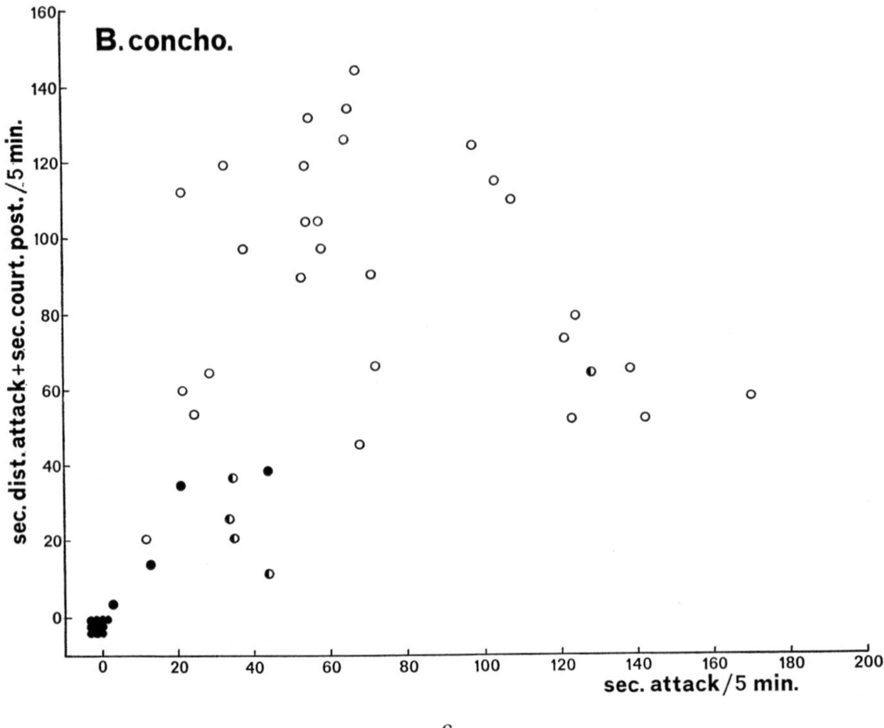

c

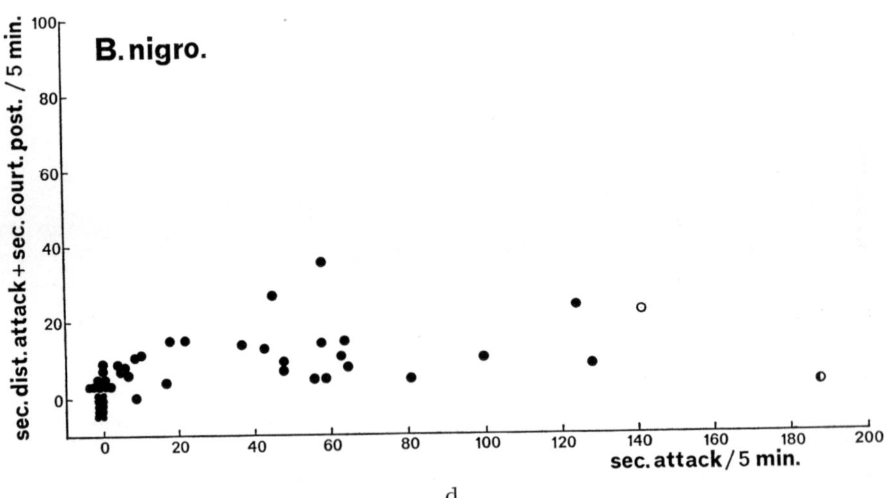

d

Legends fig. 80 sie p. 313.

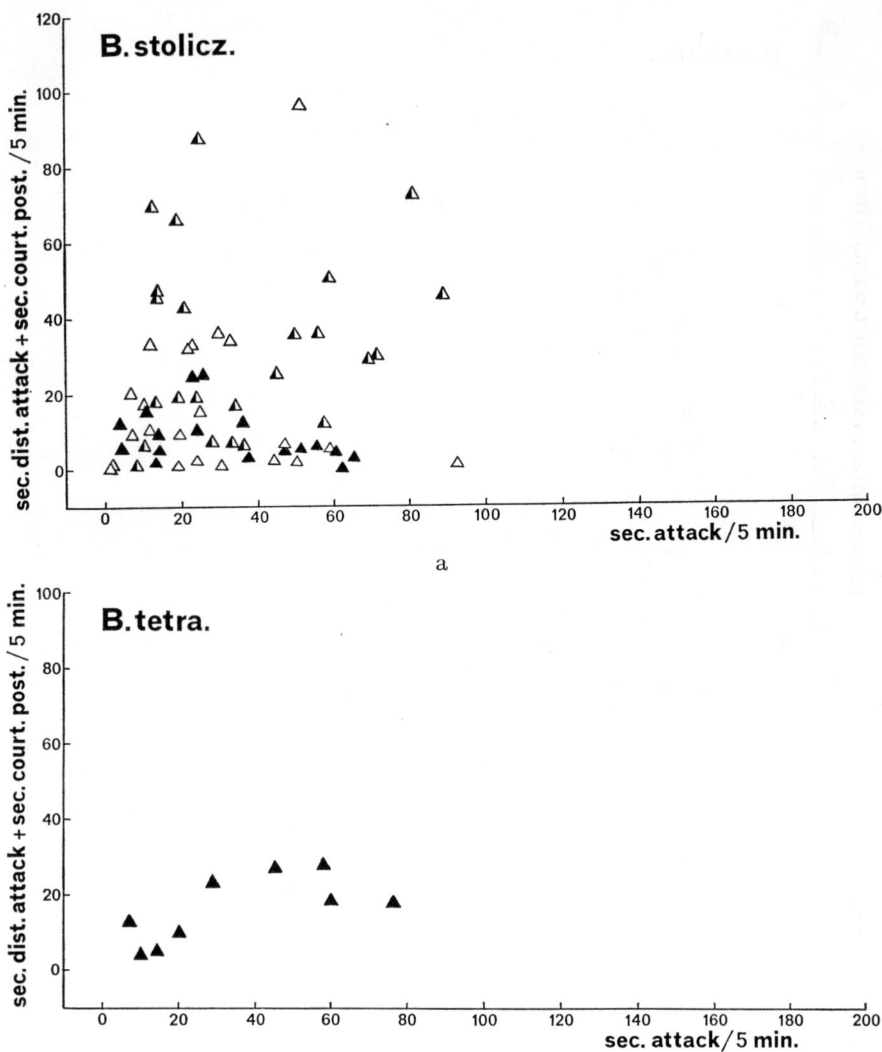

Fig. 81a-d. *B. stoliczkanus, B. tetrazona*; attacking and distance attack + courting posture scores of dominant males in 3♂♂ (triangles) and 2♂♂ (circles) groups; a.l. and n.l. series combined; periods with established dominance only.

APPENDIX 3

(*ad.* p. 252; *B. stoliczkanus, B. tetrazona*)

Figg. 81a-d represent the distance attack *plus* courting posture scores of the 3♂♂ and 2♂♂ dominants of *stoliczkanus* and *tetrazona*, plotted against the full attack scores. The same comments as given in Appendix 2 are applicable here if *'conchonius'* is replaced by *'stoliczkanus'* and *'nigrofasciatus'* by *'tetrazona'*.

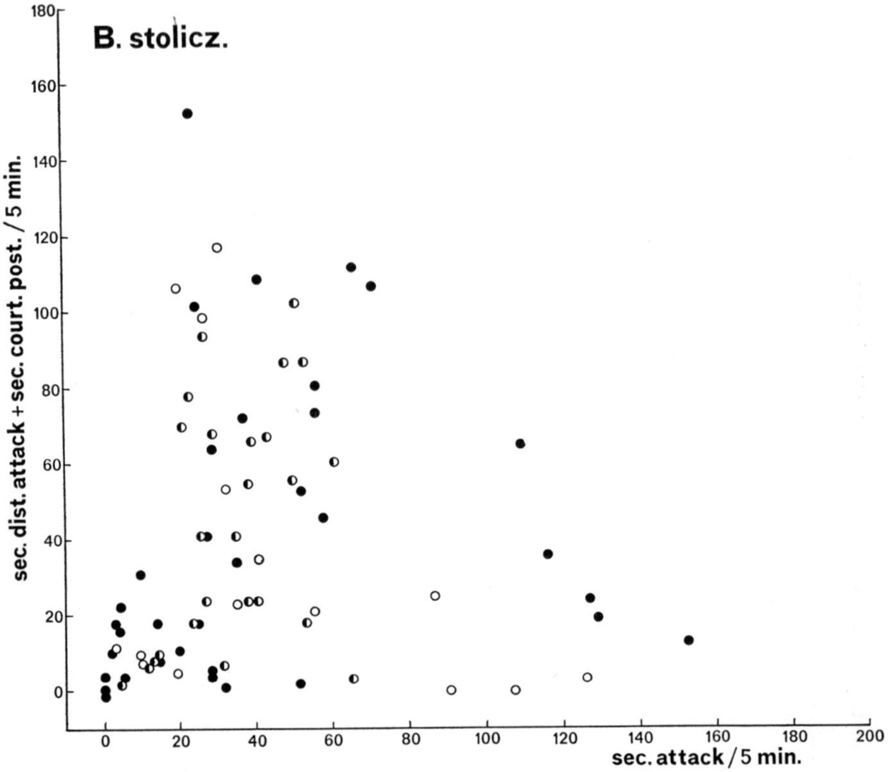

B. stolicz.

sec. dist. attack + sec. court. post. / 5 min.

180
160
140
120
100
80
60
40
20
0

0 20 40 60 80 100 120 140 160 180 200

sec. attack / 5 min.

C

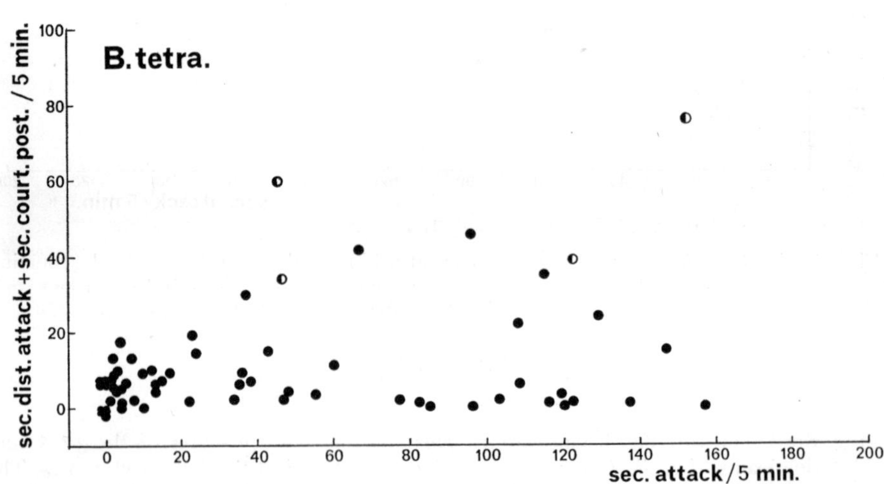

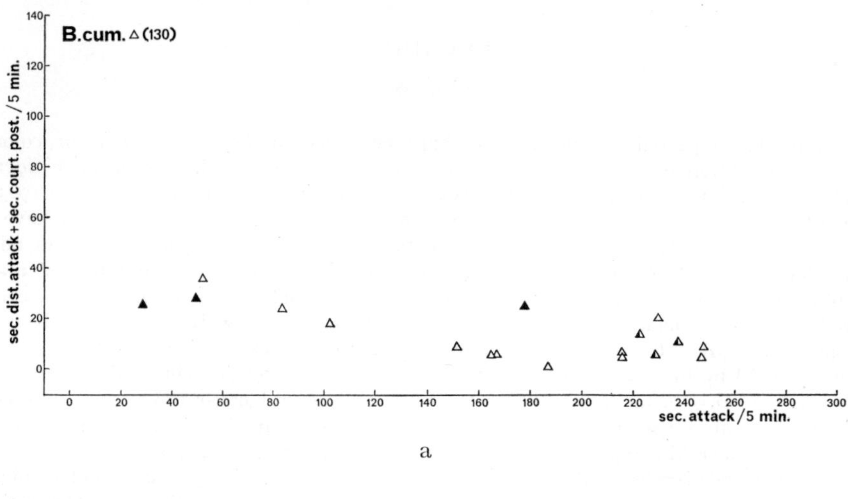

a

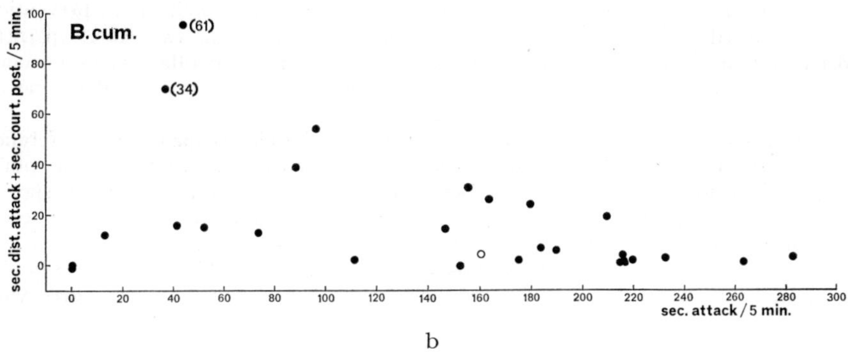

b

Fig. 82a-d. *B. cumingi*; attacking and distance attack + courting posture scores of dominant males in 3♂♂ (triangles) and 2♂♂ (circles) groups; n.l. series only; periods with established dominance only; figures in parentheses: seconds courting posture.

APPENDIX 4

(*ad* p. 284; *B. cumingi*)

Figg. 82a-b represent the distance attack *plus* courting posture scores of the dominant males of the 3♂♂ and 2♂♂ groups of *cumingi,* plotted against the full attack scores. Like in the more aggressive species *nigrofasciatus* and *tetrazona,* distance attack and courting posture scores are comparatively low; in the few cases with relatively high scores the contributions of the courting posture alone to the total scores have been indicated in the graphs.

An interesting feature not found in the graphs of the other 4 species is the negative correlation between distance attack *plus* courting posture on the one hand and full attack on the other; apparently any considerable amounts of courting may occur only in combination with comparatively low aggressive scores. This phenomenon may well be related to the rather dramatic shift in accent from aggressive towards avoidance tendencies which, in this species particularly, marks the onset of courting (see p. 106).

APPENDIX 5

(*ad* p. 299)

A remarkable partial parallel may perhaps be found in the combinations of colour patterns and behaviour of *Tilapia mossambica, T. nilotica* and *Astatotilapia strigigena* (SEITZ, 1940; BAERENDS & BAERENDS-VAN ROON, 1950; BAERENDS & BLOKZIJL, 1963). According to the observations of BAERENDS & BLOKZIJL, *T. mossambica* males have a considerably higher propensity to show courting as a primary response to an intruder than does *T. nilotica,* provided that the intruder does not carry the typically male reproductive colours. More or less the same seems to hold for *T. mossambica* as compared to *A. strigigena* (see SEITZ *l.c.* p. 48 and BAERENDS & BAERENDS-VAN ROON's comment, p. 140). All three species show some degree of sexual dichromatism which is mainly caused by the male's reproductive colours differing considerably and the female's reproductive dress being more similar to the rather neutral non-reproductive markings common to both sexes. However the most striking dichromatism is found in *T. mossambica* in which the reproductive males are entirely black with a white throat. This kind of dichromatism is reminiscent of what is found in *B. nigrofasciatus* and it might well be that the function on the part of the males is the same in both cases: prevention of being courted by other males. In fact, BAERENDS & BAERENDS-VAN ROON (p. 141) report that (pale-coloured) males may respond to the courting of other males in much the same way as do females. The problem of time-consumption may be the same as in *B. nigrofasciatus*.

On the basis of my comparison of *Barbus* species, the almost qualitative difference in dichromatism found in *T. mossambica* on the one hand and *T. nilotica* on the other would lead one to expect that also the behavioural differences between the two species are not as purely quantitative as BAERENDS & BLOKZIJL suggest. This expectation seems to be corroborated by a comparison of the most common male duelling techniques of the two species as described by BAERENDS & BLOKZIJL: carousselling in *T. mossambica* and an exchange of ramming and lateral display in *T. nilotica.* I do not agree with BAERENDS & BLOKZIJL's interpretation that the latter type of duelling would express just a higher level of aggressive motivation (relative to fear) as compared to the former type. Rather, the difference seems to lie in the degree of stability of the balance between attacking and flight (and/or in the relative influence of a third factor (such as lateral orientation or threat). I do not think it is pure coïncidence that carousselling is typical of *B. nigrofasciatus* and *T. mossambica* whereas exchange of bites is more common in such species as *B. tetrazona, B. cumingi* and in *T. nilotica.*

I made these expatiations because they illustrate an important two-fold principle of comparative functional analysis. (a) In spite of the plurality and the context-dependentness of the functional interrelations found in any one species or group of species, more or less completely parallel interrelations may be found in widely separated groups of animals (of course, ethological literature abounds of even more striking examples of this). (b) As a rule such parallels are only partial and hardly if ever allow broad generalizations such as for instance those made by NOBLE (1938) with respect to colour patterns and behaviour in several families of fishes. Apparently, every group and indeed every single species has its own unique collection of functional interrelations. (For instance, *T. mossambica* males are territorial and *B. nigrofasciatus* males are not; the whole context of mating behaviour is completely different in the two species). The latter part of the principle at least puts severe limitations to the validity or the strength of comparative functionalism. The same problem of uniqueness *vs* comparability has for decades been troubling anthropologists studying different cultures (see GOLDSCHMIDT, 1966).

SUMMARY

Chapter I.

The species involved in this study are described with an emphasis on colour patterns. Some information is given on their distributions and habitat.

Hybridization of *conchonius, nigrofasciatus, stoliczkanus* and *cumingi* yields sterile males only; therefore they are considered true species; the same is probably true of *tetrazona*.

On the basis of a comparison of ontogeny, topography and structure of the black body markings it is argued that the 5 species, together with *phutunio* and *gelius,* belong to one subgroup separate from other subgroups of the genus. This division is partly supported by (juvenile) morphological and geographical arguments. The division is supposed to reflect phylogenetic relationships.

Within the subgroup, *conchonius* and *nigrofasciatus* have many characters in common and so have *stoliczkanus* and *tetrazona*; the two pairs of species differ considerably from each other; in both pairs of species one has considerably more black colour patterns than the other. *Cumingi, phutunio* and *gelius* differ from both pairs in many characters; the latter 3 species share some of the characters in which they differ from the other 4; it is hypothesized that the 3 species are (progressively more in the order in which they are mentioned) neotenous descendants of a *nigrofasciatus*-like ancestor. It is not suggested that the classification within the subgroup reflects phylogenetic relationships.

The colour patterns and the ontogeny of the black body markings of the hybrids are described. Apart from a few patro- or matroclinic effects and apart from a few characters in which hybrids differ from both parental species, the hybrids are roughly intermediate between the parental species both in form and in colours. Reciprocal hybrids are practically identical.

Chapter II.

The behaviour elements of areproductive, agonistic and courting behaviour are listed and briefly described. A detailed form analysis suggests that the whole of agonistic behaviour results from three major tendencies: attack, flee and lateral threat and their interactions. It is argued that the interpretation of *Barbus* threat as an ambivalent posture, consisting of components of both attack and flight, is at least incomplete. An alternative hypothesis is proposed according to which the threat posture is built out of spinal reflexes similar to those described by von Holst for spinal fish. Ritualization is supposed to have occurred to mould a collection of spinal reflexes into one functional posture. It is implied in the spinal reflex hypothesis that the relative independence of threat from the attack and flight tendencies is not due to emancipation from an originally ambivalent motivation.

The pitching component of head-down threat is interpreted as an active (probably aggressively motivated) reaction against the lifting force arising from swimbladder distension; the distension itself is ascribed to autonomous activity resulting from the conflict. Dorsal and ventral rolling are interpreted as ambivalent postures arising from a conflict between the fleeing tendency and a (probably aggressive) tendency to maintain binocular fixation upon the opponent.

No arguments have been found to ascribe the various forms of attacking and fleeing behaviour to more than one major aggressive and one major fleeing tendency. At least peripherally, however, conflicts may appear between different components of the aggressive tendency on the one hand and the fleeing tendency on the other: between bite and flee, as in 'intention butts'; between attack and flee, as in tapping movements; or between binocular fixation and flee, as in rolling. Further it is suggested that in addition to these peripheral conflicts, the conflict between attack and fleeing may be more central, as in inhibited attacks. Finally, bite may also conflict with lateral orientation (threat) to produce tapping movements.

A final complication is discovered when it is attempted to range the various agonistic behaviour elements in a single series according to the different attack-flight balances that they reflect. It is argued that a rather profound difference in physiological state may underlie the differences in behaviour of inferior and 'self-actualizing' males.

Courtship is supposed to arise from three major tendencies: to attack, to flee and to mate, and their interactions. Form analysis suggests that *Barbus* male courtship is almost entirely built out of agonistic components. It is only for some components which obtain in the later phases of courtship (C curving, lateral orientation during mating attempt) that the sexual tendency may (alternatively) be supposed to express itself directly in the form of the movement. The main role of the sexual motivation may be to influence the ways in which the agonistic tendencies are expressed and to (co)-determine the sequence of courtship elements.

Courtship leaning is interpreted as a device (analogous to pitching during threat) to counteract a lifting force originating from swimbladder distension.

A comparison is drawn between the general motivational states of male courtship and inferiority.

Chapter III.

The 7 species are rather uniform at the behaviour element level. They greatly differ from each other in the relative frequencies and in the temporal and spatial arrangement of the various elements.

Section I. Reproductive behaviour elements and some general locomotory characteristics are described for each of the species (see also section II: *B. phutunio* and *B. gelius*) insofar as they deviate from the general picture offered in chapter II.

Section II. The behaviour of *conchonius, nigrofasciatus, tetrazona, stoliczkanus* and *cumingi* in large monospecific groups in large tanks is described in a semi-quantitative way. Observations of *phutunio* and *gelius* are added. Data involve: 1. temporal patterning of behaviour (a) from day to day (b) within a day; 2. spatial patterning of behaviour relative to (a) depth and (b) vegetation (c) territories; 3. interindividual patterning of reproductive behaviour; 4. other data (such as schooling behaviour and individual differences in behaviour).

Motivational aspects of the spatial and temporal patternings of reproductive behaviour are discussed. The temporal and spatial patterns of *conchonius* and *nigrofasciatus* may be understood on the basis of a small number of relatively simple assumptions. The same applies to *tetrazona, stoliczkanus* and *cumingi* except that in these species (1) the spatial patterning is further complicated by the existence of male territoriality and (2) the temporal patterning, though similar to that of the former 2 species in some respects, is less regular in others.

Conchonius and *nigrofasciatus* are very similar to one another in nearly all respects except such as may be summarized by saying that the latter species is more aggressive than the other. A similar relationship exists between *tetrazona* and *stoliczkanus,* the former of which is the more aggressive one. The two pairs of species differ greatly from each other; the most important differences comprise: group courting in the former *vs* territorial courting in the latter 2 species; regular *vs* irregular circadian rhythmicity of reproductive behaviour; 'unaggressive' courting *vs* courting interspersed with attacks at the courted female. Further differences are summarized in table 16, p. 287. *Cumingi, phutunio* and *gelius* differ from both pairs of species in many characters. In *cumingi* and *phutunio* males are territorial during reproduction, *phutunio* males most strictly so of all 7 species. *Phutunio* males were seen to use a clump of filamentous algae or a fixed location on the bottom as a kind of nest. Neither nest-*building* nor any unequivocal parental behaviour were observed. *Gelius* males are not territorial during courtship nor do they seem to form spawning aggregations. *Cumingi, phutunio* and *gelius* share some of the characters in which they differ from the other 4 species (see also table 16). The (not necessarily phylogenetic) classifications of the 7 species based upon their behaviour on the one hand and on their colour patterns on the other are essentially similar (*cf* chapter I). Some behavioural characters of *cumingi, phutunio* and *gelius* may support the hypothesis that these species are to some degree neotenous.

Data on the degree of individual variability in the various species are incomplete. So far, most pronounced individual differences in behaviour were found in *nigrofasciatus* and *tetrazona,* both in males and in females. Homosexual males, showing a definite preference for courting other males, occur in *nigrofasciatus* and possibly also in *conchonius.* Such males were not found in any of the other species.

In all species the females play a comparatively inactive role in courtship; they exhibit comparatively little aggressive behaviour as long as active males are present in the same tank. (Females of all species may fight furiously in the absence of males). In spite of this the species differ considerably in the degree of aggressive activity of the females in ambisexual groups. In such groups *conchonius* females display practically no aggressive behaviour at all and so do *nigrofasciatus* females except for a few 'notoriously' aggressive individuals. *Tetrazona, stoliczkanus* and *cumingi* females, on the other hand, often display considerable aggressiveness both towards other females and towards males. In small groups (2♂♂ 2♀♀) they may even dominate males. In the territorial species: *tetrazona, stoliczkanus, cumingi* and *phutunio,* the females must also actively go and find the males in their territories. Such behaviour is most conspicuous in *phutunio* females.

All 7 species are clearly schooling fish, at least outside reproductive periods; *tetrazona* schools much more strongly than do the other species. Schooling and reproduction are largely mutually incompatible in all species but the species do differ also on this point. *Conchonius* males in large groups (6♂♂ 6♀♀) form loose schools with considerable distances between individuals when reproductively active. It is possible that the same tendency is present in *nigrofasciatus* males. Acute conflicts between schooling and reproduction occur in *tetrazona* and *stoliczkanus* males. In the other species the transitions from the one activity pattern to the other are more smooth. Females of all species keep schooling also during reproductive periods more than do males.

Tetrazona males and females were observed to 'mob' larger fish.

Also feeding is largely incompatible with intensive reproductive behaviour in nearly all species; only *phutunio* males do feed normally when reproductively active; they do not, however, eat their 'own' eggs. In *conchonius, nigrofasciatus* and *cumingi* feeding is completely absent during periods of intensive reproductive behaviour; this is true for both males and females. In *tetrazona* and probably also in *stoliczkanus* both males and females do feed even during periods of intensive fighting and courting. It is possible that feeding is completely suppressed, also in these 2 species, when they court in well-separated territories with little interference from rivals. In the latter 5 species spawning and feeding grounds are thought to be separated spatially. More data are needed on *gelius* in order to determine the relationship between feeding and reproduction in that species; it seems possible that, like *phutunio* males, they do not eat their own eggs.

Section III. In this and the following section quantitative data are supplied in order to check and supplement the data of the foregoing section. Behaviour records are from individual fishes in small ambisexual (2♂♂ 2♀♀, section III) and all-male (2♂♂ and 3♂♂, section IV) monospecific groups of *conchonius, nigrofasciatus, tetrazona, stoliczkanus* and *cumingi* in small (60 × 35 × 35 cm) tanks.

The data from the 2 ♂♂ 2 ♀♀ groups are in accordance with the conclusions of section II on the following points:

1. Courtship of 'normal' males of all 5 species is almost exclusively directed at females. The exception is formed by the homosexual males found in *nigrofasciatus* and possibly in *conchonius.*
2. *Conchonius* and *nigrofasciatus* males (including the homosexual) direct the great majority of their aggressive behaviour at the other male. Females are attacked rarely, especially by males which spend much time courting. *Tetrazona, stoliczkanus* and *cumingi* males, on the other hand, display much aggressive behaviour towards females even during periods of intensive courtship. This is true (as far as the records reach) for both the dominant and the inferior males and for a male's behaviour towards the courted as well as towards the non-courted female.
3. *Conchonius* and *nigrofasciatus* inferior males are much less hampered both in their aggressive activities towards the dominant male and in their courting activities towards females than are the inferior males of *tetrazona* and *stoliczkanus.* In the latter 2 species also the dominant males mate rarely as compared with the dominants of the former 2 species. Both facts are in accordance with the greater 'immunity' to competitive situations in males of *conchonius* and *nigrofasciatus* as compared with the other 2 species. The almost complete subordination of *cumingi* inferior males and the low mating scores of *cumingi* dominants require a special explanation (see section IV).
4. Inferior males of *tetrazona, stoliczkanus* and *cumingi* — and to a lesser degree the females of these species — hid in corners of the tank or between vegetation in most observation periods; hiding was not observed in *conchonius* and *nigrofasciatus.* Hiding of the inferior fishes is suggestive of territorial tendencies in the dominant male. The data are in accordance with the territoriality found in *tetrazona, stoliczkanus* and *cumingi* males in the large groups. Also inferior males of *tetrazona* and *stoliczkanus* 2 ♂♂ 2 ♀♀ groups displayed vestigial territorial behaviour. The absence of territorial tendencies in *cumingi* inferior males requires a special explanation (see section IV).
5. *Tetrazona, stoliczkanus* and *cumingi* females often display considerable aggressiveness (not only towards the other female or towards the inferior male); females of the latter 2 species may become dominant in 2 ♂♂ 2 ♀♀ groups. *Conchonius* and *nigrofasciatus* females, on the other hand, are usually very placid; only some individual *nigrofasciatus* females may display considerable aggressiveness but they never become the dominant fish.
6. As fas as the sampling technique allows comparison, the following differences between the species as to male aggressiveness are found: *nigrofasciatus* males are more aggressive than *conchonius* males; *tetrazona* males are more aggressive than *stoliczkanus* males; it is not possible to compare degrees of aggressiveness of *conchonius* and *nigrofasciatus* males on the one hand and *tetrazona* and *stoliczkanus* males on the other. This is so even if the samples of the two pairs of species would be comparable; *cumingi* dominants lived up to their reputation of unrestrained agressiveness; the data of the last species cannot be directly compared with those of the other 4. The conclusions of this paragraph are more accurately checked in the next section.

Section IV. 2 ♂♂ and 3 ♂♂ monospecific groups of *conchonius, nigrofasciatus, tetrazona, stoliczkanus* and *cumingi* were established and observed according to a fixed time schedule. Observations extended over two days following establishment. Individual behaviour records are analyzed in relation to the general social situations of the groups (schooling; established dominance; inferiors free or hiding). The following inter-specific comparisons are made:

1. *Conchonius vs nigrofasciatus.* Once dominance has been established, *nigrofasciatus*

inferiors hide more than do *conchonius* inferiors. Two hypotheses may account for this difference: (1) *nigrofasciatus* males (especially when dominant) are more aggressive than *conchonius* males; (2) *nigrofasciatus* males (especially when inferior) are more afraid of each other than are *conchonius* males. Both hypotheses are tested. Complications interfering with the interpretation of the quantitative behaviour records are discussed: (a) the aggressive behaviour of the dominant and the fleeing behaviour of the inferior fishes tend to be each other's mirror image; (b) hiding by the inferior influences aggressive scores of the dominant; (c) differences between the species in agonistic scores may result from the differences in the species-specific colour patterns of the opponents as well as from species differences in motivation. The possible existence of a secondary mirror effect, between aggressive scores of inferiors and fleeing scores of dominants is discussed.

Taking into account these complications the following useful parameters remain: attacking scores of dominants on hiding inferiors; aggressive scores (attack, roll, threat) of inferiors both free and hiding; flight scores (fleeing, threat) of dominants for (hiding and free) inferiors. In addition some unquantified observations may be used. From these data various arguments are found in favour of hypothesis (1) and none that plead for hypothesis (2).

2. *Stoliczkanus vs tetrazona.* On the basis of arguments which are essentially similar to those used in the fore-going sub-section, it is concluded that *tetrazona* males are more aggressive than *stoliczkanus* males. No arguments are found which may indicate that the 2 species differ in the strength of the fleeing motivation.

3. *Conchonius-nigrofasciatus vs stoliczkanus-tetrazona.* The two pairs of species differ from each other in the behaviour of the inferior males during hiding: passive hanging in the former, tense posture and fixating the dominant in the latter 2 species. This in its turn reduces the number of parameters that may be used for comparison. The remaining parameters are: aggressive scorings of inferiors (both free and hiding) and flight scores of dominants with free inferiors.

The fleeing scores of the dominants do not indicate differences between the 4 species in fleeing motivation; the threat scores are consistent with the general impression that *conchonius* and *nigrofasciatus,* as compared with the other 2 species, are 'threat specialists'.

Somewhat paradoxical results are obtained if the aggressive scorings (attack, roll, threat) of the inferior fishes are compared across the two pairs of species. A key to an understanding of the differences between the two pairs of species may be found in the 'active' posture of hiding inferiors, in the comparatively high attacking scores of hiding inferiors and in the incidental occurrence of mouth-fight between hiding inferior and dominant in *stoliczkanus* and *tetrazona.* All three features are absent in the other 2 species; all three are indicative of marginal territoriality in the inferiors of *stoliczkanus* and *tetrazona*; they are also consistent with the general impression that *stoliczkanus* and *tetrazona,* as compared with the other 2 species, are 'biting specialists'.

It is concluded that aggressiveness in the two pairs of species is expressed in qualitatively different ways.

4. *cumingi vs* the other 4 species. Once dominance has been established *cumingi* 2♂♂ inferiors hide as much as do *nigrofasciatus* or *tetrazona* 2♂♂ inferiors; *cumingi* 3♂♂ inferiors hide even less than do *stoliczkanus* 3♂♂ inferiors. *Cumingi* dominants score more attack than do the dominants of any of the other 4 species, also if only dominants with hiding inferiors are taken into account. As the manner of hiding by the inferiors is similar to that of *conchonius* and *nigrofasciatus,* the aggressive scores of the dominants suggests that *cumingi* is even more aggressive than *nigrofasciatus.* This interpretation is corroborated by the behaviour of the

hiding inferiors but is contradicted by the behaviour of the free 3 ♂♂ inferiors. It is concluded that aggressiveness in this as compared with the other 4 species is again expressed in qualitatively different ways.

The fleeing scores of dominant *cumingi* are often higher than those of the other species. This may indicate a higher fleeing motivation in this species, but alternative interpretations are possible.

Various sets of data combine in suggesting that prolonged ambivalent states are less common in *cumingi* as compared with the other 4 species.

The fact that the 3♂♂ inferiors are so often free and yet do not show territorial tendencies may throw some light (a) on the absence of marginal territoriality also in hiding 2 ♂♂ and 3 ♂♂ and in hiding 2 ♂♂ 2 ♀♀ inferior males; (b) on the fact that *cumingi* dominants of 2♂♂ 2♀♀ groups attained comparatively low mating scores only (in spite of their comparative 'immunity' to competitive situations in large groups). Together, these facts may indicate that either the crowded condition of the small groups or the small size of the tanks *per se* interferes with the full deployment of reproductive behaviour and territorial behaviour in this species.

Chapter IV.

Chapter I and III have revealed a close concordance between classifications of the 7 species on the basis of morphological characters on the one hand and on the basis of behaviour on the other. Various hypotheses which might account for this concordance are discussed. One of them, the functional hypothesis, has been chosen as a working hypothesis for the present study. According to this hypothesis, the colour patterns and the behaviour of each of the species are products of mutual evolutionary adaptation. The relationships of the functional hypothesis to the other hypotheses and the ways in which it should be tested are discussed.

It is suggested that the colour patterns play an important role in the social interactions between conspecifics, especially during reproduction. Preliminary experiments with white, grey and black models indicate that both general darkness and dark cross-bars reduce overt attack in an opponent. In both the *conchonius-nigrofasciatus* and the *stoliczkanus-tetrazona* pair of species, the more aggressive species has more dark patterns. It is suggested that these patterns are functionally necessary in order to counter-balance the aggressive behaviour of conspecifics. *Conchonius* and *nigrofasciatus* on the one hand and *stoliczkanus* and *tetrazona* on the other differ in many behavioural *and* colour characters. It is shown that both categories of characters may be functionally understood when grouped around one central feature: group spawning in the former and territorial spawning in the latter 2 species.

The functional interrelationships of the colour patterns and behaviour of *cumingi*, *phutunio* and *gelius* cannot be understood on the basis of a simple extrapolation of the rules found for the former 4 species. Specific features, some of which are common to all 3 species, have also to be taken into account. It is concluded that no simple set of rules describes the functional interrelationships for the whole group of species (see also appendix 5).

Other than reproductive functions of the colour patterns are briefly discussed.

The question is raised what has made each of the species 'choose' for its own specific 'way of life'. It is suggested that the ultimate causes of this mainly lie in the extra-specific environment, including predators.

Some substrates on which selection may have worked to produce the present species are briefly discussed.

A few critical remarks on homologizing the behaviour elements of the different species conclude the chapter and the book.

REFERENCES

ALEXANDER, R. McN. (1959). The physical properties of the isolated swimbladder in *Cyprinidae*. — J. Exp. Biol. 36, p. 341-346.

ALFRED, E. R. (1963). Some colourful fishes of the genus *Puntius*. — Bull. Singapore Nat. Mus. 32, p. 135-142.

ANDREW, R. J. (1956). Some remarks on behaviour in conflict situations, with special reference to *Emberiza* spp. — Brit. J. Anim. Behav. 4, p. 41-45.

BAERENDS, G. P. & BAERENDS-VAN ROON, J. M. (1950). An introduction to the study of the ethology of cichlid fishes. — Behaviour Suppl. 1.

——, BROUWER, R. & WATERBOLK, H. Tj. (1955). Ethological studies on *Lebistes reticulatus* (Peters). I. An analysis of the male courtship pattern. — Behaviour 8, p. 249-334.

—— & BLOKZIJL, G. J. (1963). Gedanken über das Entstehen von Formdivergenzen zwischen homologen Signalhandlungen verwandter Arten. — Zeitschr. f. Tierpsychol. 20, p. 517-528.

BARLOW, G. W. (1961). Ethology of the Asian Teleost *Badis badis*. I. Locomotion, maintenance, aggregation and fright. — Trans. Ill. State Acad. Sci. 54, p. 175-188.

—— (1962). Ethology of the Asian Teleost *Badis badis*. III. Aggressive behaviour. — Zeitschr. f. Tierpsychol. 19, p. 29-55.

BLURTON-JONES, N. G. (1968). Observations and experiments on causation of threat displays of the great tit (*Parus major*). — Anim. Behav. Monogr. 1, p. 75-158.

BREDER, C. M. (1926). The locomotion of fishes. — Zoologica 4, p. 159-297.

CAMPBELL, G. (1970). Autonomic nervous systems; In: HOAR, W. S. & RANDALL, D. J. (eds) Fish physiology, Vol. 4. — Academic Press, New York & London.

CULLEN, E. (1957). Adaptations in the Kittiwake to cliff-nesting. — Ibis 99, p. 275-302.

DERANYAGALA, P. E. P. (1930). The Eventognathi of Ceylon. — Spolia Zeylanica 16, p. 1-42.

DIJKGRAAF, S. (1942). Swimbladder reflexes in fish. — Zeitschr. f. vergl. Physiol. 30, p. 39-66.

DUNHAM, D. W., KORTMULDER, K. & IERSEL, J. J. A. VAN (1968). Threat and appeasement in *Barbus stoliczkanus* (*Cyprinidae*). — Behaviour 30, p. 15-26.

DUYVENÉ DE WIT, J. J. (1964). Hybridization experiments in Rhodeine fishes (*Cyprinidae, Teleostei*). Crossings between female *Tanakio tanago, Rhodeus ocellatus* and *Acheilognathus limbatus* and male *Acheilognathus limbatus*. — Copeia 1964 (1), p. 156-159.

EIBL-EIBESFELDT, I. (1967). Grundriss der vergleichenden Verhaltensforschung. — Piper & Co., München.

EVANS, H. M. (1925). A contribution to the anatomy and physiology of the air-bladder and Weberian ossicles in *Cyprinidae*. — Proc. Roy. Soc. B97, p. 545-576.

FÄNGE, R. (1943). Preliminary notes on the reactions to electricical and chemical stimulation of the smooth muscles in the swimbladder of Cyprinids. — Kungl. Fysiogr. Sällskap Förhandlingar Lund 13, p. 147-150.

GOLDSCHMIDT, W. (1966). Comparative functionalism; an essay in anthropological theory. — Univ. Calif. Press, Berkeley.

HAMILTON, F. (formerly BUCHANAN) (1822). An account of the fishes found in the river Ganges and its branches. — Archibald Constable & Co., Edinburgh.

HARDY, A. C. (1958). Escape from specialization; In: HUXLY, J., HARDY, A. C. & FORD, E. B. (eds). Evolution as a process. — Allen & Unwin, London.

HINDE, R. A. (1955-56). A comparative study of the courtship of certain finches (Fringillidae). — Ibis 97, p. 706-745; 98, p. 1-23.
—— (1966). Animal behaviour; a synthesis of ethology and comparative psychology. — McGraw-Hill, London.
HOLST, E. VON (1934). Studien über Reflexe und Rhythmen beim Goldfisch. — Zeitschr. f. vergl. Physiol. 20, p. 582-599.
—— (1935). Weitere Reflexstudien an spinalen Fischen. — Zeitschr. f. vergl. Physiol. 21, p. 658-665.
—— (1950a). Die Arbeitsweise des Statolithenapparates bei Fischen. — Zeitschr. f. vergl. Physiol. 32, p. 60-120.
—— (1950b). Die Tätigkeit des Statolithenapparats im Wirbeltierlabyrinth. — Naturwiss. 37, p. 265-272.
JACOBS, W. (1940). Die Schwimmblase der Fische als Schwebeorgan. — Naturwiss. 28, p. 33-43.
KORTMULDER, K. (in prep.). A quantitative analysis of the motivations of reproductive behaviour in *Barbus conchonius*.
KRUIJT, J. P. (1964). Ontogeny of social behaviour in Burmese Red Junglefowl (*Gallus gallus spadiceus*). — Behaviour Suppl. 12.
LADIGES, W. (1951). Der Fisch in der Landschaft. — Gustav Wenzel & Sohn, Braunschweig.
LORENZ, K. (1963). Das sogenannte Böse; zur Naturgeschichte der Aggression. — Borotha-Schoeler, Wien.
—— (1964). Ritualized Fighting; In: CARTHY, J. D. & EBLING, F. J. (eds). The natural history of aggression. — Academic Press, London & New York.
MASLOW, A. H. (1968). Towards a psychology of being. — Van Nostrand Rheinhold, New York.
MENDIS, A. S. & FERNANDO, C. H. (1962). A guide to the freshwater fauna of Ceylon. — Bull. nr 12 of Fisheries Research Station, Dept. of Fisheries, Ceylon.
MORRIS, D. (1956). The feather postures of birds and the problem of the origin of social signals. — Behaviour 9, p. 75-113.
—— (1957a). "Typical intensity" and its relation to the problem of ritualization. — Behaviour 11, p. 1-12.
—— (1957b). The function and causation of courtship ceremonies; In: L'instinct dans le comportement des animaux et de l'homme. — Fondation Singer Polignac, Paris.
MÜLLER, J. (1843). Untersuchungen über die Eingeweide der Fische. — Abh. der Königl. Akad. der Wissenschaften, Berlin, p. 109-170.
MUNRO, I. S. R. (1955). The marine and freshwater fishes of Ceylon. — Canberra (Australia): Dept. of External Affairs.
NICHOLSON, A. J. (1960). The role of population dynamics in natural selection; In: TAX, S. (ed.). Evolution after Darwin. I. — Univ. of Chicago Press.
NOBLE, G. K. (1938). Sexual selection among fishes. — Biol. Rev. 13, p. 133-158.
PRECHTL, H. F. R. (1954). Reste stammesgeschichtlich alten instinktiven Verhaltens in Ausdrucksbewegungen des Menschen. — Die Umschau 20, p. 627-630.
SCHULTZ, L. P. (1957). The generic names *Barbus* and *Puntius*. — Tropical Fish Hobbyist 5, p. 14-15; 29-31.
SEITZ, A. (1940/41). Die Paarbildung bei einigen Cichliden. I. — Zeitschr. f. Tierpsychol. 4, p. 40-84.
—— (1943). Die Paarbildung bei einigen Cichliden. II. — Zeitschr. f. Tierpsychol. 5, p. 74-101.
SEVENSTER, P. (1961). A causal analysis of a displacement activity (fanning in *Gasterosteus aculeatus* L.). — Behaviour Suppl. 9.
STEEN, J. B. (1970). The swimbladder as a hydrostatic organ; In: HOAR, W. S. &

RANDALL, D. J. (eds) Fish physiology, Vol. 4. — Academic Press, New York & London.

STERBA, G. (1963). Freshwater fishes of the world. (Translated and revised by TUCKER, D. W.). — Vista Books, London.

TINBERGEN, N. (1951). The study of instinct. — Clarendon Press, Oxford.

—— (1958). Bauplan-ethologische Beobachtungen an Möwen. — Arch. Néerl. Zool. 12, Suppl. 1, p. 369-382.

—— (1959). Comparative studies of the behaviour of gulls (*Laridae*); a progress report. — Behaviour 15, p. 1-70.

WICKLER, W. (1957a). Das Kampfverhalten von *Nannostomus beckfordi aripirangensis* Meinken. — Bull. Aquatic Biol. 1, p. 13-15.

——(1957b). Das Verhalten von *Nannostomus beckfordi aripirangensis* Meinken. Mit einer Besprechung der Schwimmblasenfunktion. — Beaufortia 6, p. 203-220.

WIEPKEMA, P. R. (1961). An ethological analysis of the reproductive behaviour of the bitterling (*Rhodeus amarus* Bloch). — Arch. Néerl. Zool. 14, p. 103-199.

WILLIAMS, G. C. (1966). Adaptation and natural selection; a critique of some current evolutionary thought. — Princeton Univ. Press, New Yersey.

ZUSAMMENFASSUNG

Hauptfrage der Untersuchung ist, ob es funktionelle Zusammenhänge gibt zwischen Farbmuster und Verhalten bei 7 tropisch-asiatischen *Barbus*-Arten: *B. conchonius, B. nigrofasciatus, B. stoliczkanus, B. tetrazona, B. cumingi, B. phutunio* und *B. gelius*.

Im *ersten Kapitel* werden die Farbmuster und ihre Ontogenie eingehend beschrieben und verglichen.

Da zwischenartliche Hybriden steril sind, wird geschlossen dass es sich um richtige Arten handelt. Andererseits werden die 7 Arten gegen andere derselben Gattung als phylogenetisch zusammenhängende Gruppe abgegrenzt.

Innerhalb der Gruppe gibt es viele Ähnlichkeiten in der Farbe zwischen *conchonius* und *nigrofasciatus,* ebenfalls zwischen *stoliczkanus* und *tetrazona*; die zwei Artenpaare sind untereinander sehr verschieden. In jedem Paar hat eine Art bedeutend mehr und grössere schwarze Muster als die andere. *Cumingi, phutunio* und *gelius* weichen in vielen Merkmalen von beiden vorhergehenden Artenpaaren ab. Die 3 Arten gleichen sich in einigen der Merkmale, in denen sie von den 4 andern verschieden sind. Es wird die Hypothese gestellt dass die letzten 3 Arten mehr oder weniger neotene Deszendente eines *nigrofasciatus*-ähnlichen Vorfahren sind. Es wird nicht ohne weiteres angenommen dass die Ähnlichkeiten und Unähnlichkeiten innerhalb der Gruppe phylogenetisch begründet sind.

Im *zweiten Kapitel* werden die Bruchteile des Verhaltens, vorwiegend am Beispiel von *conchonius* oder *nigrofasciatus,* benannt und kurz beschrieben.

Aus der Formanalyse der agonistischen Elemente ergibt sich dass das Kampfverhalten der Barben in 3 Tendenzen zerlegt werden kann: Angriff, Flucht und Breitseitsimponieren.

Es wird gezeigt dass die Deutung des Breitseitsimponierens als ambivalente Bewegung wenigstens unvollständig ist. Als alternative Hypothese wird vorgeschlagen dass das Breitseitimponieren aus einem durch Ritualisierung geänderten Komplex spinaler Reflexe, wie von VON HOLST bei spinalen Fische nachgewiesen, besteht.

Die das 'Kopf-nieder-Drohen' charakterisierende Inklination wird als wahrscheinlich aggressiv motivierte Kompensationsbewegung eines von Dystonie der Schwimmblasenmuskulatur verursachten Auftriebs, die Dystonie ihrerseits als durch einen Konflikt freigesetze vegetative Erregung aufgefasst. Das beschwichtigende Rücken- oder Bauchzuwenden wird als ein Produkt der Fluchttendenz und einer wahrscheinlich aggressiven Tendenz zum zweiäugigen Fixieren gedeutet.

Es gibt wahrscheinlich nur eine aggressive und nur eine agonistische Fluchttendenz. Doch können mehrere Komponenten des aggressiven Verhaltens: Biss, Angriff und zweiäugiges Fixieren, jede für sich mit der Flucht in Konflikt treten. Auch können Konflikte sich mehr peripher oder mehr zentral abspielen.

Es wird gezeigt dass man die unterschiedenen Elemente des agonistischen Verhaltens nicht in eine einfache Folge zwischen reiner Aggression und reiner Flucht einreihen kann. Dies führt zur Voraussetzung einer tiefen physiologischen Dichotomie zwischen dem Verhalten unterwürfiger und dem Verhalten ‚selbst-aktualisierender' Tiere.

Das männliche Werben wird dem Zusammenspiel dreier Tendenzen: Angriff, Flucht und Paarung, zugeschrieben. Aus der Formanalyse geht hervor dass das Werben der *Barbus*-männchen grossenteils aus agonistischen Komponenten aufgebaut ist. Nur etliche Elemente des vorgeschrittenen Werbens lassen sich vielleicht vom reinen sexuellen Verhalten herleiten. Es wird angenommen dass die sexuelle Motivation die Form der aus agonistischen Komponenten zusammengestellten Werbe-elemente sowie die Folge der Werbe-elemente (mit)bestimmt.

Das *dritte Kapitel* zerfällt in 4 Abteilungen:

1. *Abt.* Bruchteile des reproduktiven Verhaltens der verschiedenen Arten werden beschrieben soweit sie von der generellen Beschreibung im zweiten Kapitel abweichen.

2. *Abt.* Das Verhalten der 5 Arten *conchonius, nigrofasciatus, stoliczkanus, tetrazona* und *cumingi* in grossen Gruppen in grossen Behältern wird beschrieben. Beobachtungen an *phutunio* und *gelius* werden hinzugefügt. Die zeitliche, räumliche und inter-individuelle Struktur des Verhaltens wird analysiert. Es wird gezeigt dass relativ wenige Regeln genügen zum Verstehen der räumlichen und zeitlichen Verteilungen des Verhaltens.

Conchonius und *nigrofasciatus* sind einander verhaltensmässig sehr ähnlich; nur ist die letzte Art aggressiver als die erste. Ein gleichartiger Unterschied gibt es zwischen *stoliczkanus* und *tetrazona*. Die zwei Arten-Paare sind untereinander sehr verschieden (s. Tafel 16, S. 287). *Cumingi, phutunio* und *gelius* weichen in vielen Verhaltensmerkmalen von beiden vorhergehenden Artenpaaren ab. Die 3 Arten gleichen sich verhaltensmässig in einigen der Merkmale, in denen sie von den 4 andern verschieden sind. Etliche Verhaltensmerkmale stützen vielleicht die Hypothese dass die letzten 3 Arten mehr oder weniger neoten sind.

Die Einteilungen der 7 Arten aufgrund von Farbmuster einerseits und aufgrund von Verhaltensmerkmalen andererseits stimmen weitgehend überein (vergl. Kapitel I).

Daten inbezug auf individuelle Variabilität des Verhaltens sollen noch ergänzt werden. Bis jetzt wurden die grössten Unterschiede bei *nigrofasciatus* und *tetrazona* gefunden. Homosexuelle Männchen fanden sich nur bei *nigrofasciatus* und wahrscheinlich auch bei *conchonius*.

Im allgemeinen spielen die Weibchen eine untergeordnete Rolle solange wie aktive Männchen im selben Behälter sind. Die Arten unterscheiden sich aber beträchtlich in der aggressiven Tätigkeit der Weibchen und auch in der Aktivität ihrer Rolle beim Sexualverhalten. *Conchonius* und *nigrofasciatus*-Weibchen sind in ambisexuellen Gruppen nahezu unaggressiv, wenige ‚berüchtigte' Individuen bei der letztere Art ausgenommen. *Tetrazona, stoliczkanus* und *cumingi*-Weibchen dagegen können ziemlich aggressiv sein. Auch sind die Weibchen der territorialen Arten *tetrazona, stoliczkanus, cumingi* und *phutunio* insofern aktiver dass sie die Männchen in ihrer Territorien auffinden sollen.

Alle Arten sind offenbar Schwarmfische. *Tetrazona* schwärmen aber bedeutend stärker als die anderen Arten. Auch ‚hassen' sie gemeinsam auf grössere Fische.

Schwärmen und Fortpflanzungsverhalten sind bei allen Arten ziemlich unvereinbar. Doch gibt es auch hier Artunterschiede. *Conchonius*-Männchen bilden lockere Schwärme, wenn sie in der Fortpflanzungsstimmung nicht balzen und kämpfen; dasselbe machen vielleicht auch *nigrofasciatus*-Männchen. Im Gegensatz zu *conchonius, nigrofasciatus* und *cumingi* treten bei *stoliczkanus* und *tetrazona* oft scharfe Konflikte zwischen den beiden Verhaltenssystemen auf. Weibchen aller Arten haben eine stärkere Tendenz zum Schwärmen beim Fortpflanzungsverhalten als die Männchen.

Auch die Nahrungsaufnahme ist mit der Fortpflanzungsstimmung ziemlich unvereinbar, aber *phutunio*-Männchen fressen in dieser Stimmung normal — nur nicht ihre ‚eigenen' Eier. Eine derartige Hemmung haben vielleicht auch *gelius*-Männchen und Weibchen. Bei *stoliczkanus* und *tetrazona* ist die Ausschaltung des Fressens bei der Fortpflanzung unvollständig, wenigstens wenn auch gekämpft wird. Vielleicht wird bei geringer Konkurrenz auch bei diesen Arten das Fressen eingestellt. Bei *conchonius, nigrofasciatus, stoliczkanus, tetrazona* und *cumingi* sind Nahrungsaufnahme und Fortpflanzung wahrscheinlich auch räumlich getrennt.

In der dritten und vierten Abteilung des Kapitels werden etliche Befunde der zweiten Abteilung, betreffs *conchonius, nigrofasciatus, stoliczkanus, tetrazona* und *cumingi,* anhand quantitativer Beobachtungen an kleinen (2♂♂ 2♀♀; 2♂♂, 3♀♀) Gruppen nachgeprüft.

3. *Abt.* (2♂♂ 2♀♀ Gruppen). Die folgenden Ergebnisse bestätigen jene der zweiten Abteilung.

(1) Mit Ausnahme der ‚homosexuellen' *nigrofasciatus* und *conchonius*-Männchen balzen die Männchen aller 5 Arten nahezu nur auf Weibchen.

(2) *Conchonius* und *nigrofasciatus*-Männchen (auch homosexuelle) bekämpfen fast nur andere Männchen. Weibchen werden nur selten angegriffen, besonders wenn die Männchen viel balzen, und gleichviel ob es sich um das angebaltzte oder um andere Weibchen handelt. *Stoliczkanus, tetrazona* und *cumingi* dagegen greifen samt dem angebalzten Weichchen oft an, auch während sie selbst intensiv balzen. Dominante und inferiöre Männchen verhalten sich in dieser Hinsicht gleich.

(3) Die untergeordneten *conchonius* und *nigrofasciatus*-Männchen können noch viel balzen und kämpfen auch mit dem Dominanten. Dagegen sind untergeordnete *stoliczkanus* und *tetrazona*-Männchen fast völlig in ihren aggressiven und sexuellen Tätigkeiten gehemmt.Auch können die dominanten Männchen der letzten 2 Arten nur selten paaren. Diese Tatsachen entsprechen der verhältnismässig grossen Empfindlichkeit der letzten 2 Arten für Konkurrenz. Die vollständige Unterwerfung der inferiören *cumingi*-Männchen fragt eine spezielle Erklärung (s. Abt. 4).

(4) Die unterlegenen *stoliczkanus* und *tetrazona*-Männchen und die Weibchen verstecken sich oft, unterlegene *conchonius* und *nigrofasciatus*-Männchen niemals. Dies suggeriert territoriales Verhalten der dominanten Männchen der ersten 2 Arten. Auch verteidigen die inferiören Männchen dieser 2 Arten oft ein winziges Territor in einer Ecke des Behälters. Die Abwesenheit territorialer Tendenzen in unterlegenen *cumingi*-Männchen, sowie die niedrigen Paarungsfrequenzen der dominanten, fragen eine spezielle Erklärung (s. Abt. 4).

(5) Weibchen von *tetrazona, stoliczkanus* und *cumingi* sind oft ziemlich aggressiv; Weibchen der letzten 2 Arten können auch dominant werden. *Conchonius* und *nigrofasciatus*-Weibchen dagegen sind in ambisexuellen Gruppen sehr friedlich, wenige ‚berüchtigte' Individuen der letzten Art ausgenommen; auch diese werden aber nie dominant.

(6) Soweit die vorliegenden Stichproben eine Vergleichung erlauben, finden sich die folgenden Unterschiede in der männlichen Aggressivität der Arten. *Nigrofasciatus*-Männchen sind aggressiver als *conchonius*-Männchen; *tetrazona*-Männchen sind aggressiver als *stoliczkanus*-Männchen; *cumingi*-Männchen sind ungezügelt aggressiv; Aggressionsstärken der beiden erstgenannten Artenpaare und der *cumingi*-Männchen lassen sich nicht einfach vergleichen. Die Aggressivitätsunterschiede der Arten werden in der nächsten Abteilung noch weiter geprüft.

4. *Abt.* Die Artunterschiede in der männlichen Aggressivität werden mit standarisierten Beobachtungen an 2♂♂ und 3♂♂ Gruppen näher untersucht. Die individuellen Verhaltensverzeichnisse werden analysiert mit Rücksicht auf den sozialen Zustand der Gruppe (Schwärmen, eingesetzten Dominanz, Inferiöre frei oder versteckt). Die Vergleichungen führen zu den folgenden Schlüssen:

(1) *Conchonius* vs *nigrofasciatus*. Es werden mehrere Argumente gefunden für die Hypothese dass *nigrofasciatus*-Männchen eine stärkere aggressive Motivation heben als *conchonius*-Männchen. Es gibt keine Anweisungen dass die Arten untereinander in der Stärke der Fluchtmotivation verschieden sind.

(2) *Stoliczkanus* vs *tetrazona*. Argumente werden gefunden dass die aggressive Motivation der *tetrazona*-Männchen stärker ist als bei den *stoliczkanus*-Männchen. Die Arten sind nicht nachweisbar verschieden in der Stärke der Fluchtmotivation.

(3) *Conchonius-nigrofasciatus* vs *stoliczkanus-tetrazona*. Es gibt keine Argumente dass die 4 Arten verschieden starke Fluchtmotivationen haben.

Conchonius und *nigrofasciatus* sind 'Drohspezialisten'; *stoliczkanus* und *tetrazona* dagegen ‚Beiss-spezialisten'.

Es ist unmöglich, die aggressiven Motivationen der zwei Artenpaare auf einer einfachen quantitativen Skala zu vergleichen.

(4) *Cumingi*. Die Fluchtmotivation der *cumingi*-Männchen ist vielleicht stärker als bei den andern 4 Arten; alternative Interpretationen sind aber noch möglich.

Cumingi-Männchen sind zügellos aggressiv. Die Stärke ihrer aggressiven Motivation lässt sich nicht auf einer einfachen quantitativen Skala mit jenen der andern 4 Arten vergleichen.

Verglichen mit den andern 4 Arten scheinen ambivalente Zustände bei *cumingi* relativ selten zu sein.

Die inferiören *cumingi*-Männchen zeigen keine territorialen Tendenzen, auch nicht wenn sie frei sind. Dies suggeriert dass entweder die ‚crowding' oder die geringe Grösse der Behälter in diesen Experimenten für sich, die vollkommene Entfaltung des reproduktiven Verhaltens hemmt. Dies könnte auch erklären warum die sich versteckenden Männchen in den 2♂♂ 2♀♀ und in den 2♂♂ und 3♂♂-Gruppen keine territoriale Tendenzen zeigen, somit warum die dominante Männchen der 2♂♂ 2♀♀ Gruppen nur geringe Paarungsfrequenzen leisten. (s. Abt 3).

Viertes Kapitel. Aus dem ersten und dritten Kapitel geht hervor dass die Einteilungen der 7 Arten auf Grund von Farbmuster einerseits und von Verhalten andererseits weitgehend übereinstimmen. Mehrere Hypothesen zur Erklärung dieser Konkordanz werden besprochen. Eine von diesen, die Funktionshypothese, wird zur Arbeitshypothese gewählt. Nach dieser Hypothese sind die Farbmuster und das Verhalten einer Art Produkte wechselseitiger phylogenetischen Anpassung.

Es ist wahrscheinlich dass die Farben eine wichtige Rolle in der sozialen Kommunikation spielen, zumal beim Fortpflanzungsverhalten. Provisorische Experimente mit weissen, grauen und schwarzen Attrappen weisen darauf hin dass totale Schwärze oder dunkle Vertikalstreifen die Wahrscheinlichkeit des Angriffs eines Gegners vermindern. In den beiden Artenpaaren *conchonius-nigrofasciatus* und *stoliczkanus-tetrazona* hat die aggressivere Art auch mehr schwarze Muster. Die Voraussetzung liegt nahe dass ein Individuum einer aggressiveren Art auch mehr dunkle Muster ‚braucht' um sich seine Rivalen vom Leibe zu halten. Die beiden Arten-Paare sind voneinander in vielen Farb- und Verhaltensmerkmalen verschieden. Es stellt sich heraus dass beide Kategorien von Merkmalen funktionsmässig verstanden werden können, wenn man sie bezieht auf ein zentrales Merkmal: Gruppenbalz oder Territorialität der Männchen.

Die funktionellen Beziehungen zwischen Farbmuster und Verhalten bei *cumingi*, *phutunio* und *gelius* können nicht durch einfache Extrapolation der bei den andern

4 Arten gefundenen Regeln erörtert werden. Spezifische Eigentümlichkeiten, etliche davon allen 3 Arten gemein, sollen mit in Betracht gezogen werden. Es wird geschlossen dass die funktionellen Beziehungen innerhalb der Gruppe von 7 nahverwandten Arten nicht mittels eines kleinen Satzes von einfachen Regeln verstanden werden können.

Nicht-reproduktive Funktionen der Farbmuster werden kurz besprochen.

Es wird die Frage aufgeworfen was jede Art zu der ‚Wahl' ihrer artspezifischen Farbmuster, Verhaltens- und Lebensweise veranlasst hat. Es wird angenommen dass die letzten Ursachen im artspezifischen Milieu (einschliesslich Predatoren) gefunden werden sollen.

Etliche Substrate, auf welche die Selektion gewirkt haben könnte bei der Entstehung der heutigen Arten, werden kurz besprochen.

Einige kritische Bemerkungen bei der Homologisierung von Verhaltensbruchteilen der Arten schliessen das Kapitel und das Buch.